MODERN RADIO SCIENCE 1999

International Union of Radio Science

Modern Radio Science 1987
Edited by A. L. Cullen

Modern Radio Science 1990
Edited by J. Bach Andersen

Modern Radio Science 1993
Edited by Hiroshi Matsumoto

Modern Radio Science 1996
Edited by J. Hamelin

Modern Radio Science 1999
Edited by M. A. Stuchly

Review of Radio Science 1984-1986
Edited by G. Hyde (Available from URSI Secretariat)

Review of Radio Science 1987-1989
Edited by G. Hyde (Available from URSI Secretariat)

Review of Radio Science 1990-1992
Edited by W. Ross Stone

Review of Radio Science 1993-1995
Edited by W. Ross Stone

Review of Radio Science 1996-1998
Edited by W. Ross Stone

MODERN RADIO SCIENCE 1999

Edited by

M. A. STUCHLY

Professor & Industrial Research Chair
Dept. of Electrical & Computer Engineering
University of Victoria, Victoria, BC, Canada

Published for the
International Union of Radio Science
by
Oxford University Press
1999

OXFORD
UNIVERSITY PRESS

Great Clarendon Street, Oxford OX2 6DP

Oxford University Press is a department of the University of Oxford.
It furthers the University's objective of excellence in research, scholarship,
and education by publishing worldwide in

Oxford New York

Athens Auckland Bangkok Bogotá Buenos Aires Calcutta
Cape Town Chennai Dar es Salaam Delhi Florence Hong Kong Istanbul
Karachi Kuala Lumpur Madrid Melbourne Mexico City Mumbai
Nairobi Paris São Paulo Singapore Taipei Tokyo Toronto Warsaw

with associated companies in Berlin Ibadan

Oxford is a registered trade mark of Oxford University Press
in the UK and in certain other countries

Published in the United States
by Oxford University Press Inc., New York

First published 1999

A catalogue record for this book is available from the British Library

Library of Congress Cataloging in Publication Data
(Data available)

URSI edition ISBN 0 19 856569 0
OUP edition ISBN 0 19 856570 4

Typeset by Donna Shannon, University of Victoria
Printed in Great Britain
on acid-free paper by
Bookcraft (Bath) Ltd., Midsomer Norton

Preface

Every three years, scientists and engineers from around the world and of various specialization within radio science, participate in the General Assembly of the International Radio Science Union (URSI). In today's increasing specialization in research, this year's General Assembly is a unique opportunity to gain a wider perspective on various scientific aspects of radio science.

This book presents fifteen contributions to the XXVIth URSI General Assembly in Toronto, Ontario, Canada. They range from the Public Lecture, to three General Lectures, to ten tutorials highlighting areas of research of ten URSI Commissions and one special tutorial on issues related to the frequency spectrum. Reflecting rapid developments in various areas of communications, the Public Lecture, one General Lecture and quite a few tutorials, directly or indirectly relate to radio communication. The articles assembled in this volume also reflect the scope of the research that is on the one hand quite diverse, but on the other hand clearly indicates that many mathematical tools and techniques are common. Thus, "visiting" activities of other URSI Commissions, either as represented in this book or during the scientific sessions, often can greatly benefit our own research.

I trust that the readers who attend the 1999 URSI General Assembly, as well as other scientists, will find the articles in this volume not only interesting but stimulating new ideas and perhaps scientific collaborations.

Sincere thanks go to the authors for their timely submission of manuscripts, and to Ms. Donna Shannon of University of Victoria for her expert technical editing.

Victoria, BC
April, 1999

Maria A. Stuchly

Contents

Contributions

Public Lecture

Dr. John T. Evans
COMSAT Corporation
Bethesda, MD 20817, USA

General Lectures

Dr. Louis J. Lanzerotti, Dr. David J. Thomson and Dr. Carol G. Maclennan
Bell Laboratories, Lucent Technologies, 600 Mountain Ave.,
Murray Hill, NJ 07974, USA

Dr. Hirax Hirabayashi
Institute of Space and Astronautical Science
Yoshinodai 3-1-1, Sagamihara, Kanagawa 229 8510 JAPAN

Prof. Jørgen Bach Andersen
Aalborg University
Inst. Of Electronic Systems, 9220 Aalborg, Denmark

Tutorial Lectures

Commission A
Dr. Quirino Balzano
MOTOROLA, South Florida Laboratories
8000 W Sunrise Blvd., Ft. Lauderdale, FL., 33322 USA

Commission B
Prof. Eric Michielssen[1], Prof. Yahya Rahmat-Samii[2] and Dr. Daniel S. Weile[1]
[1]Centre for Computational Electromagnetics, Dept. of Elec. & Comp. Eng.
University of Illinois at Urbana-Champaign, Urbana, IL 61801 USA
[2]Dept. of Electrical & Computer Eng., University of California at Los Angeles
Los Angeles, CA 90024-1594 USA

Commission C
Prof. Gilles Delisle[1], Dr. Khelifa Hettak[1] and Prof. Godfrey Lucas[2]
[1]Electrical & Computer Eng.,
Laval University, Ste-Foy, Quebec, G1K 7P4 Canada
[2] Faculty of Engineering, University of Western Sydney,
Kingswood, NSW 2747, Australia

Commission D
Dr. Dan Sievenpiper and Prof. Eli Yablonovitch
Dept. of Electrical & Computer Eng.., University of California at Los Angeles
Los Angeles, CA 90024-1594 USA

Commission E
Prof. David Llanwyn Jones
Physics Dept., King's College,
Strand, London, WC2R 2LS UK

Commission F
Dr. Dag T. Gjessing
Triad AS
N-2001 Lillestøm, NORWAY

Commission G
Prof. Jürgen Röttger
Max-Planck-Institut für Aeronomie
D-37191 Katlenburg-Lindau, Germany

Commission H
Dr. Llewelyn R. O. Storey
Quartier Luchene
84160 Curcuron, France

Commission J
Prof. A. Russell Taylor
Dept. of Physics & Astronomy, University of Calgary,
2500 University Dr NW, Calgary, AB, T2N 1N4 Canada

Commission K
Prof. Russel J. Reitter
The University of Texas Health Science Center at San Antonio
San Antonio, Texas 78284-7762 USA

Inter-commission Working Group
Prof. Paul DeLogne[1] and Dr. Willem Baan[2]
[1]*Université catholique de Louvain, Belgium*
[2]*Netherlands Foundation for Research in Astronomy*
Westerbork Observatory, The Netherlands

THE PAST, PRESENT AND FUTURE OF SATELLITE COMMUNICATIONS

John V. Evans

1. Introduction

The world's first commercial communications satellite "Early Bird," (later renamed INTELSAT I) was placed by NASA into a geosynchronous orbit above the Atlantic in 1965. Providing 240 voice circuits between the United States and Europe, this single satellite had a capacity almost equal to that of all of the RF/coaxial cables laid under the Atlantic up to that time. It could, moreover, be used for relaying television, which the undersea cables of that era could not. Since 1965, several hundred commercial communications satellites have been launched to provide both domestic and international communications. Used initially for completing long-distance (e.g., overseas) telephone circuits, satellites provided a means of establishing a truly global telecommunications system in which small countries for the first time had direct access to the major cities of the world. With the advent of fiber optic cables, the fraction of overseas telephone traffic carried on satellites has declined, although it still amounts to about 30 percent. Satellites today are the prime means of distributing television pictures around the globe, either to cable head ends or directly to subscribers. Satellites are also increasingly being used to provide private data networks for companies (such as banks) that have widely distributed offices. This service usually involves a central large (hub) station linked to many very small aperture terminals (VSATs).

Early development of satellite communications is traced in Section 2, and Section 3 describes current efforts to create global satellite systems providing personal communications (voice, fax, and paging) services to users equipped with cellphone-like terminal. Section 4 presents a discussion of plans announced by large (mostly U.S.) corporations to establish satellite systems to deliver multimedia services (such as Internet access, video streaming, telemedicine, and teleeducation) at high data rates to offices and homes equipped with small fixed terminals.

2. The Past

2.1 INTELSAT

Bell Telephone Laboratories built the first active repeater satellite (*Telstar I*) that received and re-transmitted simultaneously. This satellite, operating at 4 and 6 GHz, was placed into a 1,000 by 6,000-km elliptical orbit in 1962. Later the same year, the RCA-built *Relay* satellite operating at 4.2 and 1.7 GHz was placed into a similar orbit. As a result of these successful demonstrations, the U.S. Congress passed the Satellite Communications Act in 1962. This created a new private company (COMSAT) charged with exploiting satellite communications technology with the goal of improving the world's communications. In 1964, with the assistance of the U.S. State Department, an international organization called INTELSAT was created to operate such a global satellite telecommunications system. COMSAT was named U.S. Signatory to INTELSAT, as well as technical manager. This latter role ended in 1978.

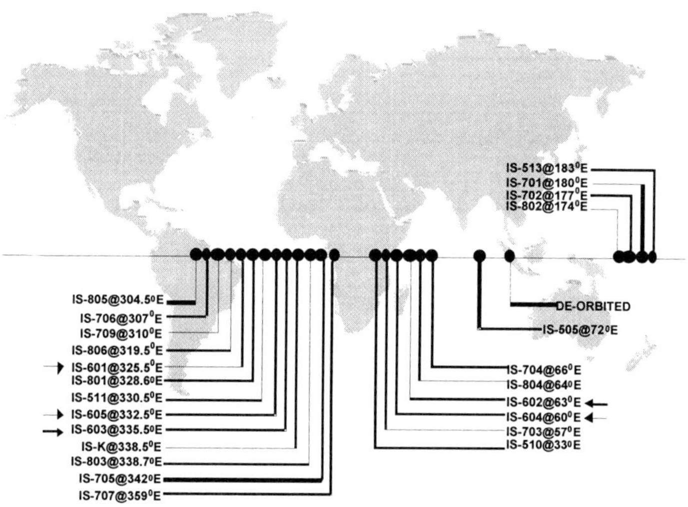

Fig. 1 Distribution of the INTELSAT satellites (circa 1998).

Figure 1 depicts the arrangement of satellites in the INTELSAT system. Satellites are placed in geostationary orbits over the major oceans. Any pair of earth stations in view of the same satellite can, in principle, establish a link, and there are now several thousand such links in the INTELSAT system. Initially, service was provided at C-band (roughly 6 GHz for the uplink frequency and 4 GHz for the downlink); to achieve added capacity, repeaters were later incorporated in the Ku-band (roughly 14-GHz uplink and 12-GHz downlink). Table 1 lists the wavelengths for bands that are available internationally (i.e., as agreed upon at the ITU) for commercial satellite communications, and which are referred to herein.

Table 1 *Wavelengths for commercial communications satellites.*

Band	**UHF**	**L-band**	**S-band (up/down)**	**C-band (up/down)**	**Ku-band (up/down)**	**Ka-band (up/down)**	**Q/V-band (up/down)**
Wavelength (cm)	~75.0	~18.6	~15.1/~13.7	~4.0/~7.5	~2.5/~2.1	~1.6/~1.0	~0.8/~0.6

An indication of the growth of the INTELSAT system can be seen by inspecting the capabilities of the individual satellites employed (Table 2). From INTELSAT I, which could handle 240 circuits, to the INTELSAT IX satellites now being built capable of handling approximately 40,000 circuits, we see a large increase in design life, mass, primary power, number of transponders, etc. The technical developments necessary to achieve the added capacity indicated by Table 2 are summarized below. Since INTELSAT was the driver for satellites of ever-greater capacity, the capabilities indicated in Table 2 were essentially "state of the art," and similar satellites were ordered from the principal manufacturers (initially Hughes, Ford Aerospace, and RCA Astro) by other systems that were inaugurated later. These included regional systems (Eutelsat and Arabsat) and domestic systems for countries such as Indonesia (Palapa), Mexico (Morelos), and Australia (Aussat).

2.2 Satellite Technology

2.2.1 System Concept

A communications satellite consists of a number of active repeaters which receive signals broadcast from one or more earth stations on the ground, amplify them and retransmit them at a new (usually lower) frequency to one or more receiving earth stations. The change in signal frequency is necessary because the amplification that must be employed to overcome path loss far exceeds the amount of isolation that could be achieved between separate transmit and receive antennas on board the satellite if these were at the same frequency.

Links can be established by any pair of earth stations that agree to receive the signals transmitted by each other (after appropriate frequency translation in the satellite). Since a geostationary satellite is visible from a large portion of the earth's surface, a large number of links can be established. The stations may in fact transmit on more than a single frequency and/or polarization, depending on the amount of traffic they carry. The most efficient use of the satellite transponder capacity is achieved by combining as many voice circuits as possible onto a single RF carrier. This is accomplished at the earth station either by assigning the voice circuits to adjacent 4 - kHz frequency channels at baseband before modulating the RF carrier, or, in the case of digital voice channels, by time-division multiplexing the bit streams into a single high-rate stream. While analog circuit combining was initially employed, the most telephone circuits have since been converted to digital operation. Initially it was the practice to keep the satellites as simple as possible and place the burden of completing the radio link on the earth station. This lead to a requirement for large earth station antennas (up to 100 feet in diameter), as shown in Figure 2. As satellite technology improved, the need for such antennas receded and today few, if any, larger than 10 meters in diameter are required.

Table 2. Details of the INTELSAT series of satellites.

Description	INTELSAT I	INTELSAT II	INTELSAT III	INTELSAT IV	INTELSAT IV a	INTELSAT V	INTELSAT V-A	INTELSAT VI	INTELSAT VII	INTELSAT VII-A	INTELSAT VIII	INTELSAT VII-A	INTELSAT IX
Number Placed in Orbit	1	3	5	7	6	8	3	5	6	3	4		TBD
Year of First Launch	1965	1967	1968	1971	1975	1980	1985	1989	1993	1995	1997	1998	2000
Number of Transponders C-Band Ku-Band	1	2	2	12	20	21 6	26 6	38 10	26 10	26 14	38 6	28 3	44 12
Maximum Marketable Capacity (in equivalent 36-MHz units)	C-band 2	C-band 8	C-band 12	C-band 12	C-band 20	C-band 37 Ku-band 12	C-band 42 Ku-band 12	C-band 64 Ku-band 24	C-Band 42 Ku-Band 20	C-band 42 Ku-band 28	C-band 64 Ku-band 12	C-band 36 Ku-band 6	C-band 76 Ku-band 20
C-Band Beam Coverages	Omni (N. hemisphere)	Omni (N. hemisphere)	Global	Global & Spot Beams	Global & Spot Beams	2 Hemi, 2 Zone and Global A	2 Hemi, 2 Zone, Global A and B, C-Spot A and B	2 Hemi, 4 Zone, Global A and B	2 Hemi, 4 Zone, Global A and B, C-Spot A and B	2 Hemi, 4 Zone, Global A, Global B, C-Spot A and B	2 Hemi, 4 Zone, Global A and B	Landmass Hemi A and Hemi B	2 Hemi, 5 Zone, Global A and B
Ku-Band Beam Coverages	None	None	None	None	None	West Spot and East Spot	West Spot and East Spot	West Spot and East Spot	Spot 1, Spot 2, Spot 3 and Enhanced Spot 2/2A	Spot 1/1X, Spot 2/2X, Spot 3 and Enhanced Spot 2/2A	Spot 1 and Spot 2	Spot 1	Spot 1 and Spot 2
Extent of Frequency Reuse/ Hemi/Zone	None	None	None	None	2-fold	4-fold (except in Cha 9)	4-fold	6-fold, with a 4-fold option on a Channel-by-Channel	4-fold with enhanced Zone connectivity	4-fold with enhanced Zone connectivity	6-fold, with a 5-fold option for the POR	2-fold in Hemi only	6-fold except in Channels (1-2) and (3-4) which is 7-fold
Prime Power W	40	75	120	400	500	1,350	1,350	2,300	4,000	5,000	5,000	5,400	8,000
Mass in Orbit (kg)	38	86	152	700	796	1,200	1,200	2,240	3,700	4,400	3,420	3,400	4,000
Design Life (yr)	1.5	3	5	7	7	7+	10+	10+	10+	10+	10+	10+	13
Launch Vehicle(s)	Thor-Delta	Thor-Delta	Thor-Delta Atlas Centaur	Atlas-Centaur	Atlas Centaur Ariane 2	Atlas Centaur Ariane 4	Atlas Centaur Ariane 4	Ariane 4 Titan	Atlas II Ariane 4	Ariane 4	Ariane 4	Atlas II	Proton Ariane 5

Fig. 2 Earth station antennas employed by COMSAT to communicate with the Atlantic satellites. These antennas are located at Etam, West Virginia, and are 105, 45, and 90 feet in diameter (as one approaches the camera). The smaller antenna operates at Ku-band, while the two larger antennas are C-band.

2.2.2 Communications payload design.

The satellite's communications payload receives the signals transmitted from the ground, amplifies them, and retransmits them (usually at a new, lower frequency). Figure 3 sketches the main elements of a payload for a Ku-band domestic satellite.

For a geostationary satellite overhead, the distance from the earth station is 35,860 km and for a satellite on the horizon the distance is 41,750 km. Signals received at a satellite from earth station antennas for these two cases experience attenuation of 162 dB and 163 dB, respectively. To overcome this, high-power transmitters (up to 12-kW) coupled to high-gain (50 to 60 dB) antennas are employed at the earth stations (Figure 2). This permits the satellite transponder to receive signals that are considerably stronger than the unwanted Johnson noise in the receiver. Typical early satellite transponders operated with an output power capability in the range 5 to 20 W. Thus, the satellite- to-earth link usually operated within lower margins than the earth-to-satellite link. More recently, however, transponder powers have been raised to as high as 120 W to permit reception for direct-to-the home TV with very small antennas (e.g., 66 cm in diameter).

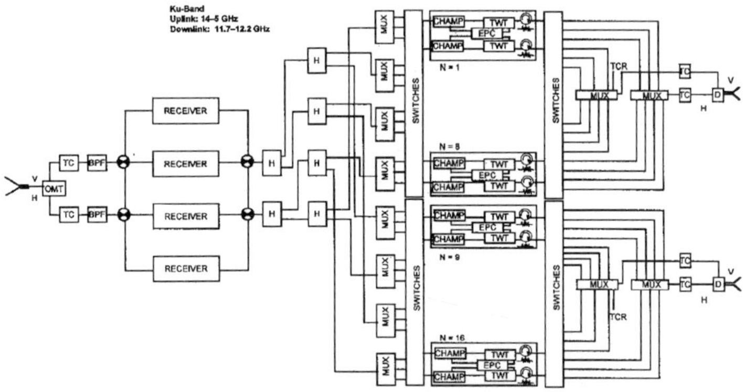

Fig. 3 Typical arrangement of payload components in a satellite. Horizontal and vertical polarized signals (H and V) are selected via an ortho mode transducer (OMT) and amplified by receivers (each of which has a spare). Signals are separated into channels via input multiplexers (MUX) and recombined by output mulitiplexers. Each channel has its own channel amplifier and travelling wave tube (TWT). The hybrids (H) and switches allow for routing around failed TWTs.

2.2.3 Channelization

The first attempt to increase communications capacity was made by increasing the bandwidth of the onboard transponders to occupy an increasing fraction of the assigned frequency band. The available satellite transmitter power then had to be shared among a larger number of RF carriers. Careful control of power levels that radiated from each earth station was also required to prevent any one carrier from capturing an undue share of the power. The increased interference between the carriers resulting from the mixing that occurs in the final amplifier, however, rendered this approach unsatisfactory; it was then necessary to employ several transponders, each designed to operate over only a portion of the assigned band. This change created a need for microwave filters with little attenuation in their passbands and steep skirts at the edges of their bands. This response was achieved by using elliptic function filters in which a series of microwave cavities are cascaded together. To achieve the proper response, it is necessary to couple energy from the first to the last cavity independently of any intervening cavities. COMSAT Laboratories developed a means for accomplishing this by exciting each cavity with two orthogonal modes and coupling these independently from cavity to cavity. This technology was critical to minimizing the guard bands needed between transponders. Figure 4 shows the response achieved using this in a modern microwave multiplexer unit. Most current satellite systems employ transponder bandwidths in the range of 36 to 54 MHz, although INTELSAT employs transponders with 72 MHz bandwidth.

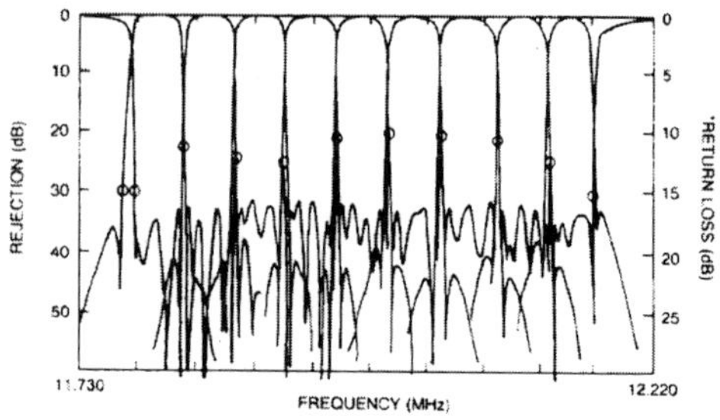

Fig. 4 The response of a set of microwave cavity filters coupled to a waveguide run to provide a means of connecting the outputs (i.e., multiplexing) a set of transponders in a way that minimizes their mutual interference and/or interaction. (Courtesy of Ford Aerospace Corporation.).

2.2.4 Antenna Technology

A second technological advance has been to use the assigned band (say 500 MHz) available at C-band several times on the same satellite. This can be accomplished by using antenna beams that illuminate separate service areas and achieve excellent isolation (i.e., low sidelobe levels) from one another. INTELSAT pioneered the application frequency reuse and extended the concept by using overlapping beams in which isolation is accomplished through orthogonal circular polarization for the two beams.

Beams are "shaped" to cover an irregular service area or footprint. To create the shape, long-focal-length reflectors (which can be illuminated by horns offset from the axis without introducing large coma lobes) are used. Each of the horns in the array must be fed by an appropriate fraction of transmitted power at the proper phase to achieve the desired flat-topped, sharp-sided response. This is accomplished using a distribution network consisting of an arrangement of coaxial power splitters and transmission lines. For domestic or regional satellites, it is simpler to achieve beam shaping with a single feed horn and distorting the reflector to achieve the desired contour.

2.2.5 Compression and Circuit Multiplication

In the terrestrial telephone network, digital transmission of voice signals was introduced to improve quality and reliability, and to reduce the need for maintenance. Typically, the analog voice signal arriving at the first switch is sampled 8,000 times per second and assigned an 8 (or 7) bit digital word that defines its amplitude and polarity. The resulting 64 (or 56) kb/s digital bit stream is then sent over the long-distance portion of the circuit. In the INTELSAT system, coders were introduced, employing adaptive pulse-code modulation which reduced this rate to 32 kb/s. Speech interpolation equipment was also introduced which could seize inactive circuits for active speakers (in a typical voice conversation, only 40 percent of circuit capacity is in use at a given time). These two developments allowed four voice channels to be created from a single 64 kb/s circuit. Standards have since been developed that achieve toll-quality voice using digital encoders operating at low as 16 and 8 kb/s. These coders capture the setting of the vocal tract and the manner in which it is being excited, and transmit this information rather than the original analog signal.

Television signals are now also transmitted digitally, following compression to remove redundancy. Standards (such as MPEG2) define how this is carried out and permit the delivery of NTSC television with good (studio) quality at rates as low as 6 to 8 Mb/s.

2.3 Mobile Communications

2.3.1 Inmarsat

Commercial use of satellites for mobile communications began with the 1976 COMSAT/ Marisat system. Ultrahigh frequency (UHF) and L-band (Table 1) satellites were positioned over the Atlantic and Pacific oceans. The U.S. Navy used the UHF capacity, while the L-band capacity was intended for a commercial service for mariners. Shipboard terminals consisted of an above-deck, 1-m-diameter antenna gimballed to remain locked on the satellite and protected from the elements by a radome, while below deck would be a telephone handset, fax, and/or teleprinter terminal (Fig. 5). Feeder links were provided at C-band via "coast" earth stations at Southbury, Connecticut, and Santa Paula, California, which were connected to the public switched network.

Fig. 5 *Shipboard Inmarsat terminal radome-enclosed antenna.*

The system became global with the addition of a third satellite over the Indian Ocean and, in 1979, was turned over to the newly formed Inmarsat international organization to manage. Inmarsat, which is headquartered in London, is a treaty organization with 81 members.

In 1991, Inmarsat deployed four Inmarsat-2 satellites constructed to its own specifications. Two, known as Atlantic East and Atlantic West (see Fig. 6), were placed over the Atlantic to handle the large amount of traffic in that region. These satellites, like their predecessors, employed a single L-band global beam for servicing mobile users. This, combined with the limited band of frequencies (28 MHz) available for this service at L-band, restricted the number of simultaneous users. In response to traffic growth, two approaches have been taken to increase the availability of a circuit. One involves terminal design and the other entails frequency reuse.

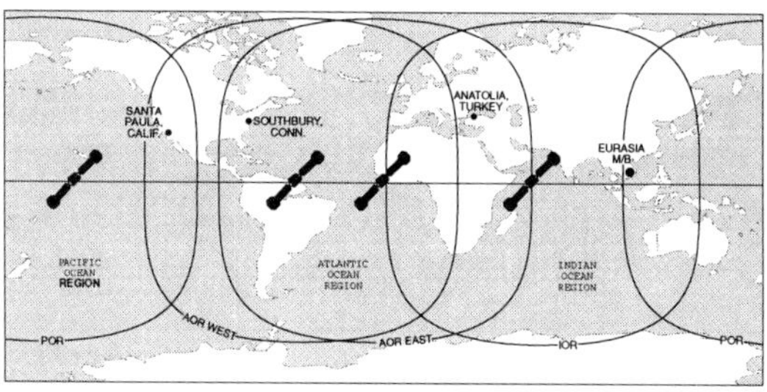

Fig. 6 *Coverage patterns of the Inmarsat-2 satellites and locations of the coast earth stations operated by COMSAT.*

The original (Inmarsat-A) terminal employed analog (FM) modulation of the L-band signals and 50-kHz channel spacing. About 20,000 such terminals are installed on vessels around the globe. To satisfy the demand for smaller terminals and to create more channels, Inmarsat introduced two new services known as Inmarsat-B, and Inmarsat-M (described in Table 3) that operate in narrower channels than Inmarsat-A. In 1996, the first two of five Inmarsat-3 satellites were launched. These satellites reuse the authorized frequencies in up to five "spot" beams, which can be selected for their coverage of land. These beams provide higher EIRP, making it possible for still-smaller terminals to operate within the system, and a "mini-M" terminal (Table 3) the size of a laptop computer was made available in January 1977. It sells for about $3,000, with a usage charge (fully terminated) of $3.00 per minute.

2.3.2 Other Land-Mobile Systems

Additional land-mobile satellite systems have been established by some countries. These all operate either at L- or S-band (Table 1), since longer wavelengths provide a larger collecting area for an omnidirectional antenna.

Table 3 *Standard services of the Inmarsat system.*

Features	Inmarsat-A	Inmarsat-B	Inmarsat-M	Mini-M
Mobile to/from Satellite, L-Band (GHz)	1.6/1.5	1.6/1.5	1.6/1.5	1.6/1.5
Fixed to/from Satellite, C-Band (GHz)	6/4	6/4	6/4	6/4
Minimum Channel Spacing (kHz)	50	20	10	5
Users per Carrier	1	1	1	1
Access Method Modulation	FDMA Companded FM	FDMA O-QPSK	FDMA O-QPSK	FDMA O-QPSK
FEC Coding Rate	N/A	3/4	3/4 (for subband bits)	2/3 (for suband bits)
Speech Coding Algorithm	N/A	Adaptive predictive coding (APC)	Improved multi-band excitation (IMBE)	Advanced multiband excitation (AMBE)
Voice Coding Rate (kb/s)	N/A	16	6.4 (incl. error correction bits)	4.8 (incl. correction bits)

In North America, the U.S. and Canada operate two satellites in what is called the MSAT system. This system can provide service to vehicles (cars, trucks, etc.) and is similar in capability to the Inmarsat-M system (Table 3) noted above. Australia has implemented an L-band system known as OPTUS, and the Japanese have an S-band system called N-Star. None of these systems are powerful enough to permit operation with small handheld terminals, and new systems proposed to provide this capability are described in Section 3.

3. The Present

Satellite technology is moving toward providing services to individual customers. The earliest advances in this direction have been satellite systems that deliver television direct to the home (DTH). These systems have benefited from the increased prime power that can now be generated on board satellites (currently 10 kW or more), allowing the operation of a sizable number (≥16) high-power (≥100 W) Ku-band transponders. Another contributor to the success of these systems, at least in the U. S., has been the development of powerful digital compression schemes that permit as many as 10 TV channels to be transmitted via a single transponder, and the picture to be recovered in a set-top box containing special purpose, very-large-scale digital integrated (VLSI) circuits manufactured at low cost.

Next to be conceived and developed were satellite systems that provide cellular-like voice service to mobile users equipped with small handheld terminals. Three such systems with global or nearly global coverage are under construction, and two more are seeking financing. Market studies have identified four potential markets: international business travelers (primarily business travelers from the developed world visiting less-developed countries), national roamers (primarily business travelers needing mobile communications in their own countries, but who travel beyond the reach of terrestrial cellular systems), national rural fixed service (an extension of the national fixed services to regions where they are presently unobtainable), and government agencies (law enforcement, fire, public safety, and other services). The designs of the global systems discussed represent different assumptions concerning the business to be attracted from these four segments. The subsections that follow discuss the five proposed systems that are proceeding. The Iridium, Globalstar, and ICO systems appear to have the best chance of being fielded, while financing for the others remains to be completed. Table 4 summarizes the parameters of these five systems.

Table 4 Proposed new global satellite PCS systems.

Parameter	Iridium	Globalstar	ICO-Global	Ellipso	Aires (ECCO)
Company	**Motorola**	**Loral/ Qualcomm**	**ICO-Global**	**Mobile Comm. Holdings, Inc.**	**Constellation Comm., Inc.**
No. of Active Satellites	66	48	10	17	46
Orbit Planes	6 circular polar (86.5°)	6 circular inclined (52°)	2 circular inclined (45°)	2 elliptical inclined (116.6°) - 1 circular equatorial	7 circular inclined - 1 circular equatorial
Orbit Altitude (km)	780	1,414	10,355	N.A. 8,060 equatorial	2,000 equatorial
Satellites per Orbit Plane	11	8	5	5 in each elliptical orbit - 7 in equatorial orbit	5 in each inclined orbit - 11 in equatorial orbit
Beams per Satellite	48	16	163	61	1
Reported Cost ($B)	4.7	2.5	4.6	0.56	1.15

3.1 Iridium

Technically, the Iridium system—proposed by Motorola and constructed by that company in conjunction with Lockheed Martin, Raytheon, and other contractors—is the most ambitious of the five to be discussed. It has been purchased and is being operated by a separate company (Iridium, Inc.), which has secured investment from many parts of the world. The design employs 66 active satellites placed in circular polar orbits at 780-km altitude. The satellites are deployed into six equispaced polar orbital planes, with 11 satellites separated equally around each orbit. Satellites in adjacent planes are staggered with respect to each other to maximize their coverage at the equator, where a user may have to access a satellite that is as low as 10° above the horizon.

Users employ small handsets operating in frequency-division-multiplexed/time-division multiple access (FDM/TDMA) fashion to access the satellite at L-band. Eight users share 45-ms transmit and 45-ms receive frames in channels with a bandwidth of 31.5 kHz, spaced 41.67 kHz. The users are synchronized so that they transmit and receive in the same time windows alternately. This approach is necessary because the three phased-array antennas onboard the satellite

(Figure 7a) are used for both transmitting and receiving. Figure 7b shows the 48 spot beam that these phased arrays form, projected onto the earth at the equator.

The Iridium system is unique in that it is designed to achieve essentially global coverage with only a small number of gateway earth stations that connect to the public switched network (in all, 11 will be built). To this end, the satellites are designed to route traffic from satellite to satellite. Each satellite employs onboard processing to demodulate each arriving TDMA burst, determine how to route it, and then retransmit it to its next destination. This can be to the ground if a gateway earth station is in view, or to one of the four nearest satellites: the one ahead or behind in the same orbital plane, or the nearest in either orbital plane to the east or west. These satellite cross-links operate at 23 GHz, while the links to the gateway earth stations are at 20 GHz. Each satellite is capable of handling as many as 1,100 simultaneous calls. The Iridium satellites are station-kept using onboard propulsion in order to overcome atmospheric drag and to have sufficient fuel for an 8-year life.

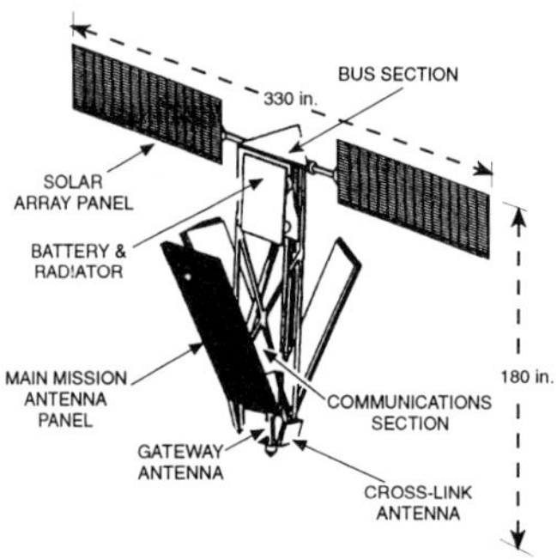

Fig. 7a. Sketch of the Iridium satellite showing the three phased-array antennas that are canti-levered off the three sides of the spacecraft bus.

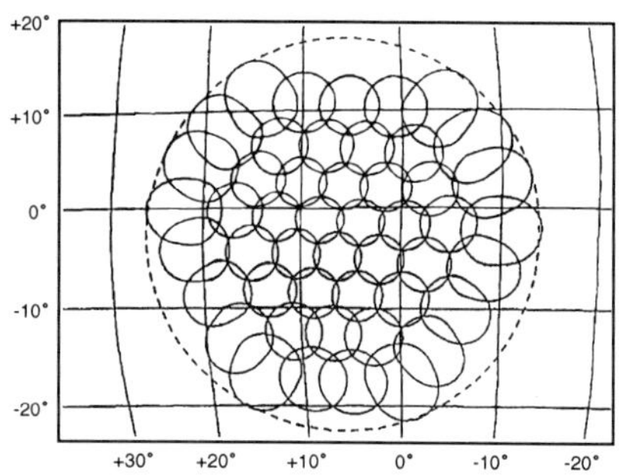

Fig. 7b. The 48 spot beams produced by the phased-array antennas shown in Figure 7a projected onto the earth at the equator.

3.2 Globalstar

The Globalstar system is being purchased by a limited partnership in which Loral and Qualcomm of the United States are principal partners. The satellites are currently being assembled in Italy (by Alenia Spazio), while Qualcomm has developed much of the ground segment. Unlike the Iridium system, which offers true global service, Globalstar's business plan calls for launching the space segment and franchising its use to partners in different countries. More than 90 such relationships have already been established.

The Globalstar system will employ 48 satellites in eight planes of six satellites each. The satellite orbits are circular, at 1,414 km and 52° inclination. The use of an inclined orbit concentrates the available satellite capacity at lower latitudes where the largest populations exist; little or no coverage is provided beyond ± 70° latitude. As can be seen in Figure 8, two or more satellites are visible (above 10° elevation) between 25° and 50° latitude at all times, and from the equator to 60° latitude 80 percent of the time. Like the Iridium satellites, the Globalstar spacecraft are three-axis-stabilized, with a mission life of 7.5 years (minimum).

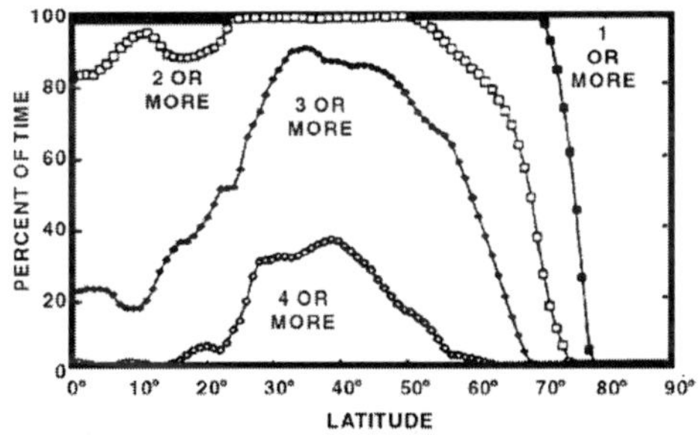

Fig. 8 *Average percentage of time that different numbers of satellites of the Globalstar constellation can be seen (above 10° elevation), as a function of latitude.*

Because the Globalstar system does not employ satellite cross-links, a subscriber gains access to the system only when a satellite in view can also be seen by a gateway earth station. Thus, service areas are within 1,000 miles of each gateway earth station. Truly global coverage would require building more than 200 earth stations, which is unlikely to happen. Thus, Globalstar is more likely to serve national roamers than international business travelers.

In contrast to Iridium, each Globalstar satellite covers a comparable area of the earth's surface with only 16 spot beams. This, together with sharing of the receive channels onboard the satellite by more users, reduces the available link margins to about 3 to 6 dB, although for a small number of users this can be increased to 11 dB. Access to and from the satellite is at L- and S-band, respectively, utilizing code-division multiple access (CDMA) in channels that are 1.25

MHz in bandwidth. Voice is encoded at 1 to 9 kb/s, depending on speaker activity. The satellites use simple "bent pipe" transponders, with the feeder links at C-band.

Since all 16 beams of all of the 48 satellites are always active, each satellite in view will pick up a subscriber's signal and retransmit it in its feeder link. By tracking all the satellites in view of a given gateway earth station, two channels can be open to the subscriber. The channel with the stronger signal can connect to the public switched network. This should mitigate blocking by buildings and provide a "soft" handoff from satellite to satellite. Globalstar hopes to have it system in operation by late 1999. Eight of the 48 satellites have been launched so far (Dec. 1998).

3.3 ICO

ICO-Global is a spin-off from Inmarsat, which owns 15 percent of the corporation. The remainder is presently owned by Inmarsat signatories, and by Hughes (the builder of the spacecraft) and TRW. ICO-Global has chosen an intermediate circular orbit for its system (10,355-km altitude), with 10 satellites arranged 5 in each of two inclined circular orbits. The inclination of the orbits is 45°—making it the lowest of the three systems. Because each satellite must serve a large portion of the earth's surface, a total of 163 spot beams will be used. Routing signals to the correct spot beam then becomes difficult with analog [e.g., surface acoustic wave (SAW)] filters, and will instead be done with a digital filter bank [which performs a fast Fourier transform (FFT) on signals arriving from the gateway earth station]. To access a given spot beam, gateway earth stations must transmit at a particular frequency.

A true TDMA scheme has been adopted for the service links, with six subscribers multiplexed into channels 25.2 kHz in width at a bit rate of 36 kb/s. However, a soft handoff (e.g., beam to beam) is not automatic, and it is difficult to exploit dual-satellite visibility.

Hughes Space and Communications Division is building the ICO satellites, and a team consisting of NEC, Ericsson, and Hughes Network Systems Division is building the ground segment. ICO hopes to have its system in operation by year 2000.

3.4 Ellipso and Aires

The Ellipso and Aries systems are both currently believed to be attempting to secure investors. Both aim to be low-cost entrants to the market. These systems would initially deploy satellites into circular orbits above the equator, and later would add satellites in additional inclined (elliptical in the case of Ellipso) orbits to cover higher latitudes.

4. The Future

4.1 Introduction

Fixed-satellite services were first offered by INTELSAT at C-band frequencies (refer to Table 1). The earth terminal antennas in the INTELSAT system were large, but have decreased in size with the launch of more powerful satellites. As other satellites providing domestic or regional fixed-satellite services were deployed, agreement had to be reached on their separation along the orbital arc. To avoid mutual interference between systems caused by an earth station with a small antenna illuminating satellites on either side of the wanted one, it was necessary to agree on the spacing between satellites using the same frequency band; this is now set at 2°.

Absence of suitable C-band orbital slots drove the construction of satellites operating at Ku-band, and most INTELSAT satellites are built with transponders operating in both bands. DTH TV broadcasting satellites also operate at Ku-band. It is now nearly impossible to secure an orbital location where satellite can be operated at C- or Ku-bands without interfering with its neighbors. This has spurred interest in systems operating at Ka-band (Table 1).

Until recently, interest in this band has been confined to experimental satellites launched by Japan, the United States, and Italy. This is because, unlike at C-band, rain greatly attenuates Ka-band signals (and to some lesser extent, Ku-band signals), making this a difficult band in which to provide satellite services. A group of private U.S. investors proposed a Ka-band satellite system. It would provide a global wideband data distribution capability known as the Calling^{sm} Network, later renamed Teledesic. This system was to employ 840 low-altitude satellites, each of which could relay to its eight nearest neighbors and provide users (who had sufficiently large terminals) access to rates up to 1.2 Gb/s.

Despite the ambitious nature of this proposal, Teledesic organization was successful in lobbying at the World Administration Radio Conference for Ka-band frequency assignments. This caused the FCC to proceed with a Notice of Inquiry offering other applicants the opportunity to seek Ka-band spectrum (and orbital locations). In May 1997, the FCC approved applications for 12 additional Ka band (geostationary) fixed-satellite systems.

Worldwide, there are believed to be over 50 proposed Ka-band projects requiring over 170 geostationary orbit locations. Most appear to be for national or regional systems, and not a great deal has been published about them. Six of the 13 licensed U.S. systems propose to offer global (as distinct from domestic) service, and these are reviewed briefly below.

In filing their FCC applications, companies must make representations to the effect that their systems will benefit the public. In so doing, almost all conceivable services have been cited as being planned. However, rain fading represents a serious problem at Ka-band and may discourage attempts to provide some of the services (such as telephony, teleeducation, or telemedicine). By far the largest market is believed to be for access to the Internet.

4.2 Proposed Multimedia Satellite Systems

Table 5 *Proposed U.S. Ka-band global satellite communications systems.*

Company	System	Orbit	Coverage	# of Satellites	Satellite Capacity (Gb/s)	Intersatellite Link	Onboard Switching	Capital Investment ($B)
Lockheed-Martin	Astrolink	GEO	Global	9	7.7	1 Gb/s	FPS	4.0
Loral	Cyberstar	GEO	Limited Global	3	4.9	1 Gb/s	BBS	1.05
Hughes	Galaxy/ Spaceway	GEO	Global	20	4.4	1 Gb/s	BBS	5.1
GE Americom	GE*Star	GEO	Limited Global	9	4.7	None	BBS	4.0
Morning Star	Morning Star	GEO	Limited Global	4	0.5	None	None	0.82
Teledesic	Teledesic	LEO	Global	840*	13.3*	1 Gb/s*	FPS*	9.0*

FPS: Fast packet switch, BBS: Baseband switch
*Original design numbers.

As noted above, a very large number of Ka-band satellite system proposals have been filed with the ITU in Geneva. In the U.S. alone, the FCC has opened two separate windows for filing for systems operating in this band, and one for systems operating at still higher frequencies (40 and 50 GHz) known as Q- and V-band. Discussion herein is limited to the six global systems that received licenses from the FCC in May 1997 (Table 5).

4.2.1 Astrolink

Lockheed Martin's proposed Astrolink system will have nine geostationary satellites in five orbit locations (Fig. 9a). Each satellite's total capacity is 7.7 Gb/s. By placing two satellites at the same location and operating them with orthogonal polarizations, a total capacity of 15 Gb/s is achieved over the Americas, Europe, and Asia (Fig. 9b). Cross-links between satellites operating at 60 GHz provide a means of routing traffic round the globe; each link has a capacity of 1 Gb/s.

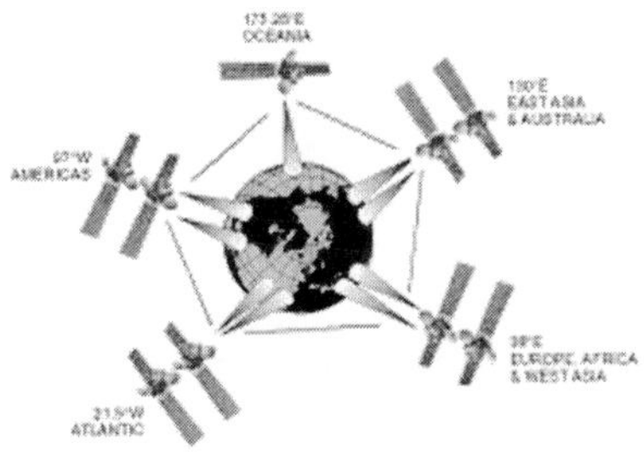

Fig. 9a *The Astrolink System constellation employs nine geostationary satellites placed in five orbit locations. Similar arrangements are planned by Hughes (for their Expressway system), GE Americom (for the GE*Star system) and Morning Star.*

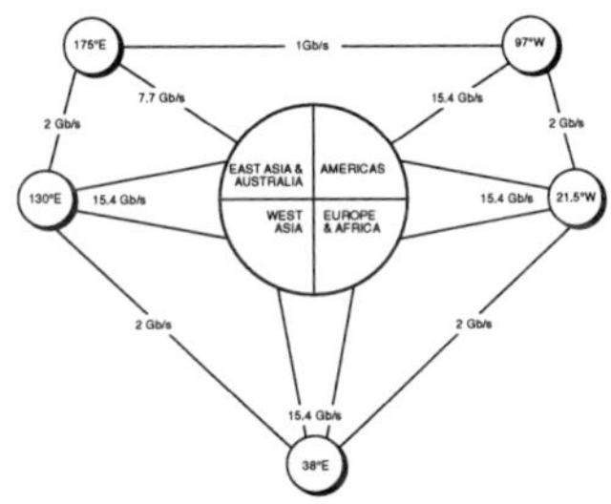

Fig. 9b *Allocation of capacity in the Astrolink system.*

The system would be built in stages, with one satellite initially located in each slot. The satellites will be large and expensive, representing a considerable financial investment.

Lockheed's original proposed design used hopping beams to provide global coverage. By adjusting the dwell time, a match could be achieved to the volume of traffic at each beam location. A subsequent redesign has the beams be fixed prior to launch in order to save weight. Traffic in the beams will be sampled and processed to yield 64 channels, each being 125 MHz wide. The channels are subjected to a frequency-division (de-multiplexing) analysis to select each carrier, which is then demodulated and decoded. A packet switch routes packets to the appropriate downlink channel, where they are multiplexed with other packets destined for the same beam, re-encoded, and modulated onto the carrier.

Astrolink proposes to equip users with terminals employing antennas of 65, 85, and 120 cm in diameter, operating at power levels ranging from 0.25 to 10.0 W, at rates in the range of 64 kb/s to 8.448 Mb/s (with larger antennas and higher powers being required for the higher rates.) These terminals would interconnect with the terrestrial switched network via gateway stations employing 2.4- or 4.5-m antennas with up to 200 W of power.

4.2.2 Cyberstar

Loral's proposed Cyberstar system represents a lower cost, less risky approach to the market. Loral proposes to launch just three geostationary satellites positioned to reach the world's largest population centers and has been awarded slots at 28°E, 105.5°E, and 115°W. The satellites were to employ multiple spot beams interconnected by an on-board processor, but the plans appear have undergone several changes. For a while Loral pursued a simpler design (Phase 1) that entailed no onboard processor, allowing very early entry to the market. In Phase 2, three larger, more powerful satellites were to be launched, each with a total capacity of 10 Gb/s and capable of providing beam-to-beam connectivity via an onboard processor. At last report, Loral appeared to be reevaluating this plan in favor of going directly to the Phase 2 system. For the moment, Loral seems intent on developing an interim service capability (called Cyberlink) with the assets it obtained from AT&T Skynet and Orion.

4.2.3 Galaxy/Spaceway

Hughes' proposed geostationary satellite system called Galaxy/Spaceway would have 21 satellites in 16 orbital locations. The Spaceway™ portion of this system resembles that proposed by Lockheed Martin for Astrolink (Fig. 9a) in that it consists of nine satellites placed in five orbital locations, with intersatellite links between four of the locations.

In the system's Spaceway™ portion, two satellites will occupy each of four orbit locations. By operating each satellite over the allowed 500 MHz of spectrum (but with different polarizations), an equivalent of 1,000 MHz of bandwidth is obtained. Each satellite supports 68 transponders, 64 which occupy 125 MHz each (for the users), and two which occupy 250 MHz (for gateways), thus achieving further frequency reuse. The transponders operate into narrow (59-dBW EIRP) and wide (52.3-dBW EIRP) spot beams that cover the land masses visible to the satellites. The intersatellite links operate at 60 GHz and have a 1 Gb/s data rate. Communications services will be provided at rates of 16 kb/s to 1.544 Mb/s via terminals with antennas in the range of 66 to 200 cm in diameter and uplink transmitters of up to 2 W. Onboard processing of arriving packets is employed, both to route traffic between beams and to merge traffic in a given transponder into a single 92-Mb/s stream.

4.2.4 GE* Star

GE American Communications, Inc. (GE Americom) proposed a system of nine geostationary satellites occupying five orbits (GE*Star). The system's satellites will produce 44 spot beams each for transmitting and receiving and operate in a fourfold frequency reuse pattern. Signals received in these beams will be separated by filters into six 24-MHz sub-bands for high-traffic regions (24 beams) and three 24 MHz sub-bands in low-traffic regions (20 beams). The 204 sub-bands are processed at base-band to recover the digital bit streams, which are routed to a downlink beam. The downlink beams would operate at 40 Mb/s, with an EIRP at beam center of 54 or 51 dBW for the high- and low-traffic regions, respectively.

GE Americom intends to work with manufacturers so that terminals are produced that are compatible with its system. The terminals' proposed antenna sizes are 75 and 150 cm, operating at 1 W achieving rates of 384 kb/s and 1.544 Mb/s, respectively. Rather than award its satellite construction contract to a potential competitor (Hughes, Lockheed Martin or Loral), GE Americom has announced that they will be built by the Harris Corporation.

4.2.5 Morning Star

The Morning Star Satellite Company, L.L.C., of Washington, D.C., has proposed a system of four geostationary satellites designed to serve parts of North America, Europe, and Asia. Like the Hughes' Galaxy proposal, Morning Star proposed using hybrid Ku/Ka-band satellites, but has been authorized to proceed only with the Ka-band portion. As originally proposed, each satellite would employ up to 10 receive spot beams operating at 30 GHz and combine their traffic into a single 20-GHz downlink beam to a gateway earth station. This station would uplink signals at 30

GHz that would be separated and transmitted via the spot beams at 12 GHz (Ku-band). Since the use of the Ku-band was not authorized, this arrangement will have to be modified

The satellites employ simple frequency-translating transponders with no onboard processing. This, together with the absence of frequency reuse, limits their capacity to 0.7 Gb/s. The design is unique among those proposed in that beam-to-beam connectivity can be provided, but requires two transmissions through the satellite (a so-called double hop). This is wasteful of satellite resources, but presumably Morning Star plans mainly to deliver movies and Internet service, both of which would originate at the gateway earth stations.

4.2.6 Teledesic

The Teledesic system has been described in numerous technical papers, but now appears to be undergoing complete redesign. As first conceived, a LEO system was chosen because very large data rates were incompatible with the delay (or latency) encountered with geostationary distances. Also, to mitigate rain fading, each satellite's service area was limited to a cone of ±40° about the nadir (i.e., subsatellite point). This (together with the low altitude chosen) brought the number of satellites needed for global coverage to 840. Each satellite in the system was capable of cross-linking with its eight nearest neighbors and employed phased array antennas to scan "cells" on the ground from which to collect and deliver traffic.

Teledesic has announced that the satellite number is to be reduced to 288 and the orbit raised to 1,400 km. The parameters given in Table 5 are for the old design. Details of the new one are unavailable, and little more can be said about this system. Teledesic probably remains the most advanced of all of the proposed systems, and with 288 satellites is likely to be the most expensive. Given that Motorola has elected to join the Teledesic project and will be technical manager, further changes to the design and/or number of satellites are probable in order to reduce capital outlay and increase the chances of financing the system.

5. Concluding Remarks

Satellite projects are enjoying a period of great interest, spurred by a number of factors. The overall telecommunications market is growing rapidly, due in part to growth in international trade, but also due to reduced prices for many services and the introduction of new services (such as cellular or PCS and Internet access). Deregulation and the opening of overseas markets offer new opportunities for telecommunications companies. Satellites can provide almost "instant" infrastructure, requiring little in the way of the civil works needed to install other systems. In addition, U.S. aerospace companies are seeking new opportunities in the civil sector now that defense orders have declined. Lockheed Martin, Loral, and Motorola appear intent on vertically integrating into the service business, as Hughes has successfully done. Finally, there is a realization that the resources (frequencies and geostationary orbital slots) available for satellite projects are quite limited. The FCC's Ka-band rule-making probably represented one of the last opportunities to lay claim to a limited resource.

Over the long term, communications satellites will be used only when they provide a clear competitive advantage. These situations include broadcasting, distributing the same information (e.g., television) to a large number of subscribers (such as schools or cable head ends), and mobile applications (ships, planes, travelers in remote places, etc.).

If some of the Ka-band systems described here are successful, then satellites may enjoy a role not heretofore theirs—providing "last mile" connections to homes and businesses for broadband data, multimedia, and related services. Success here depends critically on bringing down user terminal costs; this can be achieved only if terminals are mass-produced using specially developed chip sets for all functions. Since it will be particularly difficult to lower the price of terminals requiring tracking antennas, it seems that the LEO systems will have greater difficulty serving the consumer market. Also, timing is critical, since other technologies for wideband Internet access are being pursued in the developed countries where the largest markets are initially to be found.

Satellites will also continue to be used for the foreseeable future to deliver telephony and other public switched services to countries that are not linked by fiber optic cables, and to support private (e.g., VSAT) networks that cover a wide geographic area. However, these roles appear inadequate to justify all of the current interest in new satellite systems.

ENGINEERING ISSUES IN SPACE WEATHER

Louis J. Lanzerotti, David J. Thomson and Carol G. Maclennan

1. Introduction and Historical Notes

The earliest evidence for the impact of solar-terrestrial phenomena on technical systems appeared in the first half of the 19th century. At that time, battery-powered long wires were beginning to be installed in the U.S. and Europe to serve the demands of the fledgling telegraph industry. Unexpected changes in the powering potentials were often observed by the operators of the early systems, and the reasons for these disturbances were a puzzle to them. Accompanying the appearance of a "brilliant aurora" in March 1847, and persisting "during the whole time of its remaining visible", "strong alternating deflections" were observed in all of the galvanometers that were being used on the telegraph systems of the Midland Railroad in England [*Barlow, 1849*]. *W. H. Barlow*, an engineer for the railroad, asked for and received permission to use spare wires of several of the telegraph lines in order to conduct a dedicated investigation of the "spontaneous deflections of the telegraph needles ...." He systematically carried out this investigation over an approximately two week interval (excluding, it seems, measurements on the weekends) in May 1847 and tabulated the results in a thorough published report [*Barlow, 1849*].

Barlow's experimental investigation was conducted several months prior to a very large auroral display that occurred on November 17, 1847. Thus, he missed making direct measurements of an event which affected not only the telegraphs in England, but in Italy caused the "soft iron armatures employed in the electric telegraph between Florence and Pisa [to] remain attached to their electromagnets ... without the currents in the battery being set in action" (as reported from *M. Matteucci in Prescott [1866*]).

Any relationship of the sun to the appearances of the "spontaneous" currents that were measured on the telegraph wires was not "clarified" until the occurrence of the large white light solar flare of late August 1859 (*Carrington [1863*]; the first such flare ever recorded by astronomers). Within a day following this flare, large geomagnetic disturbances and wide-spread aurora were observed over the Earth, including at low geomagnetic latitudes in Hawaii and Rome. Telegraph systems that were in operation in the eastern United States could at times be operated without the use of batteries [*Prescott, 1866*]. The electrical currents in the wires were noted to fluctuate in association with changes in the intensity of the aurora. Nevertheless, astronomers still questioned the connection of the observed solar flare to the subsequent geomagnetic disturbances on Earth (including the disruptions of the telegraphs). Such skepticism as to the origins of the geomagnetic disturbances measured on Earth persisted throughout much of the remainder of the 19th century.

Increased development of long-distance wire-line telegraph communications in the late 19th and early 20th centuries required better characterization of extraneous currents in the wires of the systems in order to devise engineering solutions. Significant attention was devoted to the topic of Earth potentials by the electrical engineering community in those times (see review in *Lanzerotti and Gregori [1986]*). As late as the 1930's, research papers were devoted to determining the "global" distributions of these currents [*Gish, 1936*]. This seems to have followed from the observation that the daily variation (Sq) of the geomagnetic field [*Schuster, 1908*] could be reasonably characterized, and that some theoretical understanding of Sq, in the context of ionosphere currents, had begun to be obtained. What was not well recognized was the extreme complexity of the interplay between the high variability, in space and time, of the ionosphere currents and the fixed geological structures within the Earth.

Establishing an "Earth current" model that can predict currents along a given route of a long conductor (e.g., a communications cable, an electrical grid, or a pipeline) during a magnetic disturbance was then, and still remains, a difficult problem. It is even more difficult to produce a model that can be used to predict such currents following a specific solar event. Too much remains unknown about the physical processes that are involved in all of the transfer functions between a solar occurrence and the measured effects on Earth.

The measurements tabulated by *Barlow* can be used for further analysis. Plotted in Figure 1 are the hourly mean deflections observed in the galvanometers at Derby (England) that registered voltages on the unpowered wires running from that location to Rugby (England; dashed line) and to Birmingham (England; solid line) from 17 to 30 May 1847. A diurnal variation in the mean recorded galvanometer readings (which recorded the Earth potential variations between the two sets of stations) is clearly evident in the way these data are presented [*Lanzerotti, 1994*]. These may be the earliest records that contain information on the daily variations of such Earth currents.

The beginnings of wireless communications followed the electromagnetic wave research of Hertz in the 1880's and the work of *Marconi* in wireless telegraphy. Long distance telegraph "signals were received by *Mr. Marconi* at St. John's Newfoundland, from Poldhu Station, Cornwall, a distance of 1,800 miles across the Atlantic on December 12th and 13th, 1901" [Year Book, 1914]. *Marconi* received the 1909 Nobel Prize in Physics for this achievement. This work led directly to the first possibility for long distance, trans-oceanic voice communications [e.g., *Marconi, 1928*].

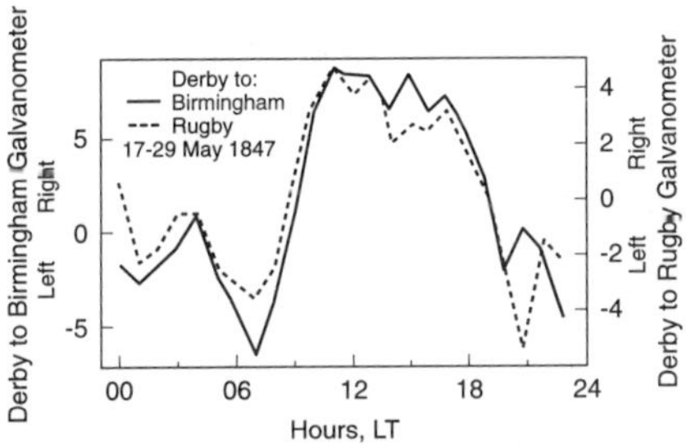

Fig. 1 *Hourly mean galvanometer deflections recorded on telegraph cables from Derby to Birmingham (solid line) and to Rugby (dashed line) in May 1847 (data taken from Barlow [1849] and analyzed).*

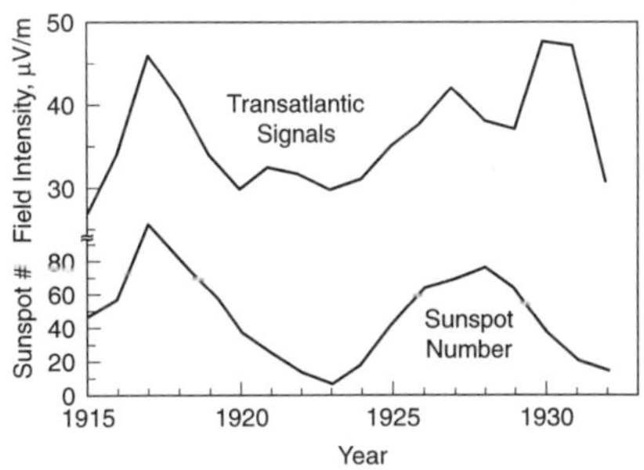

Fig. 2 *Yearly average daylight trans-Atlantic transmission signal strengths and monthly average sunspot number per year for the interval 1915-1932.*

The origins of URSI date from these earliest days of radio investigations, both for science and for practical purposes [e.g., *Senior, 1998*]. The advancement in communications provided by radio stimulated a significant need to better understand the medium that was critical in bending the radio waves around the curvature of the Earth – the ionosphere [*Appleton and Barnett, 1925*; *Breit and Tuve, 1925*]. The same ionosphere currents that could produce "spontaneous" Earth currents could also affect the reception and the fidelity of the transmitted, long-distance wireless signals (e.g., *Petersen, [1928]). Marconi [1928*] noted, with respect to wireless communications, that "... times of bad fading practically always coincide with the appearance of large sun-spots and intense aurora-boreali usually accompanied by magnetic storms ...." He also wrote that these were the "... same periods when cables and land lines experience difficulties or are thrown out of action." Such concerns have persisted throughout the twentieth century [e.g., *Gassmann, 1963*], with considerable present-day research on the ionosphere being motivated by engineering considerations similar to those encountered in the early days of trans-ocean wireless communications.

Figure 2 contains a plot (upper trace) of yearly average daylight trans-Atlantic transmission signal strengths for the years 1915 – 1932 (Figure reproduced from *Fagen [1975*], which contains historical notes on early wireless research in the Bell System). The intensities in the signal strength curve were derived by averaging the values from some 10 European stations that were broadcasting in the ~15 and 23 kHz frequency range, after reducing them to a common base (the signal from Nauen, Germany was used as the base [*Fagen, 1975*]). Plotted in the lower trace of Figure 2 are monthly average sunspot numbers per year, clearly showing an association between the two quantities. *Judson [1933*] notes that the rise in wireless signal strength in 1921 – 1922 and in 1930 – 1931 "occurred near the same phase in the sunspot cycle". This observation, made nearly 70 years ago, is probably related to the now well-recognized fact that there is a tendency for the geomagnetic activity cycle to have a double-peaked modulation. A major peak in activity can occur in the declining phase of the cycle [e.g., *Gosling et al., 1977*] and a secondary peak near the sunspot maximum [e.g., *Newton, 1948; Ohl, 1971*].

From the early telegraph, to the beginnings of the use of the ionosphere for wireless communications, to the establishment of inter-connected power grids, to the flight in recent decades of humans in space, the importance of knowledge of the solar-terrestrial environment for the successful implementations of many new technologies has continued to increase. The operations and survivability of both ground- and space-based technical systems have often encountered surprises because of natural environment effects that were unanticipated. In recent decades, as technologies have increased in sophistication as well as in miniaturization and in interconnectedness, more sophisticated understanding of the Earth's space environment has become necessary.

Some selected major disturbances of technical systems since the early 19th century are shown as a function of the sunspot cycle in Figure 3 [*Lanzerotti, 1994*]. This listing illustrates the types of disturbances that have been encountered in land-based technologies during this interval. Not until the sunspot cycle peak in the late 1930's (and the especially strong magnetic disturbances of April 16, 1938, and March 24 - 25, 1940) did the effects of Earth currents on electric power systems begin to be of major concern to the power industry [*Davidson, 1940*]. In addition to spectacular occurrences, such as complete disruptions of power grids and cable and radio communications (and, in more recent years, of spacecraft systems and sub-systems), individual smaller magnetic disturbances can also affect technologies, as is discussed in the following sections.

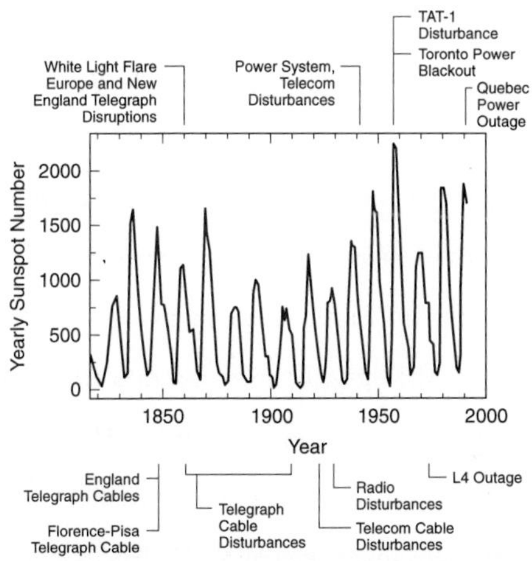

Fig. 3 *Yearly sunspot numbers, 1825-1991, with indicated times of selected major impacts of the solar-terrestrial environment on ground-based technical systems.*

2. The Solar-Terrestrial Environment

Solar emissions, including the photons (light) that are critical for sustaining life, continually affect the Earth's atmosphere and space environment [e.g., *Lanzerotti, 1992; Hargreaves, 1992*]. In particular, the hot (> 10^6 °K) solar outer atmosphere – called the solar corona – beginning about

2,000 km above the sun's visible surface (the photosphere) continually expands into the interplanetary space as an ionized gas. This gas, composed of electrons and protons with approximately 10% helium ions, is called the solar wind. At the orbit of Earth, this wind has a density of about 10 particles per cc, a temperature of about 1 keV (equivalent to about 10^7 °K) and an average speed of about 500 km/s. The solar wind gas also has magnetic fields from the sun imbedded within it. At Earth's orbit, these fields have a strength of about 10 nT (about 3,000 times less than the Earth's magnetic field at its surface).

The gases and fields on the sun's surface and in the corona vary temporally and spatially. As the schematic illustration in Figure 4 indicates, the variable activity on the sun produces disturbances in the solar wind and other solar emissions that propagate throughout the solar system. Solar emissions at visible and infrared frequencies have varied by a fraction of a percent or so over the most recent solar cycle. Changes in the sun's ultraviolet (UV) radiation over a solar cycle are much greater – about a factor of two – and are responsible for significant heating of the high upper atmosphere of Earth and for changes in the electrical conductivity of the ionosphere.

Solar flares are known to contain as much as 10^{29} joules of energy, and can accelerate electrons and protons to energies of many MeV and even hundreds of MeV at times. Intense radio emissions, easily detectable on Earth, accompany this particle acceleration. Another example of violent solar activity is the episodic injection into the interplanetary medium of large masses of coronal material, some 10^{12} kg or more at a time. These coronal mass ejection (CME) events, as they propagate away from the sun, are also capable of accelerating interplanetary particles to higher energies – perhaps many tens of MeV. The relationship of these CME events to solar phenomena such as sunspots and flares is not yet well understood. However, CMEs are now known to be important sources of disturbances of the interplanetary medium and of the space environment of Earth, even during years of low sunspot conditions.

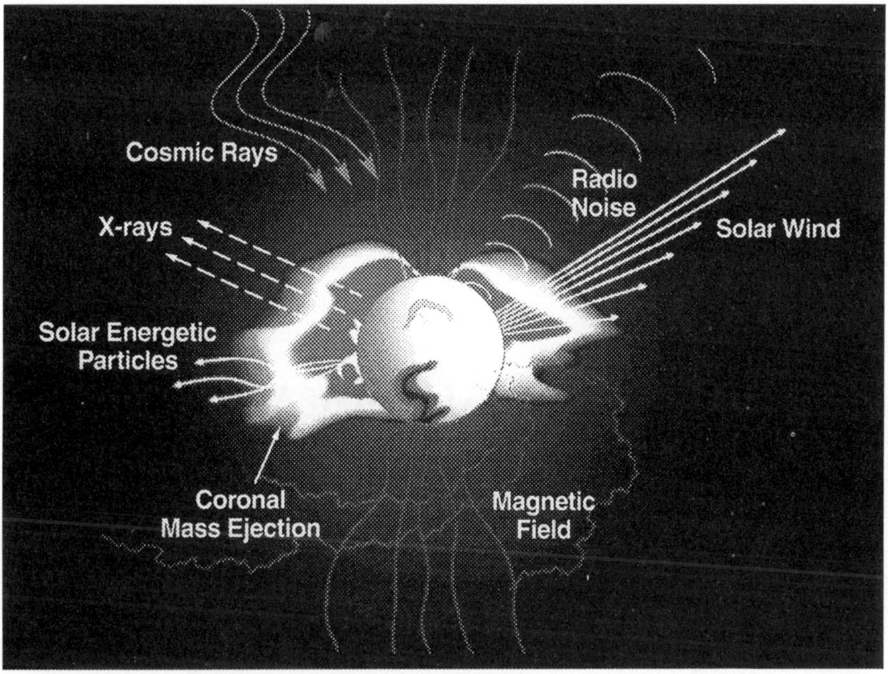

Fig. 4 *Solar emissions that fill the solar system and affect the Earth's space environment in various ways*

The time variability of several solar-produced phenomena over a one year interval are shown in Figure 5 [*Lanzerotti et al., 1997*]. These data cover 1991, during the maximum of the 22nd solar cycle. The top two panels contain daily values of the radio noise from the sun as measured at local noon by the U. S. Air Force at Sagamore Hill, Massachusetts (archived by NOAA World Data Center A). The lower frequency is in the vicinity of terrestrial and some Low Earth Orbit (LEO) wireless frequencies. The higher frequency is just below the Ka band (18 to 31 GHz) that is proposed to be used for other LEO and geosynchronous (GEO) telecommunications.

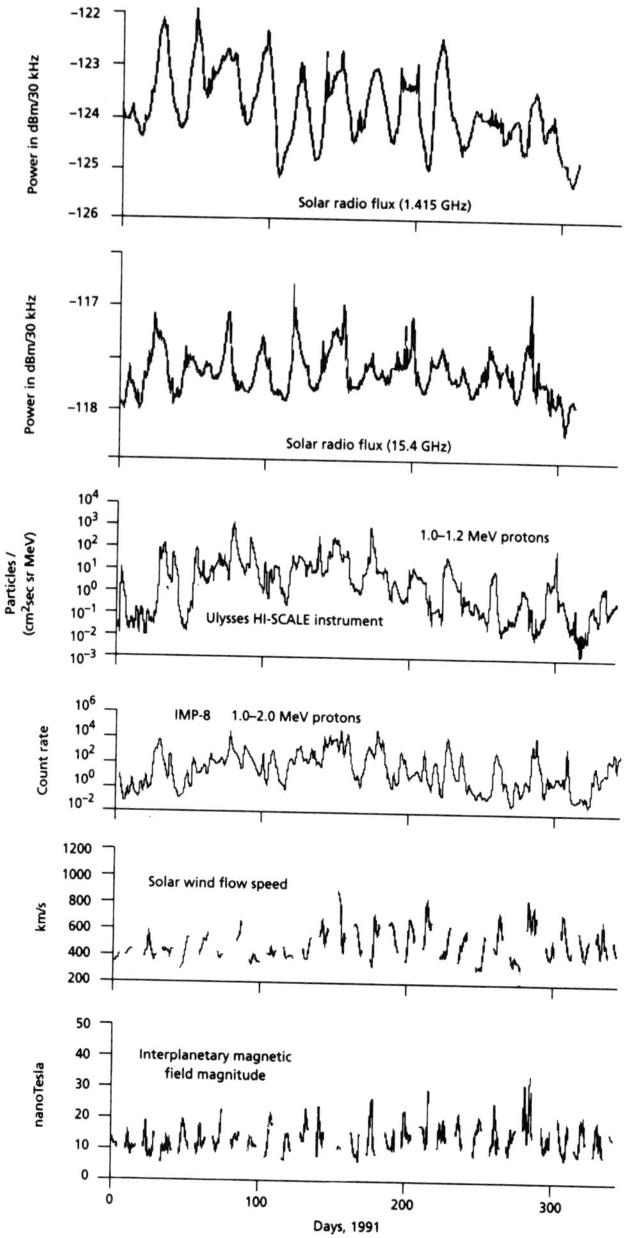

Fig. 5 *Variation with time of several solar-originating phenomena measured near Earth during the year 1991.*

The third and fourth panels from the top of Figure 5 plot the time variations of 1 MeV proton fluxes in the interplanetary medium. The third panel plots the proton fluxes measured by the HI-SCALE instrument [*Lanzerotti et al., 1992*] on the NASA/European Space Agency Ulysses spacecraft which was en route from Earth (following launch in October 1990) to Jupiter. The data in the fourth panel are proton fluxes measured on the Earth-orbiting NASA IMP-8 spacecraft. The flux levels of the charged particles that fill the solar system can vary by a factor of 10^6 or more, as shown here.

The last two panels in Figure 5 contain the velocity of the solar wind and the magnitude of the interplanetary magnetic field for the year as measured on the IMP-8 spacecraft. The gaps in data coverage occur because the orbit of IMP-8 takes it inside the Earth's magnetosphere for several days during each orbit.

Illustrated in Figure 6 are variations with time of three physical quantities in the Earth's proximate space environment (in the magnetosphere and on Earth's surface; Figure 7) during the year 1991. The top panel traces the flux of galactic cosmic rays (energies > 500 MeV) as measured on the Earth's surface at Newark, Delaware, by the Bartol Research Institute of the University of Delaware. These cosmic rays enter the solar system from the Milky Way galaxy and propagate through the magnetosphere to Earth's surface. The fluxes of these cosmic rays vary by more than 30% over the course of the year shown in the Figure, and they can change in overall baseline level by this amount or more over a solar cycle. These galactic cosmic rays are less intense in years of high solar activity since the more intense solar wind conditions during solar maximum tend to exclude the cosmic rays from the solar system.

The center panel of Figure 6 plots the fluxes of 2 MeV electrons for 1991 as measured by the NOAA GEO weather satellite GOES 6. The apparent decrease in fluxes toward the end of the year is an artifact because the instrument was losing sensitivity due to damage by space radiation (see Section 3.3). During this year, the relativistic electron fluxes at GEO varied by as much as five orders of magnitude during intervals of only a few days.

The lowest panel in Figure 6 plots the time variability of the south-north component of the Earth's magnetic field as measured at the South Pole. This magnetic field varies greatly during the year – as much as ±1000 nT. Such variability is caused by large changes in the electrical currents that flow in the ionosphere, which in turn are driven by changes in the proton and electron fluxes under the influence of the solar wind.

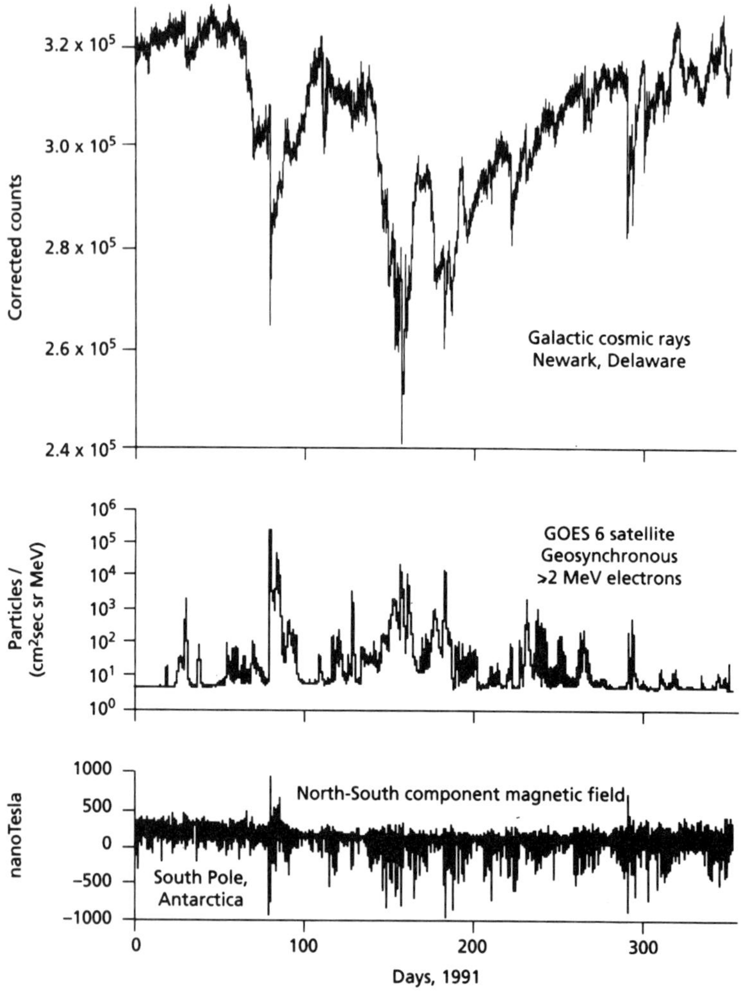

Fig. 6 *Variation with time during 1991 of several physical parameters measured in the magnetosphere and on the Earth's surface.*

3. Solar-Terrestrial Effects

Many of the present-day technological systems that must include processes and parameters from the solar-terrestrial environment in their design and/or operations are listed in Table 1. These systems are grouped into broad categories that have similar physical origins. Figure 8 illustrates these effects.

3.1 Ionosphere

3.1.1 Ionosphere and Wireless

Today, nearly a century after *Marconi's* successful trans-Atlantic telegraph transmission in 1901, the ionosphere is both a facilitator and an intruder in numerous communications applications. The military continues to make extensive use of kHz and tens of MHz communications links, and it supports active research programs into ionosphere phenomena. Police and fire emergency frequencies rely on the ionosphere for propagation, a use that can produce anomalous reception during times of high solar activity. In 1979, near the peak of the 21st solar cycle, the Orange County, California, fire department responded to a distress signal from a downed commuter aircraft only to discover – when the plane was not found – that the signal had originated from an accident site in West Virginia.

Another example of ionosphere intrusion into modern technologies occurs in the equatorial regions of the Earth where the magnetic field is approximately parallel to the horizontal layers. Here, plasma instabilities can develop, producing "bubbles" in the ionosphere's density. It is now believed that these bubbles, and accompanying plasma physics phenomena, are a prime source of scintillations of satellite to ground signals at frequencies in the 1- to 6- GHz range. Engineers at the Comsat Corporation first discovered these scintillations in the early 1970's after the initial implementation of the GEO INTELSAT satellite network. (When satellite communications were begun at these frequencies, it was not expected that the ionosphere would need to be taken into account for the provision of services.) The occurrence and intensity of these plasma physics-driven phenomena depend sensitively on local time, geomagnetic disturbance levels, and time within the solar cycle.

TABLE 1 *Impacts of Solar-Terrestrial Processes on Technology*

- Ionosphere Variations
  - Wireless signal reflection, propagation, attenuation
  - Communication satellite signal interference, scintillation
  - Interference with geophysical prospecting
  - Source for geophysical prospecting
  - Induction of electrical currents in the Earth
    - Power distribution systems
    - Long communications cables, land and ocean
    - Pipelines
- Radiation
  - Solar cell damage
  - Semiconductor device damage and failure
  - Misoperation of semiconductor devices
  - Spacecraft charging, surface and interior materials
  - Astronaut safety
  - Airline passenger safety
- Magnetic field variations
  - Attitude control of spacecraft
  - Compasses
- Solar radio bursts
  - Excess noise levels in wireless communication systems
- Micrometeoroids and artificial space debris
  - Solar cell damage
  - Damage to mirrors, surfaces, and materials
- Atmosphere
  - Low altitude satellite drag
  - Attenuation and scatter of wireless signals

Plasma processes that produce scintillations on trans-ionosphere signals have in recent years also caused considerable problems for the use of single frequency Global Positioning System (GPS) signals for precision location determination. Indeed, dual frequency GPS is now the method of choice for many applications. Much research on this phenomenon is in progress in order to explore the use of GPS signals for many purposes, including the docking of ships in harbors and the landing of aircraft. At the same time, the disturbances of GPS signals are being used to study the three-dimensional physical nature of the ionosphere using tomographic techniques.

3.1.2 Ionosphere and Earth Currents

As discussed above in the context of the early telegraph, the dynamically-changing currents in the ionosphere produce magnetic fluctuations at Earth's surface. These magnetic fluctuations in turn cause electrical currents to flow in the Earth, with the flow paths determined by the spatial heterogeneity of the conductivity structures of the Earth. The actual physical situation for inducing potentials across any given path is, in toto, a very complex phenomenon because the magnetosphere and ionosphere currents are also heterogeneous with respect to Earth's surface. The Earth currents seek out the least resistant path between two points – often that of long, grounded conductors.

Induced Earth potentials must be considered when designing long, conducting technological systems, such as pipelines (especially in the auroral regions [*Campbell, 1986; Viljanen, 1989*]), power feed equipment (PFE) for oceanic cable systems [*Lanzerotti et al., 1995*], and electrical power distribution systems [e.g., *Pirjola, 1989; Viljanen and Pirjola, 1989; Boteler et al., 1998*]. An AT&T L4 cable running from Plano, Illinois, to Cascade, Iowa, was disrupted by massive Earth potentials in 1972 [*Anderson et al., 1974*]. The entire province of Quebec suffered a power disruption for many hours during an interval of great magnetic disturbance in 1989 [*Czech et al., 1992*]. This same disturbance induced large Earth currents to flow through a multimillion-dollar transformer at a nuclear power station in southern New Jersey, causing it to burn out. In addition, more than 800 V were measured across the TAT-8 communications cable in the Atlantic Ocean at the time of the event, a potential nearly large enough to trigger an anomaly alarm [*Medford et al., 1989*].

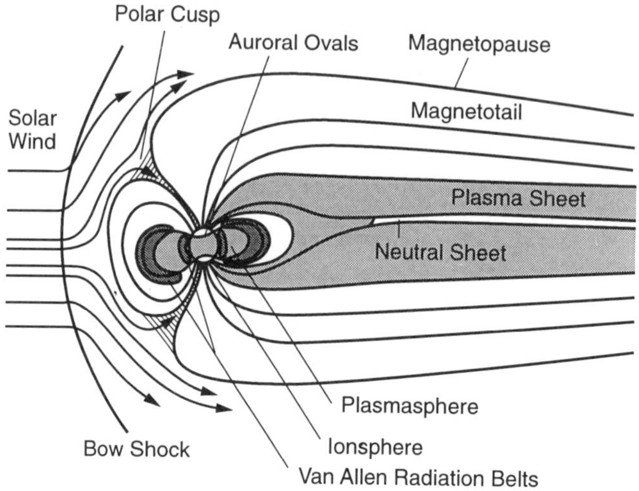

Fig. 7. The Earth's proximate space environment, the magnetosphere.

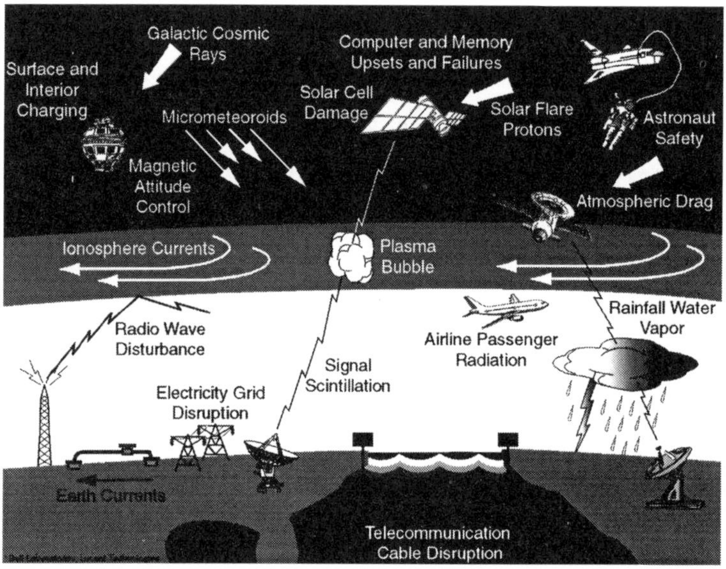

***Fig. 8.** Some of the effects of the solar-terrestrial environment on technical systems deployed on the Earth's surface and in space, and/or whose signals propagate through the space environment.*

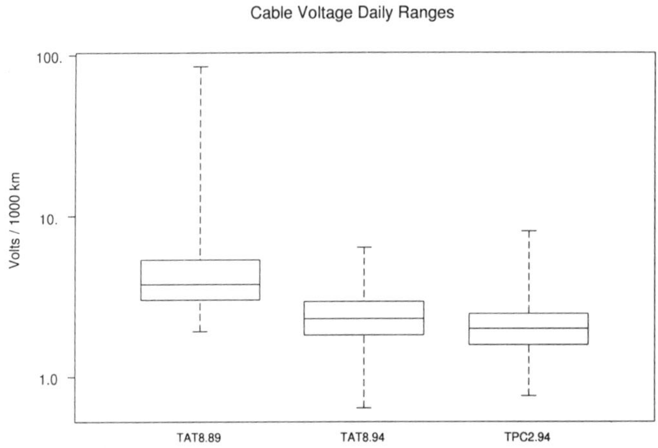

Fig. 9. *Statistical distributions of the geopotential variations measured across the TAT-8 cable during several months in 1989 (near solar maximum) and 1994 (near solar minimum), as well as across the TPC-2 cable during the same months in 1994. The center line across each box represents the median value.*

The results of statistical studies of Earth potentials along existing routes can provide some important design information for engineers. For example, the box plots in Figure 9 provide statistical information on the daily ranges of the Earth's potential, in volts per 1,000 km, across two separate cable routes in a solar maximum year (1989) and a solar minimum year (1994). TAT-8 is a cable running from Tuckerton, NJ to just off the coast of France, south of England. TPC-2 is a decommissioned Pacific cable running from Hawaii to Guam. The height of each box corresponds to the central 50% of the total ranges. In each case, the line drawn across the box represents the median. The dashed lines above and below the boxes show the upper and the lower 25% of the ranges, respectively. The results show that the median of the Earth potential for the TAT-8 cable is greater by at least a factor of two during solar maximum (designated TAT8.89 in the Figure) than during solar minimum (designated TAT8.94). Moreover, on this route, the central 50% of the ranges, as well as the upper 25%, are considerably wider during solar maximum than during solar minimum. Comparing the two rightmost boxes for the solar minimum year 1994, the median range of potentials is greater for the higher latitude cable (designated TAT8.94 in the Figure) than for the nearer-equatorial cable TPC-2 (designated TPC2.94).

3.2 Solar Radio Effects

Solar radio noise and bursts were discovered nearly sixty years ago by *Southworth [1945*] and by *Hey [1946*] in the early investigations of radar in during the Second World War. Solar radio bursts, which can have a very wide range in spectral shape and in intensity levels as measured at Earth [e.g., *Guidice and Castelli, 1975*], provided unexpected interference (jamming) of this new technology that was under rapid developement for war-time purposes [*Hey, 1973*].

There is a wide variability in the solar radio emissions at the two frequencies shown in the upper panel of Figure 5. During 1991 (the year plotted in the Figure), an analysis showed that the average noontime flux measured at 1.145 and 15.4 GHz was -162.5 and -156 dBW/(m^2 4kHz), respectively. These values are about 6 and 12 dB above the 273° K (Earth's surface temperature) thermal noise of -168.21 dBW/(m^2 4kHz). In addition, these average noontime solar fluxes on Earth are only 20 and 14 dB, respectively, below the maximum flux of -142 dBW/(m^2 4kHz) that is allowed for satellite downlinks by the ITU regulation RR2566. Because this radio noise originates at the sun in plasma physics processes in the solar photosphere and corona, and is not simple blackbody radiation, the emissions may occur in distinct frequency ranges and be neither stationary in time nor have a Gaussian distribution.

An analysis of local noontime solar radio data over the last few solar cycles (data available from the NOAA World Data Center) shows that at 10.2 cm (2.8 GHz), such signals change by about 5.3 dB peak-to-peak during the cycle, and that they have a standard deviation around the local one-year mean of about 0.7 dB [*Lanzerotti et al., 1997*]. The maximum excursion above the local mean was found to be about 3.6 dB. Therefore, it is reasonable to expect that emission levels up to two standard deviations above the 1991 levels (at least 5 or 6 dB, and perhaps as high as 7 or 8 dB) might occur at some time during the 23rd cycle. Such noise would reduce the 14 dB margin at 15.4 GHz to only about 6 or 7 dB.

A caveat is that the data shown (and analyzed) in Figure 5 are only local noontime values. Short-term variations with durations of a few seconds can be at least 35 dB higher than the noontime values [*Barron et al., 1980*]. For example, the large flare of May 23, 1967, produced a flux level measured at Earth of > 100,000 solar flux units (SFU = 10^{-22} W/(m^2 Hz)), and perhaps as high as 300,000 SFU at 1 GHz. This corresponds to –129 dBW/(m^2 4kHz), or 13 dB above the maximum limit of –142 dBW/(m^2 4kHz) noted above.

Plotted in Figure 10 are the percentages (on an hourly basis) of dropped calls averaged over all base stations in a cellular system within a major U.S. state as a function of local time during one day in 1996. There was solar flare activity during early local morning as the sun was rising at the cell sites. The east-facing antennas clearly suffered very high dropped rates just after sunrise in comparison to the other two antenna directions shown.

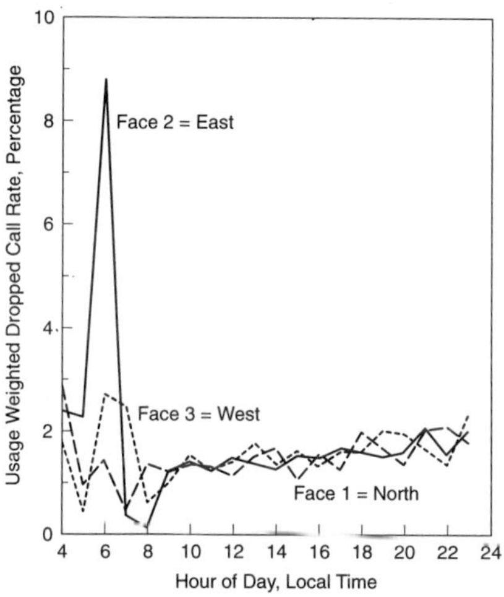

Fig. 10 *Percentages of dropped wireless calls in the cellular system of a major state in the U.S. on one day in 1996.*

3.3 Space Radiation Effects

A textbook discussion of the space environment and the implications for satellite design is contained in *Tribble [1995]*. Radiation remains a difficult environmental problem for the design and operations of space-based systems [e.g., *Shea and Smart, 1998*]. The low energy (a few eV to a few keV) particles in the magnetosphere plasma population can produce different levels of surface charging on materials that make up a spacecraft's outer envelope [*Garrett, 1981*]. If good electrical connections do not exist between the various materials, the differential charging that is possible can produce lightning-like breakdown discharges between the materials. These discharges will produce electromagnetic interference and damage to components and subsystems [e.g., *Vampola, 1987*]. The US Air Force SCATHA (P78-2) satellite was specifically designed to study radiation effects, including charging of surface materials, under a variety of conditions [e.g., *Koons, 1980; Gussenhoven and Mullen, 1983*].

The plasma populations of the magnetosphere, especially under conditions of geomagnetic activity, are highly time variable. However it is known that increases in the cross-magnetotail electric field will convect the tail plasma earthward. Under these conditions, the magnetotail plasma sheet will extend earthward to within the GEO orbit. When this occurs, the onboard spacecraft anomalies will tend to be a maximum in the midnight to dawn sector of the orbit [*Mizera, 1983*]. While records of anomalies exist, there is not an extensive body of published data on the statistical characteristics of charging on spacecraft surfaces, especially from commercial satellites that are now so important for communications. Two surface-mounted charge plate sensors were specifically flown on the AT&T (now Loral) Telstar 4 GEO satellite to monitor surface charging effects.

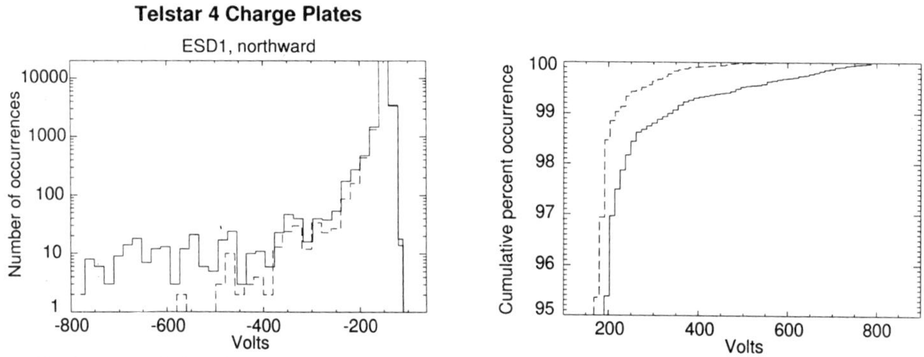

Fig. 11 *Statistical distribution of charging recorded on a charge plate sensor on the Telstar 4 spacecraft during the month of January 1997 (solid line) and for the month with data from January 10th removed (dashed line).*

Shown in Figure 11 are the statistical distributions of charging on one of the charge plates in January 1997 [*Lanzerotti et al., 1998*]. The solid lines correspond to charging data for the entire month, while the dashed lines omit data from January 10, when a very large magnetic storm occurred. Voltages as large as –800 V were recorded on the charge plate sensor during the magnetic storm, an event during which a permanent failure of the Telstar 401 spacecraft occurred (although the failure has not been attributed specifically to the space conditions).

As the central panel in Figure 6 indicates, the higher energy particles in the magnetosphere often have a very dynamic and wide range of intensities that can change by many orders of magnitude over the course of minutes, hours, and days. Generally it is prohibitively expensive (in terms of including additional spacecraft mass in lieu of additional transponders or orbit control gas, for example) to provide radiation shielding of all interior subsystems against high energy electrons. For example, the range of a 100 MeV proton in aluminum (a typical material) is ~ 40 mm; the range of a 3 MeV electron is ~ 6 mm.

These particles can penetrate deep into the interior regions of a satellite. In addition to producing transient upsets in signal and control electronics, such particles can also cause electrical charges to build up in interior insulating materials, such as those used in coaxial cables. If the charge buildup is sufficiently large, these interior materials will eventually suffer electrical breakdowns, resulting in electromagnetic interference and damage to the electronics. A number of spacecraft anomalies and even failures have occurred after many days of greatly elevated levels of relativistic electron fluxes [*Baker et al., 1987; 1994; 1996; Reeves et al., 1998*].

The data in Figure 12 plots as histograms the daily average flux of >2 MeV electrons for one month in 1998, an interval when two spacecraft failures occurred and one anomaly was noted on a NASA spacecraft (all indicated in the Figure; *Baker et al. [1998]*). After being low at the beginning of the interval, the relativistic electrons increased steadily over the next week or more, reaching a maximum just prior to the failure of the Equator-S satellite. The fluxes rose again on about May 5, and stayed elevated for more than two weeks. While the associations of the times of occurrence of the spacecraft problems with the increased electron fluxes is evident, it is still uncertain which, if any, of the failures or anomalies were specifically caused by radiation effects [*Baker et al., 1998*].

For charged particles at the very highest energies that reach the Earth's environment (the galactic cosmic rays of the upper panel of Figure 6), no realistic shielding is possible for most technological systems in space or on Earth. These very energetic particles can produce upsets and errors in spacecraft electronics as well as in computer systems on the surface of Earth [IBM, 1996]. Indeed, often DRAMs and other computer chips that are destined for use on Earth are

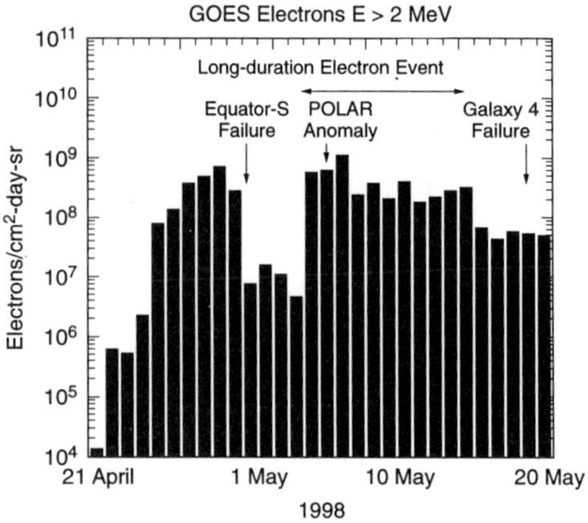

Fig. 12 Daily flux values of electrons (> 2 MeV) measured on the GOES satellite for the interval April 21 – May 20, 1998. Dates of operational problems on various spacecraft are shown [from Baker et al., 1998]

tested under cosmic ray fluxes at different altitudes on Earth, such as in Denver as compared to sea level on either coast of the U.S., in order to better understand the sensitivities of these silicon devices to realistic charged particle fluxes.

3.4 Atmosphere Effects

3.4.1 Upper Atmosphere Effects

Large changes occur in the upper atmosphere during the course of a solar cycle because of changes in heating of the atmosphere at ionosphere altitudes by solar UV components. As was noted in Section 2, the UV components from the sun change by more than a factor of two at wavelengths ≤170 nm during a solar cycle [*Hunten et al., 1991*], which is much more than the ~0.1% changes typical of the visible radiation. This heating causes the atmosphere to expand – enough to raise the "top of the atmosphere" by several hundred km during solar maximum. This greater density results in increased drag on both space debris and on spacecraft in LEO, including the space shuttle, space stations, and telecommunications satellites.

Skylab, the first United States space station, was lost due to the effects of atmospheric drag in the solar maximum period of the late 1970s because the space shuttle was not flight-ready in time to be able to be used to raise the station to a higher orbit. Some proportionate amount of orbit-control gases on LEO telecom spacecraft will be used to maintain orbit altitude against increased drag, especially because many of those now flying or in the planning stages will be in the initial part of their in-service period during the buildup to, and in, the next solar maximum [e.g., *Pickholtz, 1996*].

3.4.2 Atmosphere Water Vapor

At frequencies in the Ka band that are contemplated for high-bandwidth space-to-ground applications (as well as for point-to-point communications between ground terminals), water vapor in the neutral atmosphere is the most significant natural phenomenon that can seriously affect the signals [*Gordon and Morgan, 1993*]. However, even at GHz frequencies, it is now well recognized that the ionized channels caused by lightning strokes, and possibly even charge separations in clouds, can reflect radar signals. Lightning and cloud charging phenomena may produce as yet unrecognized noise sources for low-level wireless signals.

It would appear that, in general, the space environment itself can safely be ignored when attempting to engineer around the limitations imposed by rain and water vapor in the atmosphere. Nevertheless, it is useful to mention here that one of the more controversial areas of research (indeed, one that has been fraught with numerous false leads) involves the possible effects of magnetosphere and ionosphere (and thus the interplanetary medium) processes on terrestrial weather. After all, galactic cosmic rays produce the ionization in the atmosphere from the ionosphere down to the ground that permits the existence of the “clear weather electric field”.

Further, during the last few years, several completely unanticipated discoveries have been made related to Earth lightning. Most significant of these discoveries were measurements showing that some lightning discharges propagate upward to the ionosphere from the topside of thunderclouds. These results are convincing evidence that at times during severe weather conditions atmospheric clouds are electrically connected to the lower ionosphere [e.g., *Winckler, 1995; Sentman and Wescott, 1995*]. Questions about the overall energetics, occurrence frequencies, and many other factors (including ionosphere conductivity) remain. Nevertheless, it is now clear that atmosphere-ionosphere coupling does indeed exist, whereas in the past it was only hypothesized or dismissed out of hand.

3.5 Complications

An interesting and intriguing implication (albeit speculative to some degree) of the above discussions, as well as the listings in the Table, is that the natural phenomena that enter into reliability determinations are not independent. For example, some effects on signals propagated through the ionosphere occur during the same events when satellite anomalies occur. In addition, some satellite anomalies due to energetic particles may not be independent of some weather-related events in the atmosphere. The statistical structure of most of these natural phenomena is complicated. Statistical examinations of some interplanetary [*Thomson et al., 1995*] and solar and terrestrial data sets [*Thomson et al., 1998*] suggest that the statistics of many of these variables are often better characterized by a modal decomposition than by more conventional turbulence-based ideas. A modal decomposition implies that the extremes in a process will be worse than would be expected from processes with an "equivalent" continuous spectrum. However, processes characterized by modal decomposition would, in principal, be more predictable. Much remains to learned concerning the statistical behavior of natural phenomena that can adversely affect technical systems.

4. Summary

The solar-terrestrial environment continues to present numerous interesting and crucial engineering and science challenges for many existing and proposed technical systems that employ space for their implementation, or that can be affected by space processes. This broad overview has outlined many of the issues considered significant today. Nevertheless, it is important to recognize that the Earth's natural environment – including the sun and the space around Earth – is very dynamic. In terms of communications, many of the space-associated concepts will only begin to be realized during this new solar maximum. Based upon past history and experience with the effects of space-related phenomena on technologies, major surprises are certainly in store for some of the technical assets projected to be in use during the maximum of solar cycle 23.

References

Anderson, C. W. III, L. J. Lanzerotti, and C. G. Maclennan [1974] "Outage of the L4 System and the Geomagnetic Disturbance of 4 August 1972", *The Bell Syst.Tech.J.,* 53, 1817.

Appleton, E. V. and M. A. F. Barnett [1925] "Local reflection of wireless waves from the upper atmosphere", *Nature,* 115, 333.

Baker, D. N., R. D. Belian, P. R. Higbie, et al., [1987] "Deep Dielectric Charging Effects due to High Energy Electrons in Earth's Outer Magnetosphere", *J. Electrost.*, 20, 3.

Baker, D. N., S. Kanekal, J. B. Blake, et al., [1994] "Satellite Anomalies Linked to Electron Increase in the Magnetosphere", *Eos Trans.Am.Geophys.Union*, 75, 401.

Baker, D. N. [1996] "An Assessment of Space Environment Conditions During the Recent Anik E1 Spacecraft Operational Failure", *ISTP Newsletter*, 6, 8-29.

Baker, D. N., J. H. Allen, S. G. Kanekal, and G. D. Reeves [1998] "Disturbed Space Environment May Have Been Related to Pager Satellite Failure", *Eos, Trans.Am.Geophys.Union,* 79, 477-483.

Barlow, W. H. [1849] "On the Spontaneous Electrical Currents Observed in the Wires of the Electric Telegraph", *Phil.Trans.R. Soc.*, 61A, 61.

Barron, W. R., E. W. Cliver, D. A. Guidice, and V. L. Badillo [1980] *An Atlas of Selected Multi-Frequency Radio Bursts from the Twentieth Solar Cycle,* Air Force Geophsyics Laboratory, Space Physics Division, Project 4643, Hanscom AFB, MA.

Breit, M. A. and M. A. Tuve [1925] "A Test of the Existence of the Conducting Layer", *Nature*, 116, 357.

Boteler, D. H., R. J. Pirjola, and H. Nevanlinna [1998] "The Effects of Geomagnetic Disturbances on Electrical Systems at the Earth's Surface", *Adv.Space Res.,* 22, 17.

Campbell, W. H. [1986] "An Interpretation of Induced Electrical Currents in Long Pipelines Caused by Natural Geomagnetic Sources of the Upper Atmosphere", *Surveys Geophys.,* 8, 239.

Carrington, R. C. [1863] *Observation of the Spots on the Sun From November 9, 1853, to March 24, 186, Made at Redhill*, William and Norgate, London and Edinburgh, 167.

Czech. P., S. Chano, H. Huynh, and A. Dutil [1992] "The Hydro-Quebec System Blackout of 13 March 1989: System Response to Geomagnetic Disturbance", *Proc. EPRI Conf. Geomagnetically Induced Currents*, EPRI TR-100450, Burlingame, CA, November 8-10, 1989, 19-21.

Davidson, W. F. [1940] "The Magnetic Storm of March 24, 1940 – Effects in the Power System", *Edison Electric Inst. Bulletin*, 365.

Fagen, M. D. [1975] *A History of Engineering and Science in the Bell System*, Bell Tel. Labs., Inc., Murray Hill, NJ.

Garrett, H. B. [1981] "The Charging of Spacecraft Surfaces", *Revs.Geophys.,* 19, 577.

Gassmann, G. J., ed. [1963] *The Effects of Disturbances of Solar Origin on Communications,* AGARD Avionics Panel, Macmillan Company, New York.

Gish, O. H. [1936] "Electrical Messages from the Earth, Their Reception and Interpretation", *J.Wash.Acad.Sci.,* 26, 267.

Gordon, G. D. and W. L. Morgan [1993] *Principals of Communications Satellites,* John Wiley, New York, 178-192.

Gosling, J. T., J. R. Asbridge, and S. J. Bame [1977] "An Unusual Aspect of Solar Wind Speed Variations During Solar Cycle 20", *J.Geophys.Res.*, 82, 3311.

Gussenhoven, M. S., and E. G. Mullen, "Geosynchronous Environment for Severe Spacecraft Charging", *J.Spacecraft and Rockets,* 20, 26.

Hargreaves, J. K. [1992] *The Solar-Terrestrial Environment,* Cambridge U. Press.

Hey, J. S. [1946] "Solar Radiations in the 4 – 6 Metre Radio Wavelength Band", *Nature,* 158, 234.

Hey, J. S. [1973] *The Evolution of Radio Astronomy*, Neale Watson Academic Pub. Inc., New York.

Hunten, D. M., J.-C. Gerard, and L. M. Francois [1991] "The Atmosphere's Response to Solar Irradiation", in *The Sun in Time,* ed. C. P. Sonett, M. S. Giampapa, and M. S. Matthews, Univ. Arizona Press, Tucson, 463-497.

IBM Journal of Research and Development [1996] 40, 1-136.

Judson, E. B. [1933] "Low-Frequency Radio Receiving Measurements at the National Bureau of Standards in 1931 and 1932", *Proc.IRE,* 21, 1354.

Koons, H. [1980] "Characteristics of Electrical Discharges on the P78-2 Satellite (SCATHA)", *18th Aerospace Sciences Meeting, AIAA 80-0334*, Pasadena, CA.

Lanzerotti, L. J., and G. P. Gregori [1986] "Telluric Currents: The Natural Environment and Interactions with Man-Made Systems", in *The Earth's Electrical Environment*, National Academy Press, Washington.

Lanzerotti, L. J. [1992] "Solar-Terrestrial Physics", in *Encyclopedia of Physical Science and Technology*, Academic Press, 15, 429-439.

Lanzerotti, L. J. [1994] "Impacts of Solar-Terrestrial Processes on Technological Systems", in *Solar Terrestrial Energy Program*, ed. D. N. Baker, V. O. Papitashvili, and M. J. Teague, COSPAR Colloquium Series, 5, Pergamon Press, 547-555.

Lanzerotti, L. J., L. V. Medford, C. G. Maclennan, and D. J. Thomson [1995] "Studies of Large-Scale Earth Potentials Across Oceanic Distances", *AT&T Tech. J.,* May/June. 73-84.

Lanzerotti, L. J., D. J. Thomson, and C. G. Maclennan [1997] "Wireless at High Altitudes", *Bell Lab.Tech. J.*, 2(3), 5-9.

Lanzerotti, L. J., R. E. Gold, K. A. Anderson, et al.,[1992] "Heliosphere Instrument for Spectra, Composition, and Anisotropy at Low Energies", *Astron Astrophys. Suppl. Series,* 92, 349-363.

Lanzerotti, L. J., K. Lafleur, C. G. Maclennan, and D. W. Maurer [1998] "Geosynchronous Spacecraft Charging in January 1997", *Geophys. Res. Lett.*, 25, 2967-2970.

Lanzerotti, L. J., C. Breglia, D. W. Maurer, et al., [1998] "Studies of Spacecraft Charging on a Geosynchronous Telecommunications Satellite", *Adv. Space Res.*, 22, 79-82.

Medford, L. V., L. J. Lanzerotti, J. S. Kraus, and C. G. Maclennan [1989] "Trans-Atlantic Earth Potential Variations During the March 1989 Magnetic Storms", *Geophys. Res. Lett.,* 16, 1145-1148.

Mizera, P. F. [1983] "A Summary of Spacecraft Charging Results", *J. Spacecraft Rockets,* 20, 438-443.

Newton, H. W. [[1948] "Sudden Commencements in the Greenwich Magnetic Records (1879-1944) and Related Sunspot Data", *Mon. Not .R. Astron.Soc.*, 5, 159.

Ohl, A. I. [1971] "Physics of the 11-Year Variation of Magnetic Disturbances", *Geomag. Aeron.*, 11, 549.

Petersen, P.O. [1928] *The Propagation of Radio Waves*, Danmark Naturvidenskabelige Samfound, Copenhagen.

Pickholtz, R. L. [1996] "Communications by Means of Low Earth Orbiting Satellites", in *Modern Radio Science 1996*, ed. J. Hamlin, Oxford U. Press, 133-149.

Pirjola, R. [1989] "Geomagnetically Induced Currents in the Finnish 400 kV Power Transmission System", *Phys. Earth Planet. Inter.,* 53, 214.

Prescott, G. B. [1866] *Theory and Practice of the Electric Telegraph*, IV ed., Tichnor and Fields, Boston.

Reeves, G. D. [1998] "The Relativistic Electron Response at Geosynchronous Orbit During the January 1997 Magnetic Storm", *J. Geophys. Res.*, 103, 17559.

Schuster, A. [1908] "The Diurnal Variation of Terrestrial Magnetism", *Trans. R. Soc.,* A208, 163-204.

Senior, T. B. A. [1998] "A Brief History of URSI, I – The Early Years", *The Radio Science Bulletin,* No. 286, September 11-14.

Sentman, D. D., and E. M. Wescott [1995] "Red Sprites and Blue Jets: Thunderstorm-Excited Optical Emissions in the Stratosphere, Mesosphere, and Ionosphere", *Phys. Plasmas,* 2, 2514-2522.

Shea, M. A., and D. F. Smart [1998] "Space Weather: The Effects on Operations in Space", *Adv. Space Res.*, 22, 29.

Southworth, G. C. [1945] "Microwave Radiation From the Sun", *J. Franklin Inst.*, 239-285.

Thomson, D. J., G. G. Maclennan, and L. J. Lanzerotti [1995] "Propagation of Solar Oscillations Through the Interplanetary Medium", *Nature,* 376, 139-144.

Thomson, D. J., L. J. Lanzerotti, and C. G. Maclennan, [1998] "Low Frequency (~2.19-Day Period) Mode in Records of Interplanetary and Central England Temperature Data", *Proc. Sixth SOHO/1998 GONG Workshop*, Boston.

Tribble, A. C., *The Space Environment, Implications for Spacecraft Design*, Princeton University Press, Princeton, NJ, 1995

Vampola, A. [1987] "The Aerospace Environment at High Altitudes and its Implications for Spacecraft Charging and Communications", *J. Electrost.*, 20, 21.

Viljanen, A. [1989] "Geomagnetically Induced Currents in the Finnish Natural Gas Pipeline", *Geophysica,* 25, 135.

Viljanen, A., and R. Pirjola [1989] "Statistics on Geomagnetically-Induced Currents in the Finnish 400 kV Power System Based on Recordings of Geomagnetic Variations", *J. Geomag. Geoelectr.*, 41, 411.

Winckler, J. R. [1995] "Further Observations of Cloud-Ionosphere Electrical Discharges Above Thunderstorms", *J. Geophys. Res.,* 100, 14335-14345.

Year Book of Wireless Telegraphy and Telephony [1914] The Marconi Press Agency, Ltd., London.

SPACE-TO-GROUND INTERFEROMETRY FOR RADIO ASTRONOMY

Hirax Hirabayashi

1. Radio Astronomy

Radio Astronomy had its birth in the 1930s, when Jansky realised that a constant source of interference to radio communications was extra-terrestrial in origin. Reber furthered these studies, publishing his results in the Astrophysical Journal and thus bringing them to the attention of the astronomical community. After World War II, research in Australia and England discovered a number of celestial radio sources. As angular resolutions of several degrees were typical at the time, it was impossible to identify the radio sources with counterparts at optical wavelengths. The use of interferometric techniques -- the sea-cliff interferometer, analogous to Lloyd's mirror, in Australia, and a two element interferometer, analogous to Young's double-slit experiment, in England – enabled resolutions of ~10 arc-minutes to be achieved, and the first optical identifications of radio sources soon followed. Surprisingly, a number of the brightest radio sources were not in our own galaxy, but were found in other galaxies. By the 1950s it had been established that the radio emission was predominantly synchrotron radiation from relativistic electrons spiralling in magnetic fields.

Connected element interferometry was developed in the 1950s and early 1960s, with a central frequency standard being used to supply a reference phase to all telescopes, and with all telescope signals being combined, or `correlated', in real time. The separate developments of magnetic tape recording and stability in frequency standards led to `independent Local Oscillator' observations in the mid-1960s, the technique known today as Very Long Baseline Interferometry (VLBI). Developments since then have served to make VLBI observations more sensitive: better amplifiers, hydrogen maser frequency standards, better recording techniques, allowing larger bandwidths to be recorded, and more computer-intensive data analysis techniques.

Successful VLBI at giga-hertz frequencies (centimeter wavelengths) requires the positions of the participating telescopes to be known with centimeter precision, intercontinental VLBI quickly became a useful tool for geodecists, with continental plate motions of several centimeters per year able to be determined. Similarly, variations of up to several milliseconds in the length of the sidereal day and the apparent motion of the Earth's rotational poles around their mean position could be tracked by VLBI observations. For the remainder of this paper, however, only the astronomical applications of VLBI are considered.

One of the startling discoveries of VLBI came in the 1970s with the observation of apparent separation of components in the nuclei of active galaxies at speeds greater than the speed of light. This `super-luminal' motion was convincingly demonstrated to be real (and not an observational artefact) by the late 1970s with the introduction of VLBI imaging techniques, and is now widely accepted to be due to relativistic motion of radio-emitting plasma close to the line of sight. Such relativistic motion also results in a strong doppler boosting of the emission, with the result that superluminal sources feature prominently among the brightest, most compact radio sources known.

The VLBI technique is now well-established [*Thompson et al., 1986; Burke and Graham-Smith 1997; Zensus 1997*]. For ground VLBI experiments, practical considerations limit the longest projected baselines to about 80% of the Earth's diameter, or about 10,000 km. This, in turn, limits the highest angular resolution achievable at any given frequency by VLBI techniques, e.g. for observations at 5 GHz, corresponding to a wavelength of 6 cm, the Earth-based resolution limit is about 1 milli-arcsecond (mas). However, many galactic and extragalactic radio sources contain components that remain unresolved with the highest angular resolution available using global ground-based VLBI systems.

2. Space VLBI

The idea of improving the resolution by having at least one element of a VLBI array on an Earth-orbiting spacecraft - 'space VLBI' - has been under discussion since the first successful VLBI observations were made. The feasibility of VLBI with an orbiting element was demonstrated in a series of experiments with a communications satellite of the Tracking and Data Relay Satellite System (TDRSS). Observations using the 4.9 m antenna of a TDRSS satellite and 2--3 ground radio telescopes at 2.3 GHz and 15 GHz were made between 1986 and 1988, demonstrating that the basic concepts of space VLBI worked and that many sources have fine structure which could be observed on baselines up to two Earth diameters [*Levy et al. 1986; Linfield et al. 1990*].

These observations, and other studies in Russia, Europe, and the USA, paved the way for the development of the HALCA satellite by the Institute of Space and Astronautical Science (ISAS).

During its development, HALCA was referred to as MUSES-B as it was the second in the series of ISAS's Mu Space Engineering Satellites. The primary goals of the satellite were to test the technologies required for VLBI in space, foremost of which are the deployable main reflector of the radio telescope, stable two-way communications between the satellite and ground tracking stations including the wide-band VLBI data downlink, and accurate orbit determination. The satellite is the orbital element of the VLBI Space Observatory Programme (VSOP), a large international collaboration of space agencies and ground observatories which have combined resources to create the first dedicated space VLBI mission.

3. The HALCA Satellite

The MUSES-B satellite was carried into orbit on on 12 February 1997 on the first launch of ISAS's new M-V rocket. The solid-fuel M-V rocket has a diameter of 2.5 m, a significant increase on the 1.5 m of its predecessor, the M-3SII. The increased power of the M-V allows much heavier satellites to be placed in orbit.

After its successful launch, MUSES-B was renamed HALCA (the Highly Advanced Laboratory for Communications and Astronomy). At just under 830 kg, HALCA is the heaviest satellite to have been launched on an ISAS rocket. A breakdown of the satellite by mass is given in Table 1.

The four-stage M-V rocket placed HALCA in an orbit with an apogee height of 21,400 km above the Earth's surface and a perigee height of a little over 200 km. In the first two weeks after launch three perigee-raising maneuvers, performed by firing the Reaction Control System thrusters while the satellite was at apogee, were carried out to raise the perigee to 560 km.

The Ku-band antenna and solar panels and were deployed during HALCA first orbit. The 7 m^2 of solar panels provided 700 W after one year in orbit, and two rechargeable Ni-Cd batteries are used to power the satellite during eclipses. The main antenna and sub-reflector were not deployed until after the perigee raising had been completed. The hexagonal sub-reflector was deployed by straightening three hinged support legs, which raised the sub-reflector to a height of 3.4 m above the main reflector.

Table 1 *Satellite Mass Budget (kg)*

Satellite Mass	kg
Radio astronomy payload	25.0
Ku-band communication	30.5
Deployable antenna	256.5
GPS receiver	8.6
Electric power	115.3
S-band communication	8.3
Command and data processing	16.4
Attitude and orbit control	80.3
Reaction control system (RCS)	30.0
Launch operation	5.7
Thermal control	19.2
Structure	123.5
Wire harness and others	45.0
Satellite dry mass	764.3
RCS propellant	62.0
Total mass	826.3

The main reflector consists of a mesh of gold-coated molybdenum wire, suspended between six extendible masts and shaped by a tension-truss arrangement. Deployment of the main antenna masts took place over 26 - 27 February, and is described in more detail in *Natori et al. [1998*]. The main masts were extended for 4.8 m and locked in place, after which tensioners were extended an extra 30 cm. As shown in Table 1, the antenna system accounts for 31% of the total satellite mass, however the main reflecting mesh has a mass of only 11 kg! The main reflector is roughly hexagonal in shape, as shown in Figure 1, with a maximum tip-to-tip dimension of 10 m and with a surface area equivalent to that of an 8 m diameter reflector.

Radio waves are focussed from the main reflector to the sub-reflector and into a 2.5 m long feed horn. Ground telescopes usually use separate, optimized, feed horns for each observing band, however design constraints rendered this impossible for the satellite. HALCA is designed for observing at 1.6 GHz (18 cm), 5.0 GHz (6 cm) and 22 GHz (1.3 cm), and a separate coupler for each observing band is located near the throat of the feed horn. The antenna system is designed with a nominal efficiency (compared to an ideal reflector) of about 50% at 5 GHz, with the main losses being due to aperture illumination, spillover and blockage by the sub-reflector supports. Testing the accuracy of the antenna surface, which operates in an effectively zero-g environment in space, on Earth is a major task, but a series of tests have yielded an expected surface accuracy of 0.6 mm (rms) from the ideal surface. This surface accuracy -- outstanding as it is -- results in a lower nominal efficiency at 22 GHz than at 5 GHz. The nominal antenna efficiency at 1.6 GHz is lower due to larger spillover losses which results from the sub-reflector size being limited by the dimensions of the M-V rocket faring.

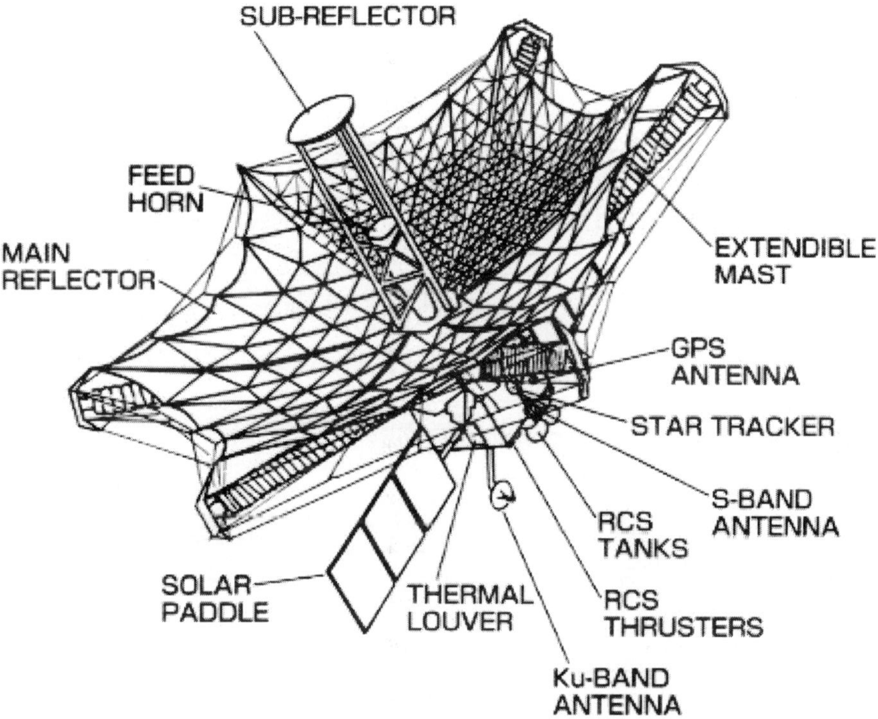

Fig. 1. *Line drawing of the HALCA satellite after deployment in orbit of the main antenna. The two solar paddles, each with three panels of solar cells, cover a total area of 7 m^2. The Reaction Control System (RCS) thrusters were used in perigee raising maneuvers but are not used in normal day-to-day operations. The satellite has one 14.2 –15.3\,GHz (Ku-band) and three 2.1 –2.3\,GHz (Sband) antennas, both bands being used for two-way communications.*

4. HALCA Science Subsystem

A block diagram of the HALCA science subsystem is shown in Figure 2. Mass and power limitations resulted in HALCA being designed to observe only Left-Circular Polarized (LCP) radiation. Quarter-wave plates combine the two orthogonal linear polarizations into LCP at the feed horn, with the resulting radiation being transmitted to the Low Noise Amplifiers (LNAs) by coaxial cables at 1.6 GHz and 5 GHz, and a waveguide in the case of the 22 GHz.

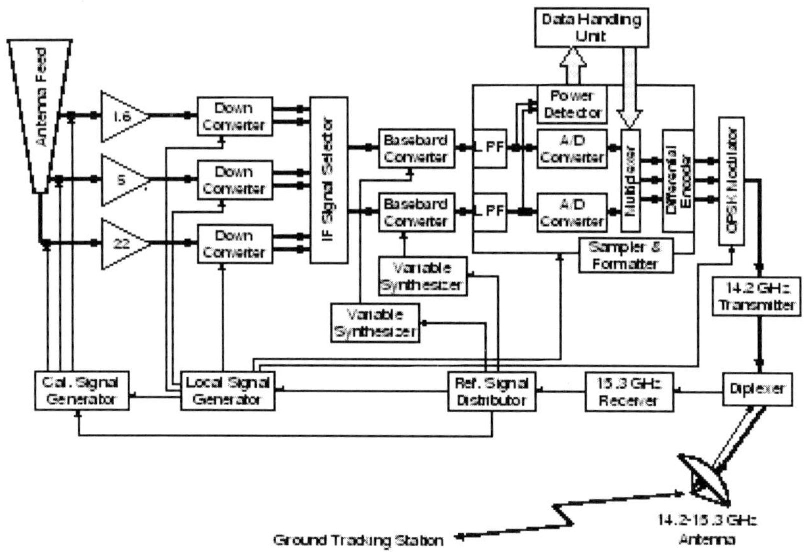

Fig. 2. Block diagram of the on-board radio-astronomical science system. The triangles represent the low noise amplifiers for the three frequency bands. After down-conversion of the Intermediate Frequency signals are selected, and coverted to baseband. After Low Pass Filtering, Analog to Digital conversion of the signals is carried out. All timing signals on thespacecraft are synthesized from an oscillator phase-locked to the 15.3 GHz reference signal transmitted by a ground tracking station.

Ground telescopes usually have the LNAs mounted as close as possible to the feed horn to avoid additional losses. For HALCA, simulations showed the base of the feed horn suffered from the largest vibrations during launch, and so it was decided to mount the LNAs on the spacecraft body. After launch it was discovered that there is a high attenuation (~10 dB) between the feed and the 22 GHz LNA, possibly caused by damage to the 22 GHz waveguide during launch. This large loss in sensitivity at 22 GHz has resulted in observing with HALCA being conducted mainly at 1.6 and 5 GHz.

The 1.6 and 5 GHz the LNAs have receiver noise temperatures of ~35–40 K and gains of ~ 35–40 dB. The physical temperatures of the LNAs vary from −24 C to +13 C, depending on the orientation of the satellite and the time since the last eclipse. The rest of the science subsystem is located on a solar panel side of the spacecraft main body, i.e. one which is not directly illuminated by the Sun, and which is therefore a good location for passive cooling of the components.

After the LNAs, the signals from each of the three radio-frequency (RF) bands are down-converted to intermediate frequency (IF), split into two paths, and sent to the IF switch. Two of the six input signals, usually from the same RF band, are then sent to two separate video (or `baseband') converters. These two channels are referred to as channel A and channel B. Each video converter is an image rejection mixer (IRM) which converts the IF signal to baseband (0–32 MHz) as well as rejecting the lower side-band image frequencies. The LO frequencies for the two video converters can be chosen in integral MHz steps and are independently set.

After the video converters, the A and B channel signals each pass through either 16 MHz or 32 MHz low pass filters, depending on which HALCA observing mode has been selected. At this stage the A and B channel signal levels are sampled by square law detectors, the 12-bit digital outputs of which are sent to the main spacecraft computer, the data handling unit (DHU), to be subsequently transmitted to the ground as part of the VLBI data frame headers. These power levels are used to determine the HALCA system temperature and to monitor the sampler power levels. The A and B channel signals are then one- or two-bit digitized, QPSK (Quadrature Phase Shift Key) encoded onto the 14.2 GHz Ku-downlink carrier, further amplified, and sent to the Ku-band link antenna for transmission to a tracking station on the ground. The data rate of the downlink signal is 128 Mbps. The steerable Ku-band antenna has a diameter of 45.5 cm and is located on a boom to extend the duration of two-way communication passes with the tracking stations.

HALCA does not have an on-board hydrogen maser frequency standard, but receives its reference signal from the ground as a 15.3 GHz tone derived from a hydrogen maser at a tracking station. This 15.3 GHz tone is received by the same Ku-band antenna that is used to transmit the

14.2 GHz downlink. The 15.3 GHz tone is sent to the Ku-band receiver, which searches for, detects, and converts the tone into the fundamental 50 MHz reference signal for the spacecraft.

HALCA system temperatures are measured using noise diodes (one for each frequency band). On/off power ratios and a model for the noise powers enables the system temperatures in both channels to be determined. Phase calibration tones every MHz throughout the observing band can be inserted during observations. Both the noise diode and phase calibration signals are injected through 25 dB couplers into the LNAs. The phase calibration tones are extracted at the tracking stations.

The gain of HALCA's radio telescope was determined in orbit by measuring the increase in detected power during observations of the strong radio sources of known flux density Cygnus A and Cassiopeia A. The Cygnus A measurements resulted in values of 0.0043 K Jy^{-1} at 1.6 GHz and 0.0062 K Jy^{-1} at 5 GHz, where a flux density of 1 jansky (Jy) corresponds to 10^{-26} Wm^{-2} Hz^{-1}. For an antenna with a total physical area of 50 m^2 the gain corresponds to an antenna efficiency of 24% at 1.6 GHz and 33% at 5 GHz.

Typical measured values of the system temperature are 75 K at 1.6 GHz and 95 K at 5 GHz, with a ~20% increase in system temperature observed during perigee passes due to terrestrial interference.

5. VSOP Observations

During VSOP observations, a reference phase, derived from a hydrogen-maser--based frequency standard at one of the tracking stations, is uplinked to HALCA at 15.3 GHz. HALCA downlinks the astronomical data in real time to the tracking station at 14.2 GHz where it is recorded. Five dedicated 15 GHz-band tracking stations are used to support VSOP observations: at Usuda (Japan), operated by ISAS; Green Bank (USA), operated by the U.S. National Radio Astronomy Observatory (NRAO); and Tidbinbilla (Australia), Goldstone (USA), and Madrid (Spain), operated by NASA/JPL Deep Space Network (DSN). Range-rate data from this ground telemetry network is used by ISAS and JPL (Jet Propulsion Laboratory) orbit determination teams for providing pre-observation 'predicted' orbits for the tracking stations and post-observation 'reconstructed' orbits for use in correlation.

A number of ground radio telescopes have committed time for simultaneous observations with VSOP. Commitments of ground radio telescope time for the mission have been negotiated through the Global VLBI Working Group (GVWG), formed under URSI Commission J. Participating VLBI networks include the Very Long Baseline Array and the European VLBI Network, Asia-Pacific Telescope, DSN 70 m antennas, and individual telescopes.

HALCA's elliptical orbit makes it interferometric observations possible on baselines up to 30,000 km -- three times longer than achievable from ground-based observations. The maximum baselines are obtained when observing sources located near the normal to the orbital plane, however, due to the Earth's oblateness, HALCA's orbit precesses with a period of ~1 year and so good imaging capability over the whole sky can be obtained during the life of the mission.

The standard recording mode for VSOP observations has two channels of 16 MHz bandwidth with 2-bit sampling, producing a data stream of 128 Mbps. Three different recording formats are used, VSOPT, VLBA/MkIV, and S2. Observations involving ground telescopes with VLBA or MkIV recorders are usually correlated at the 20-station VLBA correlator in Socorro, operated by NRAO. Observations using predominantly ground telescopes with S2 recorders are usually correlated at the 6-station S2 correlator in Penticton, operated by the Dominion Radio Astrophysical Observatory and the Canadian Space Agency. Observations with ground telescopes with VSOPT recorders are most often correlated at the 10-station VSOP correlator in Mitaka, operated by Japan's National Astronomical Observatory. When an observation requires mixed-format recording, translation to the format appropriate for the designated correlator is undertaken at Mitaka.

Observations are classified into two main categories; the General Observing Time Program and the mission-led Survey Program. General Observing Time observations last typically 10 hours. Each pair of telescopes forms a two-element interferometer, and a wide range of baseline lengths and orientations needs to be sampled during the observation to enable high fidelity images to be obtained. General Observing Time proposals from astronomers worldwide are reviewed based on their scientific merits by an international Scientific Review Committee, and scheduled by the VSOP Science Operations Group. The VSOP World Wide Web site (http://www.vsop.isas.ac.jp/) contains additional details about observing Announcements of Opportunity as well as much other information about the mission, including images obtained from VSOP observations. General Observing Time observations have been made of radio galaxies, OH masers and pulsars, but most observations are made of quasars and BL Lacs. VSOP Survey Program observations are made of Active Galactic Nuclei at 5 GHz and are generally of shorter duration and made with fewer telescopes than the General Observing Time observations. Nevertheless, the Survey will provide a large complete sample of sub-milliarcsecond radio structures for statistical studies.

Although the 22 GHz observing system was damaged during launch, a large outburst of the water vapor maser in the Orion KL nebula enabled a VSOP test observation at 22 GHz to be carried out. Interferometric fringes were detected on baselines between ground radio telescopes and HALCA for short projected baselines.

An example of an image derived from a VSOP observation is shown in Figure 3. This image was made from an observation at 5 GHz of the quasar 0212+735 [*Hirabayashi et al., 1998*]. A redshift of 2.367 was measured for this quasar, placing it at a distance of 12 billion light-years (assuming a Hubble's constant of 75 km s^{-1} Mpc^{-1}, where 1 mega-parsec (Mpc) = 3.26 million light-years, and a cosmological deceleration constant of 0.5). This image was obtained from a 14 hr observation on 6 September 1997 with the ten telescopes of the Very Long Baseline Array. Typical of most radio-loud extragalactic sources, the source has a compact bright `core' and an extended milli-arcsecond scale 'jet' on one side of the core. The standard physical model for this structure is that the core contains a super-massive black hole and that symmetric relativistic jets, powered by accretion of material onto a disk surrounding the black hole, are ejected from this core. In most cases only the jet moving towards us is seen on the milli-arc-second scale due to substantial Doppler boosting.

The core of 0212 + 735 is the bright feature on the right-hand side of the image. Comparison of Figure 3 with previous ground-only images indicates that components at 2.1 and 6.3 mas from the core can be identified with components observed over the last 15 years. It had previously been reported that the further of the two from the core was moving away from the core with an apparently superluminal speed, and combining a 1985 epoch and the VSOP image indicates the closer of the two components to the core is moving with a similar speed [*Hirabayashi et al., 1998*]. It is natural to assume the jet extends to the feature ~14 mas from the core, though any such connection is too weak and/or too resolved to be detected. This bright feature (on the left-hand side of the image) was also evident 15 years ago but it does not appear to have moved over the intervening period, although it is difficult to discern motions of such extended features.

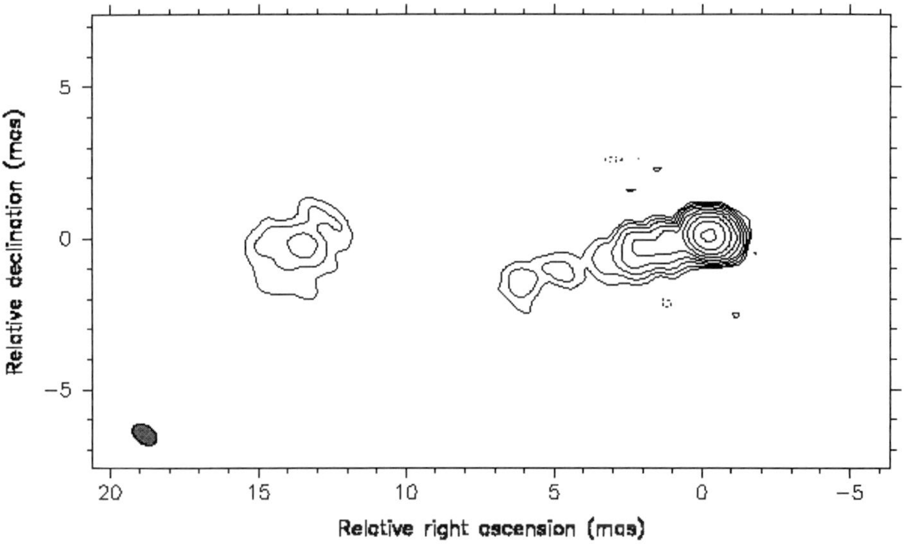

Fig. 3. *A 5 GHz image of the quasar 0212+735 made from an observation on 6 September 1997 with HALCA (using the Green Bank and Goldstone tracking stations) and the ten-element Very Long Baseline Array. The core of the quasar is the bright feature at the right-hand end of the image. The beam synthesized by the array of telescopes (including HALCA) is 0.9 mas x 0.6 mas (FWHM) at a position angle of 57.0 degrees, and is shown at the bottom left of the image. The noise level in the image is 0.9 mJy per beam (rms) and the controus levels are –1, 1, 2, 4, 8, 16, 32, 64, 128 and 256 times 5 mJy per beam, with the negative (noise) contours shown as dashed lines.*

6. The Future of Space VLBI

HALCA is expected to have a lifetime of about 5 years, limited in part by degradation of the solar panels by cosmic rays. Several missions are being prepared or planned to follow the VSOP project. The Russian RadioAstron [*Kardashev, 1997*] is planned for a higher orbit than HALCA providing higher resolution at frequencies up to 22 GHz. As a successor to HALCA, a 10 m class antenna in a slightly higher orbit than HALCA, with a range of frequencies possibly up to 43 GHz is being considered in Japan [*Hirabayashi, 1999*]. A U.S.-led mission, ARISE, is pursuing the use of a 25 m class inflatable antenna for observations up to 86 GHz [*Ulvestad and Linfield, 1998*]. All three missions will allow higher angular resolution studies of celestial radio sources in order to study how they are powered and how they, and the Universe, are evolving.

Acknowledgements

The VSOP Project is led by the Institute of Space and Astronautical Science, with significant contributions from National Astronomical Observatory of Japan, the Jet Propulsion Laboratory, the U.S. National Radio Astronomy Observatory, the Canadian Dominion Radio Astrophysical Observatory, the Australia Telescope National Facility, the European VLBI Network, the Joint Institute for VLBI in Europe, and the directors and staff of many of the world's radio observatories. Philip Edwards and the MRS editors are thanked for their help in preparing this manuscript.

References

Burke, B. F. and F. Graham-Smith [1997], An Introduction to Radio Astronomy, Cambridge, Cambridge University Press.

Hirabayashi, H., H. Hirosawa, H. Kobayashi, et al., [1998], "Overview and Initial Results of the Very Long Baseline Interferometry Space Observatory Programme," *Science*, 281, 1825-1829.

Hirabayashi, H. [1999], "The Japanese Space VLBI Mission After VSOP," *Advances in Space Research*, in press.

Kardashev, N. S. [1997], "RadioAstron - a Radio Telescope Much Greater than the Earth" *Experimental Astronomy*, 7, 329-343.

Levy, G. S., R. P. Linfield, J. S. Ulvestad, et al. [1986], "Very Long Baseline Interferometric Observations Made with an Orbiting Radio Telescope," *Science*, 234, 187-189.

Linfield, R. P., G. S. Levy, C. D. Edwards, et al. [1990], "15 GHz Space VLBI Observations Using an Antenna on a TDRSS Satellite," *Astrophysical J.*, 358, 350-358.

Natori, M. C., T. Takano, A. Ohnishi, et al. [1998], "In-Orbit Deployment of a High Precision Mesh Antenna System for Space VLBI," in 39th AIAA/ASME/ASCE/AHS/ASC Structures, *Structural Dynamics and Materials Conference*, paper AIAA-98-1837.

Thompson, A. R., J. M. Moran, and G. W. Swenson [1986], Interferometry and Synthesis in *Radio Astronomy*, New York, Wiley.

Ulvestad, J. S. and R. P. Linfield [1998], "ARISE - A Proposed Future Space VLBI Mission," in J. A. Zensus, G. B. Taylor, and J. M. Wrobel (eds.), IAU Colloquium 164: Radio Emission from Galactic and Extragalactic Compact Sources, *ASP Conference Series* 144, 397-398.

Zensus, J. A. [1997], "Parsec-Scale Jets in Extragalactic Radio Sources," *Annual Review Astronomy Astrophysics*, 35, 607-636.

FUTURE GENERATIONS OF MOBILE COMMUNICATIONS SYSTEMS - THE SCIENTIFIC ASPECTS

Jørgen Bach Andersen

1. The ITU Vision

Communication via radio is at the heart of URSI′ s scientific activities, embracing many URSI Commissions. In the last few years it has also experienced an unprecedented growth globally with hundreds of millions of users of mobile telephony. There is no reason to believe that this trend will stop, it will more likely gradually change into other more sophisticated uses of the wireless possibilities. It is the purpose of this overview to look a little into the future and take note of some possibilities, advantages and disadvantages, basic constraints versus practical and economic constraints. Looking 5 - 10 years ahead is an almost impossible task and the result will probably be full of mistakes, since many important decisions will be taken from an economic, regulatory or other non-scientific aspect. However, it is an interesting exercise to probe the limits of our scientific barriers to progress, with the hope of giving an overview to those not familiar with the subject.

The forecasters are highly influenced by the rapid growth of the Internet, where browsing and downloading of large files, on-line reception of radio broadcasts and TV-news and a multitude of other possibilities abound. The thinking is that the public will pay for similar uses over a wireless connection, or maybe more likely that there will be a number of professional uses justifying the additional investments in terminals, infrastructure etc.

Whatever the outcome, the convergence between distributed computing, data communications and telecommunications in general is rapidly taking place, and wireless will have its share of the future IT scenarios. The future scenario envisaged by ITU in Fig.1 tells the story succinctly, in a few years there will be more mobile terminals than fixed.

Considering the huge capital invested in the infrastructure of 2nd generation systems it is likely that there will be gradual transitions to more advanced services, a gradual extension of the bandwidth like in GSM phase 2+, where systems like HCCSD (High Speed Circuit Switched Data services),

GPRS (Generalised Packet Radio Service), and EDGE (Enhanced Data rate for GSM Evolution) will be introduced in the near future (Fig. 2).

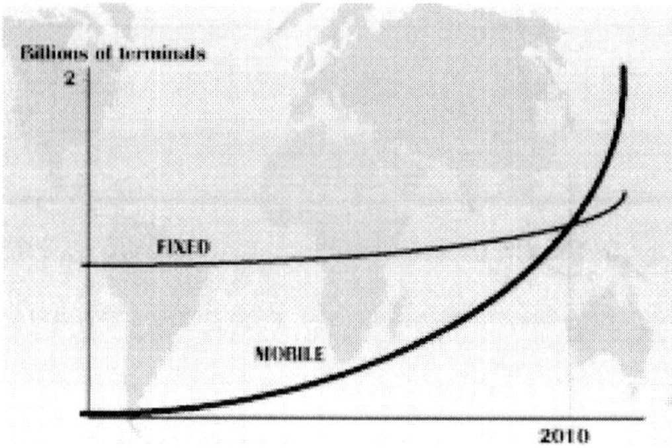

Fig. 1 *Future developments in terminals (ITU)*

The present paper is focussing on the terrestrial systems, but it is clear that there is, and will be more in the future, an important satellite component more or less integrated with the terrestrial systems, the so-called GMPCS (Global Mobile Personal Communicationsby Satellite). The capacity is limited for obvious reasons, the total number of voice channels is quoted to be of the order of 200.000 for all competing systems (ICO, Globalstar, Iridium), but in contrast to the terrestrial systems they will be truly global.

The new uses are dominated by much higher data rates than available today over a mobile phone, and it is this bandwidth expansion coupled with an increasing share of non-voice traffic that will dominate future systems. The mobility is the important by-product of a wireless system, but mobility may also just mean freely moveable systems, not necessarily at high velocities. At the time of writing this paper several national and regional agencies have submitted proposals to ITU for what is now called IMT-2000 (International Mobile Telecommunications by 2000), where the intent is that there should exist a global standard for third generation systems. Such negotiations have many non-scientific aspects like backwards compatibility to existing systems and intellectual property rights, and the outcome is uncertain at the moment, but we shall briefly review the present state of affairs later.

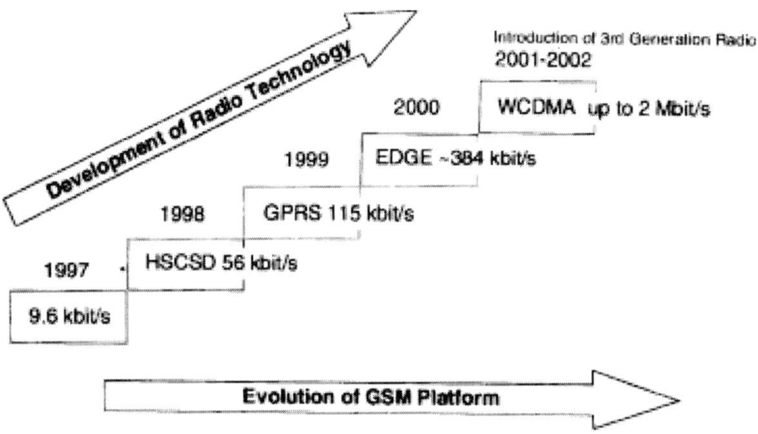

***Fig.** 2 Evolution of GSM towards 3rd generation*

One possible definition of the social scenario of the future is the following:

The user boundary condition

The high frequency electromagnetic spectrum (from 1 to 60 GHz) should be utilised efficiently for the provision of high data rates (from 10kb/s to 150 Mb/s) to many simultaneous, communicating users with portable equipment (with speeds from 0 to 500 km/h) over a certain geographical area (from a single room to the globe).

It is obvious that not all conditions can be satisfied simultaneously, but it sets the scene for what the ultimate goal is. It is a unique technical situation due to the openness and vulnerability of the sharing of spectrum, different from broadcasting, since everybody is both transmitting and receiving, and limited by the bottleneck of politically allocated pieces of spectrum. The main objectives that ITU has arrived at for the 3rd generation terrestrial systems are

- Full coverage and mobility for 144 kb/s, preferably 384 kb/s
- Limited coverage and mobility for 2 Mb/s
- High spectrum efficiency compared to existing systems

- High flexibility to introduce new services
- A fast and efficient packet-access scheme

2. Modulating and Accessing the Wide Bandwidth

During 1998 there has been a general focus on spread spectrum systems globally for 3rd generation systems, although some details are not yet settled (December 1998). The access systems are usually hybrid with FDMA, TDMA, and CDMA components. It is a good guess at the moment that some type of CDMA (Code Division Multiple Access) is a likely candidate for the 3rd generation systems, since it is supported by major industries. We shall briefly outline the air interface [*Ojanperä and Prasad, 1998*] without going into specific details of the various proposals. Some of the more scientific aspects will be dealt with later.

The major advantage of the spread spectrum systems, where CDMA is one example, is that it is well suited to multi-user systems, since other users or interferers appear as noise, if the codes are orthogonal. Thus bandwidth extension is used to make a power efficient system, and not really to accommodate high data rates, but the total bandwidth allocated is also a compromise of many factors, including regulatory. The bandwidth allocation is 5 MHz, although there are plans for extension to 10, 20 MHz later for more advanced services. The chip rates are a few Mchips/s (the details depend on the specific proposal), so indeed the spreading gains are marginal for the highest data dates.

Multi rate design is important for the flexibility so that different quality of service can be provided in a spectrum-efficient manner. The multiple rates may be time multiplexed or code multiplexed and can relate to different services. Adaptive systems which send a high data rate when a channel is good, and a low data rate when it is bad, will improve the overall spectral efficiency, but the gain depends on the acceptable delay and the additional cost of feeding back the information on the state of the channel. In general coherent detection is foreseen with user-dedicated pilots, and I and Q (complex) multiplexing for data and control channels. The additional pilot channel facilitates the use of adaptive antennas in the downlink. Power control is essential for CDMA systems with frequencies as high as 1600 Hz.

The main proposals involve an FDD (frequency division duplex) concept with transmit and receive frequencies separated in paired bands, with 1920-1980 MHz in one band and the other in the 2110-2170 MHz band. It has also been suggested to use unpaired bands, more suitable for unregulated applications, where the mode is TDD (time division duplex), i.e. a fraction of the time burst is used for transmission, the other part for reception. This allows for an asymmetrical mode, where more slots are allocated for downlink than for uplink in a typical situation, where a user is browsing the Internet. The TDD mode might still use code spreading for individual users.

Table 1 *Key parameters of 3rd generation IMT-2000 cellular system as of December 1998*

Spectrum allocation (WARC-92)	1885-2025 and 2110-2220 MHz
For terrestrial use	1885-1980, 2010-2025, 2110-2170 MHz
FDD use (UMTS)	2 * 60 MHz
TDD use	35 MHz
CDMA channel bandwidths	5 MHz (10, 15, 20 MHz)
User data rates (variable)	144-384- 2000 kb/s
Spreading and data modulation	complex BPSK, QPSK
Power control, open and fast closed loop	1600 Hz
Handover	soft handover, interfrequency handover

The modulation is QPSK or BPSK for the CDMA systems. It should be mentioned that there are also TDMA schemes in the bag of proposals for ITU, and they may use offset quadrature amplitude modulation (Q-O-QAM) which provides high data rates and good spectral efficiency, such that 384 kb/s may be sent over a channel with a 200 kHz channel spacing with reasonable range and capacity.

3. Describing the Mobile Channel - The Electromagnetic Environment

From a radio science point of view it is the mobile channel that makes the wireless communication different from the wired case. Understanding it, taking proper measures against it or utilizing it, and modelling it for simulation purposes or for planning purposes are all important endeavours. In general the theoretical description of the mobile channel in terms of Bello′ s functions is well understood. The different environments have path loss dependencies and second order moments in terms of delay spread, Doppler spread and angular spread as seen from a base station, and at least for the 2GHz region a number of experimental results are available. What is much harder and where work remains to be done, is to know in advance, to predict, the wideband properties for different environments. In general ray-tracing tools are adequate (but time consuming) for predicting the total power, but not very successful in describing the impulse response. The reason is probably that the urban and indoor environments require many reflections, diffractions and scatterings to adequately describe the situation, and these are either too time consuming to apply, or the necessary information is not available.

Most outdoor above rooftop prediction results refer to 2D models, where the wave is supposed to arrive via the shortest direct path involving a number of reflections and diffractions. Planning tools most often use parametric models where the parameters are derived from measurements, often such models are 20 dB or more wrong, so the situation is not satisfactory. It would be valuable if power prediction was within a few dB averaged over the fast fading based on geographical data base information, either by a deterministic or a stochastic approach, where in the latter case the environment is described in stochastic terms. Such research is ongoing, and the results would be valuable for the position of and frequency allocation of base stations. The maximum capacity in a region is interference limited, so controlling the interference levels is important. A few 3D models are in progress, but most of them are very complex. An interesting new development is a pure stochastic approach to urban propagation [*Blaunstein, 1998*].

Antennas are in integral part of the propagation environment serving the usual antenna purpose of delivering the power efficiently, but they also serve to increase the capacity of a multi-user network by sectoring a cell, or in a more modern fashion electronically adapt its radiation pattern or element weights, if an array, to continuously maximize the wanted signal over the interfering signals, a so-called smart antenna. We shall expect this trend to continue, so the antenna community will specialize in making small, efficient, wideband antenna elements with a wide antenna pattern, possibly dual polarized, i.e. classical antenna virtues. For the mobile parts the trend will probably be the same, a number of small integrated elements serving as a basis for an adaptive array. The traditional pattern concepts will have less importance, since an antenna in a random medium will be independent of the radiation pattern.

4. Taking Care of the Path Loss -the Coverage Problem

Neglecting for a while the multi-user problem, which creates co-channel interference, a receiver must have a certain energy per bit in order to guarantee a certain bit-error-ratio, and using Shannon's equation

$$\frac{C}{W} = \log_2(1 + \frac{S}{WN_0}) \quad \text{b/s/Hz} \tag{1}$$

where C is the capacity, W the bandwidth, and S the signal power, it follows that in order to keep a desired spectral efficiency (C/W) then power density reaching the receiver must increase proportional with increase in bandwidth for a Gaussian noise limited system. In the high data rate region, $C/W > 1$ it follows that the required signal power increases exponentially with C/W, so there is a limit to how far it is possible in this direction to satisfy the demands of spectral efficiency, or getting enough data through the bottleneck. In the power limited region, $C/W<1$, recent advances in Turbo coding pushes the required SNR/bit towards 1 dB but with R/W around 0.5, where R is the realized symbol rate. The ratio W/C may also be called the spreading factor for a spread spectrum system, and large spreading factors alleviate the need for large powers, but at the expense of bandwidth, which is one expensive factor. Realizing that higher order modulations (like *QAM*) is one way of increasing the bit rate these possibilities are under further study. From a battery powered portable terminal this leads to problems with the output power and linearity problems in the transmitter, but let us assume that signal processing (although that also requires power) takes care of the latter problem, we are still left with output power limitations. In order to appreciate the fundamental limitations a closer look at the path loss is required.

For a terrestrial system the propagation path adds its own well-known peculiarities of multipath propagation and fading. For a narrowband system without antenna diversity a rather large power margin is used in order to guarantee a certain quality in spite of fading. This seems to be a waste of power and in future systems we should strive to eliminate this power margin. The use of a high bandwidth in itself solves part of the problem if there is sufficient delay spread, since frequency diversity diminishes the fading. Also, there are many good reasons for having multiple antennas in both ends of a link taking advantage of polarization and other types of antenna diversity. Concentrating on the mean power transmit, receive diversity or both may be applied to give large average antenna gains and virtually no fading [*Andersen, 1999b*] through a random medium. Adaptive arrays at both ends will help solve the power problem.

Comparison between GSM and W-CDMA shows that for same output power CDMA for a speech service will tolerate a few dB higher path loss than a GSM speech service, which can be translated into improved coverage [*Dahlman et al., 1998*].

The most obvious way of getting around the power problem is by reducing the path loss, i.e. by reducing the distance between transmitters and receivers. This is also a current trend in 2nd generation systems with smaller and smaller cells, partly motivated by the higher density of users in urban areas. We shall expect future base stations to be obnoxious and plenty, in certain areas like present day lamp posts, so there will always be line-of-sight and a short distance. The fundamental power problem has then been turned into an infrastructureproblem, i.e. an economic problem instead of a law of nature. The IMT-2000 proposals reflect the path loss problem in the sense that the highest data rate of 2Mb/s is only available in limited areas, while for general coverage only 144 kb/s should be guaranteed.

For wireless portable devices driven by batteries the power problem is important, and a number of ways of reducing the transmit power are being studied. They include integrated circuit technology issues and protocol ways of only using power when it is needed. An overview on energy management may be found in [*Zorzi, 1998*].

If the bandwidth problem is solved by going to millimetre wave frequencies, where GHz bandwidths are possible, atmospheric attenuation limits the power exponentially in contrast to the power law dependency at lower frequencies. The output power of solid state devices tends also to decrease with carrier frequency, aggravating the problem of increasing the carrier frequency. Directive, and possibly adaptive, antennas and line-of-sight seem to be the only way to achieve the required power at the terminal.

5. Signal Processing Aspects

With the digital systems more and more emphasis is put on the signal processing, especially since the possibilities keep progressing in accordance with Moore's law. In the beginning much emphasis was on the source coding, and certainly great results have been achieved in terms of speech and image coding, which is part of the 2nd generation systems. More recently it is becoming more and more difficult to separate the signal processing from the information and communication theory. The signal processing plays now a major role in the communication aspects of channel estimation, blind processing, and multi sensor processing, so they deserve a few separate sections. Some recent results may be found in *Li et al. [1998]*.

5.1 Estimating and Predicting the Channel

It is likely that 3rd generation systems will use training sequences and/or pilot signals to keep updating the major characteristics of the time-varying channel, i.e., the receiver uses known information to estimate the instantaneous channel transfer function for a certain instant of time, and assumes that this information is approximately valid until the next pilot or training sequence. For high Doppler frequencies this is only approximately valid, and some sort of interpolation may be required. Inserting training sequences is in a way a waste of valuable bits, if they can be avoided. It is one of successes of signal processing to have come up with ways of so-called blind channel estimation, where the channel properties are estimated from basic symbol properties, not from known symbols. These properties could be of many types, like constant magnitude modulations, finite alphabets, or most simply from a sampling of a constant symbol period signal, where the sampling may be in time or space, or both. Some possibilities are indicated in Figure 4 [*Boss et al., 1998*].

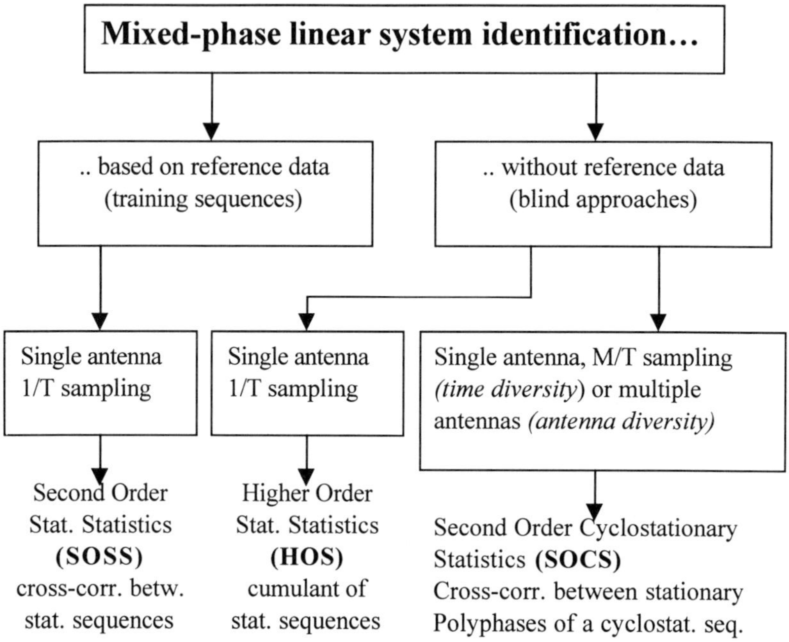

Fig. 4 *Principal ways of identification of linear system excited by a stationary random sequence [Bos et al., 1998].*

For symbol rate sampling the resulting sequence is stationary, and second order statistics are inadequate for identification of complete channel information, including the phase. Thus either higher order statistics, using higher order cumulants, or oversampling is needed, and in general the results are encouraging. It seems however that the present plans for 3rd generation systems do not include blind channel estimation, reflecting a certain caution in industry. Considering cochannel interference the blind techniques may still be of value for suppressing noise-like unwanted users, although blind approaches cannot distinguish interferers and the wanted signal, unless they have different codes.

The fast channel variations are due to the phase mixing of the signals coming from fairly constant scatterers or reflectors in the environment. If these scattering signals are identified it is in principle possible to predict the channel transfer function for future samples. Such predictions would be of value for channel estimates, receiver design, and antenna diversity. Preliminary results have been obtained by *Hwang et al.* [*1998*] and *Eyceos et al.* [*1998*], and some recent results obtained by *Andersen et al.* [*1999a*] are shown in Figure 5. The first 100 samples are used for identification of the Doppler spectrum, and the latter samples are predicted. Note that the fast fading can be predicted a few fades into the future.

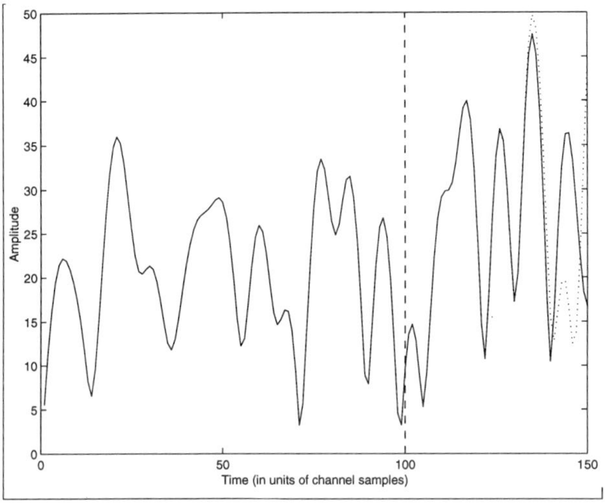

Fig. 5 *Prediction of future channel variations. Left part: The observable channel envelope (100 samples). Right part: Predicted future (dotted) and actual future (solid).*

5.2 The Multi-user Problem

For the multi-user CDMA systems in the 3rd generation the other users serve as interferers and the total capacity is limited even after effective power control has made everybody equal as seen at the base station. The total throughput in a cell for M users is theoretically limited to [*Viterbi, 1998*]

$$MR_b = 1.4W \text{ bits / s} \quad (2)$$

with W as the total bandwidth, and where the interferers are considered as Gaussian noise.

Among advanced methods multi-user detection [*Verdú, 1998*] - also known as multi-user demodulation or interference cancellation - has developed in the mid 1980'ies to become one of the most active research fields in mobile communications. Multi-user detection exploits the inherent structure of the received signals of the active users in order to cancel or mitigate the interference among these signals before separately detecting the information data of each user.

The optimum multi-user receiver in a Gaussian channel has been derived and its performance theoretically analyzed [*Verdú, 1986*]. This scheme achieves a dramatic performance gain compared to the conventional single-user receiver at the cost of a computational complexity which grows exponentially with the number of users. For that reason, investigations have focused on the identification of multi-user detectors of tractable complexity which still allow for a significant performance enhancement over the conventional receiver.

Worth mentioning among the many merits of multi-user detection is the significant reduction of the required precision in the control of the power transmitted by the mobile stations. Moreover, cancellation of the interference among the active users prior to individual data detection leads to a significant performance gain. This gain translates into a reduction of the transmitted power at the mobile stations and therefore into an increased battery life. Overall, multi-user detection enables a transfer of the processing complexity from the mobile stations to the base station.

However, the question whether cost-effective multi-user detection schemes exist which allow for a substantial performance increase is still open. Indeed most theoretical and simulation-based investigations performed nowadays rely on very simplifying, somewhat academic assumptions -- especially regarding the radio channel -- which prove to be far away from the situation encountered in the real world. Nevertheless the appreciable gain potentially achievable by using multi-user detection is still driving the considerable research activity which is currently being conducted in this field. It has also had an impact on one of the IMT-2000 proposals, since the TD-CDMA part of the UMTS proposal assumes multi-user detection and reaches twice the number of users as the W-CDMA part [*Baier et al., 1998*].

Multi-element antennas at the base station, known as smart antennas, may also play a significant role in reducing the interference level, either by nulling strong interferers or placing them into a deep fade (essentially the same thing), or by general reduction of side lobe level in the case of a uniform distribution of interferers.

5.3 High Spectral Efficiencies - the Antenna Solution

Antenna diversity is a well-known mechanism for improving the reception of fading signals, but even with transmitting it is possible to have the benefits of diversity, if the signals transmitted by the antennas are coded differently. In case of a reciprocal channel where the propagation channel may be known at the transmitter, joint transmit-receive diversity is possible with a large gain of two antennas radiating into a scattering medium.

The possibility of high spectral efficiencies (bits/s/Hz) for one channel with Gaussian noise is a well explored subject. It has recently been realised that the heavily scattered channel (large angular spreads) effectively may be considered as a number of independent, parallel channels [*Foschini ,1996*]. The reason is that the environment becomes part of the antenna and only a small antenna separation leads to a different set of channels. By suitable channel estimation of the N M paths existing between M transmit antennas and N receive antennas the coupled paths may be decoupled with an approximate channel capacity for fixed signal-noise-ratio P ($P>>1$) [*Andersen, 1999b*]

$$C = N\log_2(\frac{M}{N}P) \tag{3}$$

with the result that the theoretical capacity grows linearly with the number of elements for M/N fixed. For the case of unknown channels at the transmitter the power is divided by M instead of N giving the asymptotic capacity for N large

$$C_{unknown} = N\log_2(P) \tag{4}$$

6. Software Radio

As was apparent from section 5 signal processing will play a dominant role in future terminals and base stations. A full extrapolation of the concept leads to the software radio, a term coined to express the idea that terminals could essentially consist of a battery, a few antennas and a computer. The vision sees the terminal as reconfigurable, such that the software is not fixed once and for all, but downloadable. In this way the terminal might sense its radio environment and choose whether it is a CDMA terminal or a TDMA terminal with given parameters, or any other system related parameters.

The vision may not be realized for various reasons, but some aspects will no doubt be introduced. Contrary to the name, software radio is a lot about the hardware. A generic transceiver must have linear transmitter requirements, a wide operational bandwidth, say from 100 MHz to 2.3

GHz, a wide channel bandwidth of the order 5 MHz, and a high power efficiency, wideband amplifiers, upconversion mixers, and high rejection analogue filters. The analogue-to-digital conversion is a major technical limitation.

7. Local Area Networks (LANs), Wireless Broadband and Short Range Radio

The previous discussion has been centered around cellular systems where the cells could range from kilometres to tens of metres. Wireless communications is, however, much more than cellular communications. Ad hoc networks created between computers are already commercial products, working in the 5 and 17 GHz regions. Wireless local loops make an important impact in an area totally dominated by wires, namely the last mile connection to the subscriber. Roll-out costs are small compared to digging wires or fibres, and it is fast. Broadband Wireless Access is a similar technique in the 40 GHz region bringing high data rates to consumers. Both point-to-point and point-to-multipoint links are important, and it should not be forgotten, that communication between base stations in a cellular system is popular, especially for those operators who do not have their own wired network, so radio is all pervasive.

The latest innovation is a low-cost, radio based cable replacement technology, called Bluetooth. It is in the unlicensed ISM band at 2.4 GHz, and other technical features are peak data rates of 1 Mb/s, fast frequency hopping over 75 hop channels, and small output power. The basic idea is that many consumer products, incl. computers, in the future will contain a Bluetooth radio, so a laptop could be connected to a nearby printer, a mobile phone, a headset, another laptop, all via radio. The range will be small, maybe a few tens of metres, but radio will have replaced cables.

8. The Biological Problem?

The bioelectromagnetic problem of having transmitters near the human head is expected to diminish in future generations with lower output power due to short distance transmissions, and the use of handheld devices with visual information and a simple wire connection to the ear. The connection to the ear might also be via a very low power radio connection as mentioned above.

Nevertheless, still for many years there will be hundreds of millions of users using traditional handsets with antennas near the head, and much attention is being paid now through many research projects around the world. Assuming that amplitude modulation of the power is an active biological agent, as it is generally assumed, how do 3rd systems relate to the biological problem?

A full rate CDMA transmitter will mot have the characteristic, periodic burst of a TDMA transmitter, but as described earlier there is a fast power control of small steps with frequencies of hundreds of Hz, the actual power variation depending on the speed of the user and the environment [*Pedersen and Andersen, 1999*]. The types of modulation planned are no longer constant power, as is the case in GSM, but a BPSK or QPSK signal, the power of which will vary with the chip rate of several MHz. This is probably too high a frequency to be biologically active, but this is unknown.

9. Conclusion

Some specifications of 3rd generation mobile communication systems have been decided by ITU at the time of the URSI General Assembly. No matter what the decision is, there will be a large number of challenging problems for radio scientists for many years to come. In this overview the emphasis has been on the air interface and the signal processing, but equally interesting problems arise on the network side, with mobility management and frequency control, and with protocol issues in general, considering that everything will be data, most likely packet switched. The satellite link has only been briefly mentioned, but it will play its necessary, although limited role.

The overall trend is one of bandwidth demand and an all pervasive use of radio, especially in the lower microwave bands, but with increasing tendencies of involvement of millimetre wave frequencies. The spectral efficiency per area will mainly be achieved by a diminishment of cellular sizes, and for many uses the propagation distances will be only tens of metres. We may thus conclude that there are no technical or scientific barriers to a wideband, global connectivity of people and machines, only economic!

References

Andersen, J.B., J. Jensen, S.H.Jensen, F. Frederiksen [1999a], "Prediction of Fast Fading Based on Past Measurements", Aalborg University, February 1999.

Andersen, J.B. [1999b], "New Aspects of Multiple Antennas for Spectral Efficiency", COMMSPHERE'99, Toulouse, January 1999.

Baier, P.W., T.Weber, M.Weckerle [1998], "Spreading Techniques, A Far Reaching Technology", in *CDMA Techniques for Third Generation Mobile Systems*, F. Swarts, P. van Rooyan, I. Oppermann, M.P. Lötter (eds.), Kluwer Academic Publishers, 1998.

Blaunstein, N. [1998], "Parametric Model of UHF/L-Wave Propagation in City with Randomly Distributed Buildings", *Proc. IEEE Antennas Propag. Symp.*, Atlanta, June 1998, 1684-1687.

Boss, D., K-D Kammeyer, T. Petermann [1998], "Is Blind Channel Estimation Feasible in Mobile Communications Systems? A Study Based on GSM", *IEEE J. Selected Areas in Communications,* 16 (8), October 1998, 1479-1492.

Dahlman, E., B. Gudmundson, M. Nilsson, J. Sköld [1998], "AUMTS/IMT-2000 based on Wideband CDMA", *IEEE Communications Magazine*, September 1998, 70-80.

Eyceos, T., A.Duel-Hallen, H.Hallen [1998], "Deterministic Channel Modelling and Long Range Prediction of Fast Fading Mobile Radio Channels", *IEEE Communication Letters,* 2 (9), September 1998, 254-256.

Foschini, G.J., [1996], "Layered space-time architecture for wireless communicationin a fading environment when using multi-element antennas", *Bell Labs Technical J.*, Autumn 1996, 41-59.

Hwang, J-K., J.H.Winters, "Sinusoidal Modeling and Prediction of Fast Fading Processes", *Globecom'98*, Sydney, 667-671.

Li, Y.G., K.J.R.Liu, J.H.Winters, and J.B.Andersen (Eds.), "Special Issues on Signal Processing for Wireless Communications, I and II", *IEEE J. Selected Areas in Communications,* 16, (8 – 9), October-November 1998.

Ojanperä, R. Prasad [1998], "An Overview of Air Interface Multiple Access for IMT-2000/UMTS", *IEEE Communications Mag.*, September 1998, 82-95.

Pedersen, G. F., J.B.Andersen [1999], "RF & ELF Exposure from Cellular Phone Handsets - TDMA and CDMA Systems", *Radiation Protection Dosimetry,* to be published.

Verdú, S., [1986], "Minimum Probability of Error for Asynchronous Gaussian Multiple-Access Channels", *IEEE Trans. Information Theory*, vol. IT-32, 85-96, Jan. 1986.

Verdú, S. [1998], *Multiuser Detection*, Cambridge University Press, 1998.

Viterbi, A.J., [1998],"Four laws of nature and society: The governing principles of digital wireless communications networks", in H.V. Poor and G.W. Wornell (Eds.), *Wireless Communications-Signal Processing Perspectives,* Prentice Hall, 1998.

Zorzi, M. (Ed.) [1998], "Energy Management in Personal Communications and Mobile Computing", *IEEE Personal Communications*, 5 (3), June 1998.

EM METROLOGY ISSUES IN WIRELESS COMMUNICATIONS

Quirino Balzano

1. Introduction

The wireless communication band of devices of most commonly used by the public extends from about 27 MHz to 2.2 GHz. Recent advances in the technology of "intelligent" magnetic card reading from a distance of several feet have extended the frequency of wireless personal communications down to 0.12 MHz. The intelligent magnetic cards have some embedded data storage; these cards are energized by the magnetic field when approaching a card reader. This perturbation initiates a wireless transaction of relevant information from the magnetic card to the card reader. The traditional modulation technology for wireless portable communication has been Frequency Modulation (FM); few products used Single Side Band (SSB), which entails linear amplification of the carrier. The explosive demand of channels for cellular telephony has caused the application of more sophisticated modulation technologies (pulsed and linear amplitude modulation). The subscriber units of the newer cell phone technologies produce EM fields in the appropriate RF communication band and low frequency magnetic fields in the modulation baseband (11-217 MHz). The sources of these low frequency magnetic fields can interfere with proper functions of other electronic equipment in their immediate vicinity. From the above considerations, it is clear that the EM metrology challenges in wireless communication technologies extend from few Hz to several GHz.

At all frequencies, the metrology challenge arises from the need of evaluating the E- fields in the immediate vicinity of their sources. The E-fields must be averaged either over a small volume (e.g. 1 cm^3 or 10 cm^3). Similarly H-fields need to be measured which have a sphere of influence with an extremely small radius (e.g. 10-12 cm) with respect to the wavelength of the low frequency (<300 Hz) signal.

The technology challenges arise from concerns about human exposure to external EM fields. For this reason the metrology discussion will be subdivided in two parts: RF dosimetry and localized measurements of extremely low frequency (ELF) magnetic fields. A brief discussion of low frequency current density measurements will follow as a corollary to the human exposure to intense, localized magnetic fields.

2. RF Dosimetry

2.1 General Considerations

RF dosimetry has developed very rapidly in the last 10 years over the frequency band of cellular telephony (800-2200 MHz), in response to regulatory requirements [*FCC, 1997, CENELEC, 1996, ACA, 1998*]. Communication products of other frequency bands (<800 MHz) also are facing regulatory exposure restrictions and present difficult metrology problems. The challenge of precise dosimetric measurements at the frequencies of communication can be reduced to two basic difficult steps: 1) dielectrometry; 2) probe physical dimensions and calibration. By "dielectrometry" it is meant not only the correct measurement of the dielectric constant and conductivity of tissues, but also the determination of appropriate chemical mixtures to simulate the dielectric characteristics of human tissues in a given frequency band. Dielectrometry will be discussed first because it presents interesting, open questions at frequencies below 70 MHz.

2.2 Properties of Tissue Simulating Materials in Low Frequency Band

In the frequency band below 70 MHz, the cell membrane capacitance is not short circuited, so biological tissues, which are organized structures of cells, have a dielectric constant higher than water, which is the main component of intracellular and extracellular fluids. The RF losses in tissues are caused by dielectric relaxation of biomolecules (proteins, fat and water) and by collisions of free moving ions (K^+, Na^+, and Cl^-) in the extracellular fluids.

The main difficulty of finding the correct material simulating the dielectric properties of human tissue arises from the necessity that such material be in liquid form. All present safety regulations [*FCC, 1997, CENELEC, 1996, ACA, 1998, ARIB, 1998*] require the evaluation of the maximum value of the E-field amplitude square in a small volume (either 1 or 10 cm^3) of simulated tissue in the immediate vicinity of an RF source (e.g. a portable radio). Since there is a tight reactive coupling between the radio device and the human body, the exposure assessment must be conducted by fixing the position of the RF source on to the simulated body and searching for the maximum value of the E-field in the artificial tissue. If E-field sensors are used in the dosimetric assay, the tissue simulating material must be liquid in order to enable the free movement of the probing device. From Fig. 1, the dielectric constant of the tissue simulating mixture must reach value as high as 20,000 to be of use in the frequency bands down to 0.1 MHz, where RF dosimetry of products may be regulated.

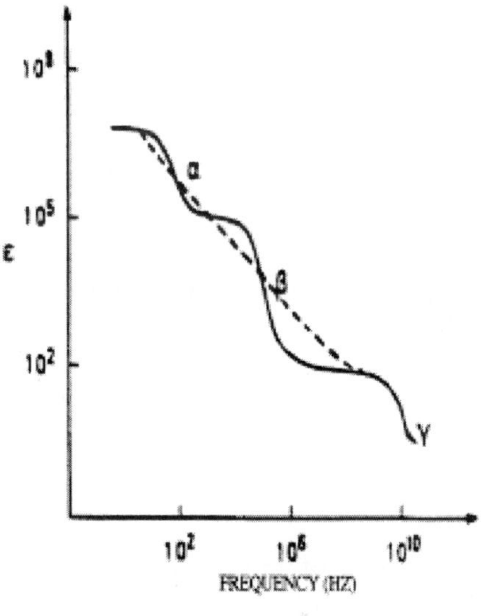

Fig. 1 *Dielectric Constant of Tissue*

While it is not difficult to construct solid materials with the appropriate dielectric characteristics, there is practically no available liquid of desired connotation. It may require some painstaking tests of emulsions of water with proteins or fats to achieve the desired dielectric constant range in a liquid. A stable, microscopically homogeneous and isotropic suspension of non-soluble materials in water is a challenging effort for accurate dosimetric measurements at the low end of the RF spectrum.

One can conceive of other solutions to skirt the liquid dielectric simulation difficulties. Near RF sources, the deposition of energy in tissue rich in water happens mainly by magnetic coupling and in fatty tissues by electric coupling [*Balzano et al, 1977*]. For magnetic coupling, the conductivity net distribution constant is the major contributing factor to dissipation. Therefore, it is possible to compose human equivalent phantoms with fatty tissue simulating mixtures near the tips of antennas and mixtures simulating just the conductivity of human tissues near the high current section of the RF sources. There is no simple criterion for the transition between the two different materials of the composite phantom. From measurements of near field electric and magnetic energy density it is possible to find the boundaries of the spatial prevalence of the needed human tissue equivalent, but the geometry of the transition that best simulates the human presence remains to be investigated.

2.3 Probe Dimensions and Calibration in Low Frequency Band

In the frequency band below 30 MHz available probes cease to have sufficient sensitivity if the sensors are about 1.5 - 2 mm long with an energy density averaging volume of 0.125 to 0.25 cm^3 [*Balzano et al, 1979*]. The sensor's length must be substantially increased to perform dosimetric measurements in the frequency band down to 100 kHz, the lower limit of applicability of specific absorption rate (SAR) to the quantification of exposure of humans to RF electromagnetic energy [*IEEE, 1991*].

Simple square-law extrapolation of probe dimension indicates that the sensor linear dimensions must be larger than 1.5 cm, giving energy averaging volumes greater than 1 cm^3. All national and international standards for the safe exposure of humans to RF energy require SAR volume averaging over 1 cm^3 or 10 cm^3. These volume-averaging limits may not be achievable with the technologies used so far in detecting RF electric energy in simulated body tissues. A new generation of semiconductor devices may increase the sensitivity of available detectors to reduce the linear dimension of the sensors. Another possible solution is to devise perturbation methods of measurement to evaluate the SAR over a volume smaller than the probe physical dimensions. Such methods are available for single E-field sensors [*Novotny, 1998*], but have not been developed for tri-axial probes. The issue of probe response anisotropy compounds the difficulty of extending perturbation techniques to tri-axial sensors. If a single axis probe is used, the measurement of volume averaged SAR by perturbation techniques becomes unacceptably long and possibly imprecise.

The calibration of E-field probes at low frequencies also presents interesting challenges. Given the long wavelength of the EM fields, it is difficult to confine the energy in relatively small volumes to achieve either a substantial temperature increase (> 0.1 °C) or large energy density, as it is possible in higher frequency bands [*Balzano et al, 1995, Meier et al, 1996*]. A recent attempt to measure SAR has required the development of special loop antennas with very high Q [*Kanda et al, 1998*]. To achieve sufficient energy density in the simulated tissue it is necessary to generate power levels approaching 100 W, which makes the measurement process more difficult and possibly less precise than the case when lower amounts of RF power are required. New methods of EM energy dosimetry and densitometry in lossy materials are needed for SAR calibration in the band below 10 MHz. The need of confining long wavelength energy in a comparatively small volume of lossy material needs to be resolved in the frequency band from 100 kHz to 10 MHz.

2.4 Dosimetry in Cellular Phone Frequency Bands

The dosimetric methods in the band 800-2200 MHz are reaching a level of maturity and sophistication that permit the establishment of international standards for the measurement procedures [*CENELEC, 1996, IEEE, 1998*]. There are still open issues related to the geometry and the composition of the most appropriate phantom to evaluate the upper limit of the exposure of the users of a given cellular phone model.

At present, the mixture more commonly used to fill a simulated human head and upper torso has the average dielectric constant and conductivity of the white and gray cerebral tissue. While this represents a reasonable approach, some experimental evidence of its validity should be collected. In addition, there should be experimental confirmation that the dielectric properties of the perfused skull bone are similar to those of the gray cerebral tissue. Only computational models are available to show that the homogeneous phantom just described is indeed representative of the absorption by the human head of RF energy emitted by a cellular phone placed in its immediate vicinity [*Hornbach et al, 1996*].

There is also a question about the thickness of the pinna when a phone is placed against an ear. This apparently trivial parameter actually is a determinant factor in dosimetric measurements, because it represents, in some cases, the maximum distance between RF currents and the simulated tissue. A change of only 1-mm of the thickness (5 to 6 mm) can cause 30% variations in some SAR measurements. The IEEE Standards Coordinating Committee 34 has agreed to use models with 6 mm ear space [*IEEE, 1998*].

The appropriateness of homogeneous phantoms in representing the upper limit of the RF absorption in human head tissue exposed to a cellular phone may be resolved by experimental measurements of RF attenuation in freshly deceased, blood perfused cadavers. Using balanced dipole antennas as RF sources and MRI scans of the decedents' head anatomy, it should be possible to obtain precise data on the dielectric properties of perfused human tissue and the attenuation of RF energy in the various layers of skin or bone and brain. Decedents of different sex and age should be evaluated to measure possible variations of dielectric properties of tissue in humans. The issue of the statistical exposure of the users of cellular phones is emerging as an important EM metrology challenge as discussed in the following section.

2.5 RF Dosimetry for the Epidemiology of Users of Wireless Devices

There are several dosimetric hurdles that must be overcome to perform a scientific assessment of the RF exposure of the cellular phone users for epidemiological studies. The epidemiological investigation is in the initial phases [*Cardis, 1998, Rothman et al, 1996*], but it is expected to produce several studies in the next few years. It is essential for the correct outcome of the studies that the exposure of the population be evaluated with some degree of precision to avoid misclassification.

Epidemiological studies of cellular telephone users require information to identify the device, the user, call duration, and to estimate exposure. For cellular phones the following parameters can or should be determined, recorded, and made available to epidemiological investigators in a usable form:

1. Portable phone model identification which defines the antenna type, electrical and geometric features of the phone, frequency band of operation, and channel access technology, e.g. FM, CDMA, or TDMA.
2. User identification, in case of multiple users of one device.
3. Duration of conversation and RF amplifier power levels during conversation.
4. Position of the phone with respect to the head of the user.
5. Antenna configuration (extended or retracted).
6. Electromagnetic coupling of the antenna to the user obtained by measuring antenna reflection/coefficient.

Requirement 1 can be satisfied by adding more digits to the phone's Electronic Serial Number (ESN). This requires international agreements among operators and manufacturers of cellular equipment.

Requirement 2 can be implemented by requiring user-specific security codes similar to those used to deter cloning. An alternative is the use of a user-specific phone card. Questionnaires may also be useful to supplement and, within well-known limitations, help to validate these methods.

Requirement 3 needs the development of some software to extract the information from the cellular infrastructure system. Call duration information is extracted for billing purposes and power level information exists in principle because the base station commands the power of the mobile phone, although presently the data are not retained.

Requirement 4 is probably the most difficult to obtain. Relative head and phone positions are variable between and among users, but so far such position shifts remain not quantified. For a given phone, the exposure of specific organs can change over one or two orders of magnitude by tilting the phone away from contact with the face.

Requirement 5 is relatively easy to satisfy by simply detecting the position of a switch. Only the initial position of the antenna, and possible changes during RF transmission, need to be recorded.

Requirement 6 can be represented by the antenna reflection coefficient, which is a measure of the mismatch of antenna, and amplifier impedance's which occurs because of coupling between the portable wireless device and the user's body. The degree of mismatch and reflection vary with proximity and anatomical relationship of the phone and user, thereby permitting, in principal, real-time information on the amount of energy coupled to the body. Because of the mismatch, power is reflected back into the phone and is not radiated by the antenna. Reflected power may be a substantial fraction (as much as half of the power output of the phone's RF amplifier). To obtain the antenna reflection coefficient a miniature network analyzer must be incorporated in the phone circuitry. This is technologically feasible and a design of such device is available and is being manufactured on an experimental basis.

Data acquired within individual user devices would be telemetered to the base station along with other signals exchanged during cellular communications. Together with data acquired at the base station, such as call duration, all data pertaining to a user would be collated by computer methods operating in parallel with usual billing operations and assembled in a format specified by a research epidemiologist. A cell phone system equipped to satisfy the foregoing requirements can be considered "a dosimeter".

Portable dispatch transceivers are commonly held in front of the face during use. Unlike cellular phones, portable radios operate at a constant nominal power level during RF transmission. The devices are rated (10, 10, 80), that is; the radio is transmitting and receiving 10% of the time and on stand by during 80% of the time, or (5, 5, 90). There is no RF transmission during receive and stand-by mode of operation.

Dosimetric studies of dispatch radios [*Balzano et al, 1978a, 1978b, 1978c, Fujimoto and Jones, 1994*] have shown that exposure of the head of the user changes substantially with the distance of the radio case from the mouth. Another important exposure parameter is the tilt of the axis of the transceiver with respect to the normal use position (antenna parallel to the body axis). Users hold the device in a variety of positions with a statistical distribution that should be quantified. If the transceiver is held with the antenna in horizontal position (parallel to the ground) there is minimal exposure of the eyes and the brain, which are more exposed to RF if the antenna is vertical.

The exposure of the user varies with duration of use of the device. Even the highest capacity batteries tend to slump after about 45 minutes of RF transmission. The 45 minutes of activity represent approximately a total of 90 minutes of two-way conversation, which is unusual, but possible for a person in a supervisory position at a construction site or in an emergency situation.

From the above considerations the determination of dose for the users of portable transceiver can be achieved with the quantification of the following parameters:

1. Duration of RF transmission.
2. Distance of the radio case from the mouth of the user.
3. Tilt angle of the device during use.
4. Transceiver frequency band, output power, and antenna type.

To the effect of dosimetry for epidemiology, parameters 1, 2 and 4 are not difficult to quantify using the methods devised for cellular telephones. The third parameter is difficult to quantify correctly and presents technological challenges of no easy solution for the creation of "dosimeters" for transceiver users.

3. Extremely Low Frequency (ELF) Magnetic Field Metrology

3.1 General Considerations

The pulsed channel signaling in cellular telephones technology causes the battery of the devices to be drained in synchrony with the pulsed transmit mode. This event causes low frequency (ELF band < 300 Hz) currents in the power supply of the phone electronic circuitry. The battery supply lines of cellular telephones on PC boards are designed for the efficient distribution of B+ voltage to critical components, so they may not have a parallel return path in their immediate vicinity. The direct current pulses on these lines are of the order of one ampere (1 A). At very short distances of the order of few millimeters, it is possible to find magnetic fields approaching one gauss (1 G). These fields can produce electromagnetic interference with a variety of electronic personal devices including hearing aids and heart pacemakers.

In magnetic "smart" card technology there is an increasing need to activate and read the devices at a distance varying from a few feet to a few dozen feet. The coils generating the magnetic fields at frequencies of about 100 kHz carry currents of the order of several amperes. The partial body exposure of attendants of these devices can reach values above proposed national or international standard even if averaged over the whole body of a person. In these conditions it is necessary to measure either SAR or current density in an equivalent human phantom. The description of the metrology challenges presented by the necessity of measuring magnetic fields or induced currents at low frequencies will be discussed in the next two sections.

3.2 Measurement of ELF Magnetic Fields

Instrumentation to measure 50-60 Hz magnetic fields with sufficient spatial resolution (less than 10 cm^3) is commercially available. There are no available instruments to cover the band from 2 Hz to about 1 kHz with a spatial resolution of about 10 cm^3. Such instruments, if available, would substantially simplify and improve the measurement of interference and compatibility between cellular phones and medical electronic devices in their immediate vicinity. At present, the detection and measurement of interference phenomena is performed by observing the performance of a device placed near a cellular phone. The procedure is awkward because it requires checking a variety of devices near a variety of cellular phones instead of measuring the susceptibility to magnetic fields from standard sources near the device under test. This last procedure, similar to those used in standard electromagnetic compatibility (EMC) measurements, relieves the measurements from the necessity of testing a variety of different potential sources of interference. Although the methods of EMC measurements using cellular phones have made substantial progress in the last few years [*Grant and Schligel, 1998*], the characterization of these devices as a low frequency magnetic field source remains an open metrology question.

3.3 Magnetic Partial Body Exposure and Current Density

The evaluation of partial body exposure to magnetic fields presents some interesting corollary measure of body current density, which has not yet been completely resolved. Current safety standards for the human exposure to magnetic fields [*FCC, 1996, ICNIRP, 1998, IEEE, 1991*] all call for the measurement of current density in the body. The peak current density is estimated by measuring the total ground currents from the exposed individual. If the exposure is mainly localized in the trunk of a person, there might be minimal ground currents, while the RF current density at the periphery of the torso may be relatively intense.

At present, there are no commercially available instruments to measure current density in a simulated human in the frequency band below 1 MHz [*IEEE, 1992*]. The development of such technology would be a welcome event not only for the magnetic card reading industry but also for other industrial concerns.

4. Conclusions

The paper has presented a broad sweep over the metrology issues facing the wireless communication technologies of the near future. The biggest challenges remain in the measure of accurate field intensity in volumes confined by dimensions very small with respect to the wavelength of the EM energy under evaluation.

References

ACA. Australian Communication Authority, Draft Technical Standard, "Specific Absorption Rate Test Method Using Phantom Models of Human Anatomy", Nov. 1998.

ARIB. Version 1.0, "SAR Estimation for Cellular Phones", Japan, January. 27, 1998.

Balzano Q. et al., "Energy Deposition in Biological Tissue near Portable Radio Transmitters at VHF and UHF", *27th Conf. of IEEE VT-4*, Orlando Florida, 25-39, March 16-18, 1977.

Balzano Q. et al. "Heating of Biological Tissue in the Induction Field of VHF Portable Transmitters", *IEEE Trans Vehicular Tech.*, 27, 51-56, 1978a.

Balzano Q. et al. "A Comparison Between the Energy Deposition in Portable Radio Operators at 900 MHz and 450 MHz", *Proc. 28th Conf. IEEE Vehicular Tech.,* Denver, CO, 46-55, March 22-24, 1978b.

Balzano Q. et al. "Energy Deposition on Simulated Operators of 800 MHz Portable Transmitters", *IEEE Trans. Vehicular Tech.*, 27, 174-181, 1978c.

Balzano Q. et al., "An Attempt to Evaluate the Exposure of Operators at 30 MHz", *Proc. 29th IEEE Vehicular Tech. Conf.*, Arlington Heights, IL, 187-188, March 27-30, 1979.

Balzano Q. et al., "Electromagnetic Energy Exposure of Simulated Users of Portable Cellular Phones", *IEEE Trans. Vehicular Tech.*, 44, 390-403,1995.

Cardis E. et al. "International Case Control Study of Adult Brain, Head and Neck Tumors", Phase 1, Feasibility Study, International Agency for Research on Cancer, Lyon, France, 1998.

CENELEC PRES-59005, "Consideration for Evaluation of Human Exposure to EM Fields from Mobile Telecommunication Equipment in the frequency range 30 MHz - 6 GHz". 1996.

FCC Order 96-326 "Guidelines for Evaluating the Environmental Effects of Radiofrequency Radiation", August 1, 1996.

FCC OET Bulletin 65, Supplement C Edition 97-01, "Additional Information for Evaluating Compliance of Mobile and Portable Devices with FCC limits for Human Exposure to Radiofrequency Emissions", December, 1997.

Fujimoto K. and Jones J. R. (Editors), "Mobile Antenna Systems Handbook", Artech House, Norwood MA. Chapter 4, 251-270, 1994.

Grant H. and Schligel R. "In Vitro Study of the Interaction of Wireless Phones with Cardiac Pacemakers", University of Oklahoma, EMC Report 1998-2, June 1998.

Hornbach V. et al. "The Dependence of EM Energy Absorption Upon Human Head Modeling at 900 MHz", *IEEE Trans. Microwave Theory Tech.*, 44, 1865-1872, 1996.

ICNIRP "Guidelines for Limiting Exposure to time-varying Electric, Magnetic and Electromagnetic Fields (up to 300 GHz)", *Health Physics*, 74, 494-522, 1998.

IEEE, "Standard for Safety Levels with Respect to Human Exposure to RF Electromagnetic Fields, 3 kHz-300 GHz", IEEE C95.1-1991.

IEEE "Recommended Practice for the Measurement of Potentially Hazardous Electromagnetic Fields RF and Microwave", IEEE Std. C95.3-1991, Section 5.4, 71, August 21, 1992.

IEEE SCC34, "Electromagnetic Energy Product Performance Safety", SC2, WG1 Experimental Dosimetry. (Draft) Dec. 1998.

Kanda M. et al., "Measurement of Localized SAR for Contactless Smart Card Reader operating in the HF Band", *IEEE Trans. Electromagnetic Compatibility*, 40 (4), November 1998 (in press).

Meier K. et al., "Broad Band Calibration of E-field Probes in Lossy Media", *IEEE Trans. Microwave Theory Tech.*, 44, 1954-1962, October 1996.

Novotny D., National Institute of Standards and Technology, Boulder, Colorado, (Personal Communication), 1998.

Rothman K. et al. "Overall Mortality of Cellular Telephone Customers", *Epidemiology*, 7, 303-304, 1996.

ELECTROMAGNETIC SYSTEM DESIGN USING GENETIC ALGORITHMS

Eric Michielssen, Yahya Rahmat-Samii and Daniel S. Weile

1. Introduction

Genetic algorithms (GAs), in their simplest incarnation, are numerical algorithms inspired by the theory of evolution, first proposed by Charles Darwin in the nineteenth century [*Darwin, 1985*]. Though GAs have numerous uses in machine learning and biological research, this study is confined to their use as tools for function optimization [*Holland, 1975;Goldberg, 1989*]. Even restricted to this realm, GAs have many appealing features that have made them into popular design tools in electromagnetics and other fields of engineering:

- GAs tend to find global or strong local optima of a given function in their search domain.
- GAs can optimize multimodal and discontinuous functions without the need for any gradient information whatsoever.
- GAs can search spaces involving both continuous and discrete parameters.

These beneficial GA qualities are derived from their novel nature: Unlike most standard optimization algorithms, GAs operate on an entire population of encoded candidate solutions (called chromosomes) to search for optima [*Holland, 1975;Goldberg, 1989*]. Moreover, GAs are innately stochastic and owe their convergence solely to the Darwinian maxim of "survival of the fittest" (rather than to some mathematical property of the function being optimized).

Electromagnetic design represents a useful and difficult realm for GA applications. Though the physical laws that govern electromagnetic devices are well understood, the analysis of such devices is often very time-consuming, and their design is still an art. Through the careful construction and application of GAs, however, much of the guesswork can be removed from synthesis. This chapter, then, can be viewed as a guide to this careful application of GAs.

The most basic form of a GA, often dubbed the Simple GA (SGA), applies a suite of three different operators to the optimization of a population of design candidates [*Goldberg, 1989*]:

- Selection, which implements the GA analog of fitness-based differential survival,

- Crossover, which hybridizes surviving designs to (one hopes) produce better designs, and
- Mutation, which randomly perturbs designs to ensure lasting populational diversity.

These three operators (described in more detail in the next section) have been used to design a profusion of electromagnetic devices, ranging from frequency-selective surfaces to microwave filters to wire antennas *et cetera* [*Haupt, 1995; Johnson and Rahmat-Samii, 1997; Weile and Michielssen, 1997*]. The application of GAs to electromagnetic device design continues today unabated.

Despite past successes and the glorification of GAs in the electromagnetics literature, the application of GAs to electromagnetic optimization problems does have some drawbacks. For instance, the convergence of GAs is notoriously slow, so their use in many cases has been restricted to problems involving easily computable objective functions. In spite of the claims of robustness made on behalf of GAs, for some problems their performance has been shown to be highly dependent upon the design of the three genetic operators. Finally, because GAs are based on evolutionary theory, they implicitly assume that not only the hybridization effected by crossover results in the inheritance of traits, but also that there exists a strong correlation of these traits to survivability of the chromosome.

Fortunately, advances in GA research are beginning to solve these problems through the increasing application of advanced operator GAs (AOGAs) [*Weile and Michielssen, 1997*]. AOGAs modify the three basic operators (or add to them) to create algorithms that may be more reliable, faster, and physically rooted to ensure that inheritance and survival are firmly related. One AOGA, the Pareto GA, is discussed in detail toward the end of this chapter.

The rest of this chapter will proceed as follows. Section 2 presents the simple, three-operator GA, and some of the theory associated with it. In addition, several common AOGA extensions are discussed. This material is thus intended not only to explain GA mechanics, but also to elucidate the principles behind their operation so that GA practitioners can diagnose and repair common problems. Section 3 then relates a history of the application of GAs to problems in electromagnetic device design and gives two examples of current GA applications. Finally, section 4 concludes the chapter.

2. Genetic Algorithms

This introduction to GAs is divided into three sections. Section 2.1 presents a tutorial on SGAs. Section 2.2 highlights the principal aspects of basic GA theory as developed by Professor John Holland, the inventor of GAs. Section 2.3 discusses commonly used AOGAs.

2.1 A Tutorial on Simple Genetic Algorithms

Consider a physical synthesis problem that requires the selection of a set of design parameters to optimize a given objective function. GAs differ from the majority of classical deterministic and stochastic optimization schemes that could be used to solve this problem in two respects. First, GAs typically do not operate directly on the design parameters themselves; rather, they act on a coded representation of the parameters. Second, contrary to most classical optimization methods that iteratively refine a single solution candidate to arrive at a suitable design, GAs optimize entire populations of candidates simultaneously. Other evolutionary optimization schemes—e.g., population based incremental learning and (μ,λ) evolution strategies—share some of these properties with GAs [*Bäck and Schwefel, 1995*].

The above notions are formalized through the following definitions. Consider a device that is fully described by N_{par} parameters arranged in a vector $\mathbf{x}=\{x_i | i=1,\ldots,N_{par}\}$. Knowledge of $\mathbf{x}$ permits the evaluation of the objective function $f(\mathbf{x})$, which indicates the worth of a design. It is assumed that the x_i take on either real or discrete values, and that $f(\mathbf{x})$ needs to be maximized. As mentioned above, the GA does not operate directly on $\mathbf{x}$ but on a discrete representation or chromosome $\mathbf{p}=\{g_i | i=1,\ldots,N_{par}\}$ of $\mathbf{x}$, each parameter x_i being described by a gene g_i. Each gene g_i in turn consists of a set of N_{allele}^i alleles that are selected from a finite alphabet and that together decode to a unique x_i as illustrated -

$$
\begin{array}{ll}
\mathbf{x}=\{ & x_1 \quad\quad x_2 \quad \cdots \quad x_i \quad \cdots \quad x_{N_{par}} \quad \} \\
 & \uparrow \quad\quad \uparrow \quad\quad\quad \uparrow \quad\quad\quad \uparrow \\
\mathbf{p}=\Bigg\{ & \overbrace{a_1 a_2 \ldots a_{N_{allele}^1}}^{g_1} \overbrace{a_{\tilde{N}_{allele}^2+1} a_{\tilde{N}_{allele}^2+1} \ldots a_{\tilde{N}_{allele}^2+N_{allele}^2}}^{g_2} \ldots \overbrace{a_{\tilde{N}_{allele}^i+1} a_{\tilde{N}_{allele}^i+2} \ldots a_{\tilde{N}_{allele}^i+N_{allele}^i}}^{g_i} \ldots\ldots \overbrace{a_{\tilde{N}_{allele}^{N_{par}}+N_{allele}^{N_{par}}}}^{g_{N_{par}}} \Bigg\}
\end{array} \quad (1)
$$

In Eqn. (1) $\tilde{N}^i_{allele} = \sum_{j=1} N^j_{allele}, i = 2,\ldots,N_{par}$; also, in what follows $N_{allele} = \sum_{i=1} N^i_{allele}$ denotes the total number of alleles in $\mathbf{p}$. In the SGA, alleles are selected from the binary set $\{0,1\}$. If x_i is a discrete parameter then N^i_{allele} should be chosen such that $2^{N^i_{allele}}$ equals or exceeds the number of values x_i can assume (In the latter case, several allelic representations will map to the same x_i.) If x_i represents a real parameter allowed to vary in the range $\left[x_i^{\min}, x_i^{\max}\right]$, N^i_{allele} should be picked large enough to ensure that it can be decoded with tolerable quantization error from its binary representation using

$$x_i = x_i^{\min} + \frac{\left(x_i^{\max} - x_i^{\min}\right)}{2^{N^i_{allele}} - 1} \sum_{j=1}^{j=N^j_{allele}} a_{\tilde{N}^i_{allele}+j} 2^{j-1}. \tag{2}$$

To illustrate the above concepts, consider the problem of synthesizing a planarly layered absorber. The synthesis problem consists in choosing the optimal material for each layer from a predefined and finite set of available materials, and determining the thickness for each layer. The objective function to be maximized can be defined as the minimum absorption of the multilayer over a given set of incident angles and polarizations and a frequency band of interest. Assume, for the sake of simplicity, that the absorber is composed of three layers, and that each layer is composed of a material chosen from eight available materials. Moreover, assume that the maximum thickness of each layer is one centimeter, and that fabrication tolerances do not exceed one hundredth of a centimeter. Then, a chromosome suitable for describing the absorber consists of six genes. The first three genes describe material choices for each layer and contain three bits each, and the remaining genes describe layer thicknesses and contain seven bits each. The total chromosome length is thirty.

As pointed out before, GAs do not limit themselves to the iterative refinement of a single coded design candidate; instead the SGA simultaneously acts upon a set of candidates or *population* $\overline{\mathbf{p}} = \left\{\mathbf{p}(i) \middle| i = 1,\ldots,N_{pop}\right\}$, where N_{pop} is the *population size* (Fig. 1). Starting from an initial population $\overline{\mathbf{p}}^0$, the SGA iteratively constructs populations $\overline{\mathbf{p}}^k, k = 1,\ldots,N_{gen}$, with N_{gen} denoting the total number of SGA *generations*. Subsequent generations are constructed by iteratively acting upon $\overline{\mathbf{p}}^0$ with a set of *genetic operators*. The operators that induce the transition $\overline{\mathbf{p}}^k \rightarrow \overline{\mathbf{p}}^{k+1}$ are guided solely by knowledge of the vector of objective function values

$\mathbf{f}^k = \left\{ f\left(\mathbf{x}\left(\mathbf{p}^k(i)\right)\right) \middle| i = 1,\ldots,N_{pop} \right\}$ and induce changes in the genetic makeup of the population leading to a $\overline{\mathbf{p}}^{k+1}$ comprising individuals that are, on average, "better adapted to their environment" than those in $\overline{\mathbf{p}}^k$, i.e., they are characterized by higher objective function values. This change is effected by the three operators mentioned in the introduction, viz., selection (S), crossover (C), and mutation (M), the actions of which are described next.

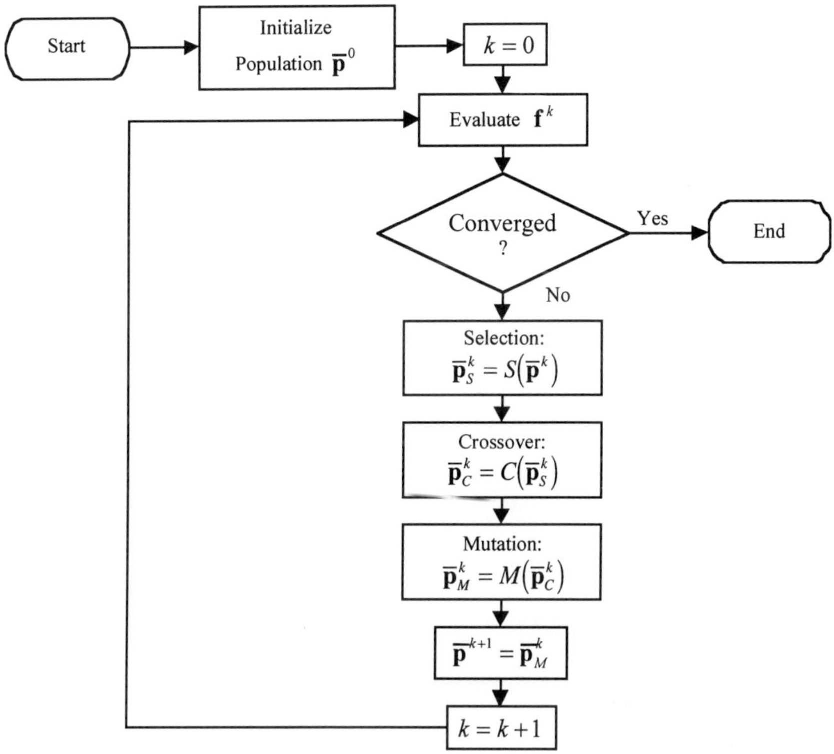

Fig. 1. *Flowchart of the Simple GA.*

The *selection operator* S implements the principle of survival of the fittest [*Holland, 1975; Goldberg, 1989*]. Acting on $\overline{\mathbf{p}}^k$, S produces a new population $\overline{\mathbf{p}}_S^k = S(\overline{\mathbf{p}}^k)$ again of size N_{pop} that is, on average, populated by the better-fit individuals present in $\overline{\mathbf{p}}^k$. Many schemes for selecting these better-fit individuals are available. Here *tournament selection* is discussed, because it is the most widely applied selection mechanism in engineering applications, because it is amenable to theoretical analysis (see Section 2.2), and because it can be tuned to favor any combination of exploration and exploitation [*Goldberg and Deb, 1991*]. Using tournament selection, the population $\overline{\mathbf{p}}_S^k$ is constructed by organizing N_{pop} *tournaments* among N_{tour} randomly selected members in $\overline{\mathbf{p}}^k$, and by placing the winner of each tournament (i.e., the participating chromosome with the highest objective function value) into $\overline{\mathbf{p}}_S^k$. The parameter N_{tour} is referred to as the *tournament size*. Symbolically,

$$\overline{\mathbf{p}}_S^k = \bigcup_{i=1}^{N_{pop}} s\left(\underbrace{ch\left(\overline{\mathbf{p}}^k\right), ch\left(\overline{\mathbf{p}}^k\right), \cdots, ch\left(\overline{\mathbf{p}}^k\right)}_{N_{tour}\text{ designs}}\right) \tag{3}$$

where the operator ch chooses a random design from the population on which it acts, and the chromosome selection operator s chooses the best member from the N_{tour} designs considered on the basis of their fitness. If $N_{tour} = 2$, the above scheme is known as *binary tournament selection.*

The *crossover operator* C mimics natural procreation [*Holland, 1975; Goldberg, 1989*]. Specifically, C acts upon the population $\overline{\mathbf{p}}_S^k$ by mating its members, thereby creating a new population $\overline{\mathbf{p}}_C^k = C\left(\overline{\mathbf{p}}_S^k\right)$. The action of the crossover operator C can be codified as

$$\overline{\mathbf{p}}_C^k = \bigcup_{i=1}^{N_{pop}/2} c\left(ch\left(\overline{\mathbf{p}}_S^k\right), ch\left(\overline{\mathbf{p}}_S^k\right)\right) \tag{4}$$

where the chromosome crossover operator c selects a random crossover allele $1 \le N_{cross} < N_{allele}$ and swaps genetic information following allele $a_{N_{cross}}$ between the two chromosomes upon which it acts with probability P_{cross}. Symbolically, letting the two chromosomes to be crossed be given by $\mathbf{p}^l = \left\{a_1^l, a_2^l, \cdots, a_{N_{cross}}^l, a_{N_{cross}+1}^l, \cdots, a_{N_{allele}}^l\right\}$, $l = 1,2$, the chromosome crossover operator can be expressed as

$$\begin{aligned} c\left(\mathbf{p}^1, \mathbf{p}^2\right) &= \left\{\hat{\mathbf{p}}^1, \hat{\mathbf{p}}^2\right\} \quad \text{with probability } p_{cross} \\ &= \left\{\mathbf{p}^1, \mathbf{p}^2\right\} \quad \text{with probability } 1 - p_{cross} \end{aligned}, \tag{5}$$

where the chromosomes $\left\{\hat{\mathbf{p}}^1, \hat{\mathbf{p}}^2\right\}$ are

$$\begin{aligned} \hat{\mathbf{p}}^1 &= \left\{a_1^1, a_2^1, \cdots, a_{N_{cross}}^1, a_{N_{cross}+1}^2, \cdots, a_{N_{allele}}^2\right\} \\ \hat{\mathbf{p}}^2 &= \left\{a_1^2, a_2^2, \cdots, a_{N_{cross}}^2, a_{N_{cross}+1}^1, \cdots, a_{N_{allele}}^1\right\} \end{aligned}. \tag{6}$$

Finally, the *mutation operator* M again generates a population of size N_{pop}, denoted $\overline{\mathbf{p}}_M^k = M(\overline{\mathbf{p}}_C^k)$, by introducing small random changes into $\overline{\mathbf{p}}_C^k$ [*Holland, 1975; Goldberg, 1989*]. Specifically, M randomly flips each bit in the chromosomes of $\overline{\mathbf{p}}_C^k$ with mutation probability p_{mut}. The action of M can be represented in operator form as

$$\overline{\mathbf{p}}_M^k = \bigcup_{i=1}^{N_{pop}} m\left(\overline{\mathbf{p}}_C^k(i)\right) \tag{7}$$

where the chromosome mutation operator m acts upon a design as follows

$$\begin{aligned} & m\left(\left\{a_1, a_2, \cdots, a_{N_{all}}\right\}\right) = \left\{\tilde{a}_1, \tilde{a}_2, \cdots, \tilde{a}_{N_{all}}\right\} \\ & \tilde{a}_j = a_j \qquad \text{with probability } 1 - p_{mut}. \\ & \tilde{a}_j = \text{NOT}\left(a_j\right) \qquad \text{with probability } p_{mut} \end{aligned} \tag{8}$$

The authors emphasize that mutation by no means constitutes the principal search mechanism of GAs [*Goldberg, 1989*]. In contrast to the sequence of selection and crossover operators that attempts to generate offspring with a genetic makeup superior to that of their parents, M blindly changes population members. Nonetheless, the mutation operator is invariably included in the SGA operator set to mask some shortcomings associated with the use of the other operators alone. Specifically, the mutation operator assists in maintaining genetic diversity and prevents the SGA from converging prematurely [*Goldberg, 1989*].

The population $\overline{\mathbf{p}}_M^k$ serves as the starting point for the next SGA iteration, i.e., $\overline{\mathbf{p}}^{k+1} = M\left(C\left(S\left(\overline{\mathbf{p}}^k\right)\right)\right)$. However, before reapplying the sequence of selection, crossover, and mutation operators, the optimal individual in $\overline{\mathbf{p}}^{k+1}$ is identified and its performance is compared to the design criteria. If it satisfies these criteria, the SGA has successfully completed its task, and the optimization process is halted. In contrast, if after many iterations the optimal individual still does not meet the design specifications, the search may be halted. The decision to stop a search before finding a design meeting the specifications is often based on population statistics: if the average population fitness is close to that of the optimal population member, little progress can be expected from continued iteration, and the SGA should be stopped. On the other hand, if the optimal design far outperforms the population average, further iterations are likely to produce gains, and the SGA should be allowed to continue for a set number of generations.

2.2 The Theory of Simple Genetic Algorithms

Although the SGA's population structure and operators are simple to describe and implement, constructing a theory of their operation has proven extremely difficult. Many early studies on the topic attempted to apply powerful mathematical techniques (notably Markov chain analysis) to describe GA convergence and guide the selection of GA parameters [*Horn, 1993*]. Typically, these studies did not elucidate much as the models they constructed were either too pessimistic to describe the many instances of GA success or too complicated to provide any guidance whatsoever to the GA practitioner. Today, GA theory consists primarily of a curious blend of stochastic models and approximations that are fused to form a picture of the inner workings of the GA. Though such models do not represent a theory that is as quantitative as GA practitioners would like, they do provide guidance in determining whether a given problem will pose difficulties to a GA and how to choose GA parameters, e.g., population size and crossover probabilities. The theory presented here is based on the concept of *schemata*, formulated by the GA's inventor Professor John Holland [*Holland, 1975*].

In GA theory, the word *schema* denotes both a subset of all possible chromosomes with identical alleles in several different loci, and the similarity template used to signify this set [*Holland, 1975; Goldberg, 1989*]. When a schema is considered as a set, its members are called *examples* of the schema. As similarity templates, schemata are denoted as strings of the same length as the chromosomes they represent, with elements chosen from the alphabet {0,1,*}. In this notation, a chromosome is a member of a schema if it has zeros in all loci where the similarity template has zeros, and ones in all loci where the similarity template has ones. The character "*" in a given locus represents a *don't care* where schema examples may have either allele. In a given schema, the positions occupied by either 0 or 1 are called the *fixed positions* of the schema.

For example, consider a five-bit chromosome. The schema $H_1 = *1*0*$ has two fixed positions and contains the eight chromosomes 01000, 01001, 01100, 01101, 11000, 11001, 11100, and 11101. Similarly, the schema $H_2 = *0111$ has four fixed positions and represents the two chromosomes 00111 and 10111. For a chromosome of length N_{allele}, there exist $3^{N_{allele}}$ schemata.

Schemata are important in GA theory because they provide a formal way of both describing common traits and quantifying the evolving populational representation of these traits [*Holland,*

1975; Goldberg, 1989]. The growth or decline of a schema's representation in a GA population is influenced by three attributes: order, defining length, and fitness [*Holland, 1975; Goldberg, 1989*].

The *order* $o(H)$ of a schema H is its number of fixed bits. The schema H_1 mentioned above has order 2, whereas H_2 has order 4. Given a chromosome of length N_{allele}, the number of members of a schema H is $2^{N_{allele}-o(H)}$. Thus, a schema of high order represents a more specific trait than one of low order in that there exist fewer examples of the high-order schema. The *defining length* of a schema H, symbolized by $\delta(H)$, is the difference between the loci of its first and last fixed position. Therefore, the two example schemata given above have defining lengths $\delta(H_1)=2$ and $\delta(H_2)=3$. Finally, the concept of schema statistics is introduced. Given a schema H, the schema mean $\bar{f}(H)$ and standard deviation $\sigma(H)$ are defined in the natural manner [*Holland, 1975; Goldberg, 1989*]:

$$\begin{aligned} \bar{f}(H) &= \frac{1}{2^{o(H)}} \sum_{\mathbf{p}\in H} f(\mathbf{p}), \\ \sigma(H) &= \sqrt{\frac{1}{2^{o(H)}} \sum_{\mathbf{p}\in H} \left[\bar{f}(H) - f(\mathbf{p})\right]^2} \end{aligned} \tag{9}$$

Also, let $\bar{f}*$ and $\sigma*$ denote $\bar{f}(*,*,\cdots,*)$ and $\sigma(*,*,\cdots,*)$, respectively. In what follows it is assumed that (i) schema averages and standard deviations calculated based only on chromosomes in a given population closely comply with Eqns. (9), (ii) that populational averages and standard deviations are near $f*$ and $\sigma*$, (iii) that the distributions of $f(H)$ are approximately Gaussian, and (iv) that binary tournament selection is used. Assumptions (i) and (ii) hold true in the early stages of the GA optimization provided that large populations are used; hence, the model derived hereafter only accurately describes initial GA iterations. Assumption (iii) relies on long chromosomes.

With these definitions in place, the action of a GA on schemata can be quantified. Let $m_{\bar{\mathbf{p}}^k}(H)$ denote the number of examples of schema H in $\bar{\mathbf{p}}^k$. The number of representatives of H in $\bar{\mathbf{p}}_S^k$ is on average given by

$$m_{\bar{\mathbf{p}}_S^k}(H) = \chi\, m_{\bar{\mathbf{p}}^k}(H) \tag{10}$$

where

$$\begin{aligned} \chi &= 2\,N\left(\frac{f(H)-f^*}{\sqrt{\sigma(H)^2+\sigma^{*2}}}\right) \\ N(x) &= \frac{1}{\sqrt{2\pi}}\int_{-\infty}^{x} e^{-t^2/2}dt \end{aligned} \tag{11}$$

Indeed, if the number of tournaments pairing two representatives of H are assumed negligible, members of H will participate in $2\,m_{\mathbf{p}^k}(H)$ tournaments on average. The probability of a member of H winning this tournament is $\chi/2$ because both schema and populational quantities are assumed to be Gaussian and the difference of two such quantities again adheres to Gaussian statistics. Next, a bound for $m_{\mathbf{p}_C^k}(H)$, the number of examples of H in $\overline{\mathbf{p}}_C^k$, is constructed. Assume that every crossover involving a chromosome in H, in which the crossover point is chosen between the first and last fixed positions of H, results in children that are no longer examples of H. Moreover, assume that no crossover actually produces more examples of H (These are conservative assumptions.) Then the probability of destroying a given occurrence of H in $\overline{\mathbf{p}}_S^k$ is given by

$$p_{dest} = \frac{\delta(H)}{N_{allele}-1}\,p_{cross} \tag{12}$$

as this is the likelihood of disrupting H. Combining Eqns. (10) with (12) yields

$$m_{\overline{\mathbf{p}}_C^k}(H) \ge m_{\overline{\mathbf{p}}_S^k}(H)\left(1-p_{dest}\right) = \chi\; m_{\overline{\mathbf{p}}^k}(H)\left[1-\frac{\delta(H)}{N_{allele}-1}\,p_{cross}\right]. \tag{13}$$

Finally, mutation is applied to produce $\overline{\mathbf{p}}_M^k = \overline{\mathbf{p}}^{k+1}$ from $\overline{\mathbf{p}}_C^k$. Again, assuming that mutation produces no additional examples of H, the probability that a given representative of H survives mutation is the probability that none of its fixed positions are mutated. This probability may be estimated as

$$p_{surv} = \left(1-p_{mut}\right)^{o(H)} \approx 1-p_{mut}\,o(H) \tag{14}$$

since typically $p_{mut} \ll 1$. Using (13) and (14) and ignoring all higher-order terms, the representation of H in generation $k+1$ is bounded by

$$m_{\overline{\mathbf{p}}^{k+1}}(H) = m_{\overline{\mathbf{p}}_M^k}(H) \ge \chi\; m_{\overline{\mathbf{p}}^k}(H)\left[1-p_{cross}\frac{\delta(H)}{N_{allele}-1}-p_{mut}\,o(H)\right]. \tag{15}$$

This inequality is known as the *schema theorem*, or the *fundamental* theorem *of genetic algorithms* [*Holland, 1975; Goldberg, 1989*]. In essence, this theorem means that in the absence of crossover and mutation, schema represented in above average chromosomes receive exponentially increasing representation in subsequent generations.

Many interesting conclusions about GA behavior can be drawn from this theorem. The first, and most important, is that above-average schemata with low order and small defining length (called *building blocks*) will receive increasing representation in subsequent generations [*Goldberg, 1989*]. It has therefore been conjectured that GAs will operate most successfully when the most important contributions to the fitness function come from low-order, short-defining-length schemata. This hypothesis has become extremely important in GA research and is supported by much experimental data [*Goldberg, 1989*]. Because of its centrality in the theory of GAs, it has been dubbed the building block hypothesis, and much ongoing research is dedicated to understanding how to exploit it to construct GAs that work better than the SGA.

The schema theorem can also be of some guidance when designing chromosomes. Specifically, because low-order schemata are less likely to be disrupted by crossover, related genes should be close to one another in the chromosome. For instance, the example of chromosome construction for a three-layer absorber problem presented in the last section is probably not optimal. A better organization of the chromosome would contain the same genes, but the ordering of these genes would alternate between material choice and thickness, with each such pair of genes fully describing a single layer of the absorber. Because the material choice and thickness for a given layer are intimately related, the schema theorem predicts that this second organization would result in faster convergence.

While the schema theorem is of some help in chromosome design, it fails to give any guidance to the choice of crossover rate, mutation rate, population size, or selection strategy. Indeed, the derivation of the schema theorem pessimistically assumes that crossover always destroys good schemata when it disrupts them and that mutation never results in creation of a new example of a building block. To choose these parameters, the readers is referred to the theories of mixing and population sizing [*Goldberg et al., 1991; Thierens and Goldberg, 1993; Harik et al., 1996*].

2.3 Extensions to the Simple Genetic Algorithm and Advanced Operator Genetic Algorithms

Analogous to the processes that inspired them, GAs have evolved since John Holland first proposed them in the 1960s [*Holland, 1975*]. As in biological evolution, GAs have shed less useful operators such as inversion [*Goldberg, 1989*], and only the fittest operators have survived to be incorporated in modern algorithms. Today, many different techniques are used in genetic optimization [*Weile and Michielssen, 1997*], and each has its "niche." Much current GA research is engaged in finding robust GAs, which, like human beings in nature, are intelligent enough to adapt to many different environments [*Goldberg, 1998*]. Although the SGA has proven useful for many electromagnetics problems [*Haupt, 1994; Haupt, 1995*], it can stall or fail outright when applied to problems involving high computational loads, or requiring more information than a simple optimum. Fortunately, however, the GA is a very malleable algorithm—with a little imagination and a lot of experimentation, it may be twisted to optimize various pathological objective functions.

Perhaps the most common deviation from the SGA found in the electromagnetics literature are small changes to the representations and operators. Real-coded chromosomes have proven useful in some instances as they mimic the search space better than do binary codings [*Goldberg, 1991; Michielssen et al., 1992*]. There are also various modifications to selection, such as the roulette-wheel and ranking selection operators, that find use in many applications. Crossover can be modified to hybridize chromosomes at multiple sites or to exchange each allele with a give probability. In real-coded GAs, both the crossover and mutation operators need to be reconstructed as they act on real genes.

Besides these "first-order" GA improvements, other schemes exist that completely overhaul the SGA construct to extend GA applicability or to improve GA robustness and efficiency. Although these topics are interesting enough in their own right to merit entire studies, we only mention them briefly here:

- *Sharing* is an operator that parallels the competition for resources in environmental niches [*Goldberg and Richardson, 1987; Goldberg, 1989*], and is generally used to induce the GA to find multiple optima for a given problem. Sharing has also been shown empirically to help the GA search more thoroughly, even if only a single optimum is desired [*Goldberg, 1995; Weile and Michielssen, 1996; Weile et al., 1996; Weile and Michielssen, 1997*].
- *Pareto GAs* are GAs for multiobjective optimization [*Fonseca and Fleming, 1993; Srinivas and Deb, 1995; Weile and Michielssen, 1996; Weile et al., 1996*]. These will be discussed in more detail in Section 3.3.

- The GA can be hybridized with local search techniques in many ways. Local optimization may be used inside the GA to help its search, or to refine optima at the end of a GA run [*Goldberg, 1989*]. More recent examples of hybridization in GAs include the use of the *hybrid GA*, which uses a GA to optimize in a subset of the search space and a local optimizer to search the complement of this subset [*Weile and Michielssen, 1999; Weile and Michielssen, 1999*].
- The GA can be applied in a hierarchical fashion to spend more time optimizing along important dimensions of the underlying domain. This algorithm has been dubbed the *community GA* [*Weile and Michielssen, 1999; Weile et al., 1996; Weile and Michielssen, 1997*].

Additionally, GA-inspired algorithms that do not use any population (such as population-based incremental learning [*Horrell and duToit, 1996*] and the compact GA [*Harik et al., 1997*]) can be used to solve easy problems quickly or to give an initial indication of a problem's GA-hardness.

3. Selected Applications in Electromagnetics

GAs were first applied to electromagnetics (EM) optimization in 1992 [*Michielssen et al., 1992*]. Ever since their introduction, they have gained popularity steadily, and the number of GA papers appearing in the EM literature increases every year. Section 3.1 provides a broad-based review of GA applications in EM. Sections 3.2 and 3.3 highlight the GA-based design of Luneburg lenses and log-periodic monopole arrays, respectively.

3.1 An Overview of Genetic Algorithm Applications in Electromagnetics

This section refers to publications on the use of GAs in electromagnetics that appeared after January 1997. For applications predating this, the reader is referred to several earlier review papers published on this topic [*Haupt, 1995; Johnson and Rahmat-Samii, 1997; Weile and Michielssen, 1997*].

Devices that have been successfully designed by GAs include (but are not limited to) wire antennas [*Altschuler and Linden, 1997; Altschuler and Linden, 1997; Jones and Joines, 1997; Linden, 1997; Yegin and Martin, 1997; Weile et al., 1996*], antenna arrays [*Haupt, 1997; Treyer, Weile et al., 1997; Yan and Lu, 1997; Ares, Villanueva et al., 1997; Marcano,. 1997; Shimizu, 1994*], printed antennas and antenna systems [*Delabie et al., 1997; Johnson and Rahmat-Samii, 1997; Reineix et al., 1997; Sarabandi, 1997; Edwards et al., 1998; Cohen 1997; Himdi and*

Daniel, Jul. 1997; Johnson and Rahmat-Samii, 1997; Sarabandi and Li,. 1997; Johnson and Rahmat-Samii, 1994; Eclercy et al., 1997], microwave absorbers [*Weile and Michielssen, 1997*], strip gratings and frequency selective surfaces [*Treyer et al., 1997; Weile and Michielssen, 1997; Aygün et al., 1997; Mittra, Jul. 1997; Weile and Michielssen, 1997*], layered and dielectric filtering structures [*Zuffada, et al., Jul. 1997*], microwave devices [*Weile and Michielssen, 1997*], and electro- and magnetostatic machinery [*Mohammed and Üler, 1997; Rong et al., 1997; Sim et al., 1997; Vasconcelos et al., 1997; Wurtz et al., 1997; Mohammed, 1997*]. Beyond the design of devices, however, GAs have been successfully applied to nondestructive testing and evaluation, inverse scattering problems, resonance extraction, scattering center analysis, mesh generation, and the design of absorbing boundary conditions for finite-difference frequency domain computations. In the section that follows, many of the techniques used in the papers referred to above are demonstrated.

3.2 Design of Luneburg Lens Antennas Using Genetic Algorithms

A Luneburg lens antenna [*Luneberg, 1944*] transforms radiation from a point source into a plane wave and vice versa. An ideal Luneburg lens consists of a spherically symmetric dielectric sphere with a relative permittivity that varies continuously from 2 at the sphere's center to 1 at the sphere's surface according to $\varepsilon_r = 2-(r/a)^2$, where a is the radius of the sphere (Fig. 2).

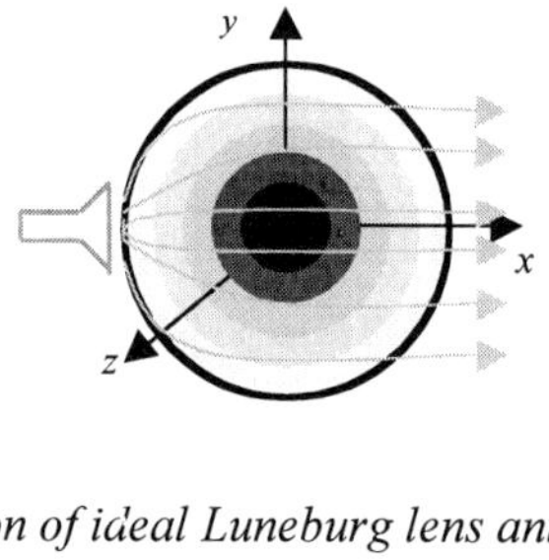

Fig. 2. *Action of ideal Luneburg lens antenna with continuous permittivity $\varepsilon_r = 2-(r/a)^2$.*

In practice, however, the Luneburg lens is constructed from a finite number of spherical shells. While the performance of such a structure is inferior to that of a true Luneburg lens, its performance can be improved by optimizing the thickness and the choice of the material of each shell. From a practical and engineering viewpoint, it is advantageous to use as small a number of shells as possible.

In this section, a novel procedure for synthesizing a 3-D Luneburg lens antenna is presented. Due to the excessive computational time involved in analyzing the 3-D Luneburg lens, the optimization process is first performed on a 2-D structure. The results are then used to design the 3-D lens. The 2-D lens is analyzed using a mode matching technique and then a GA is applied to simultaneously determine the optimal material choice for each shell and its thickness. The presented technique assumes that the lens diameter and number of shells are fixed. The GA determines the optimum structure to improve the performance of Luneburg lens antenna.

Consider a concentrically layered dielectric cylinder, that is illuminated by a plane wave as shown in Fig. 3. The total electric field in the presence of dielectric cylinder can be written as

$$\mathbf{E} = \mathbf{E}^i + \mathbf{E}^s \tag{16}$$

where $\mathbf{E}^i$ and $\mathbf{E}^s$ are the incident and scattered fields, respectively.

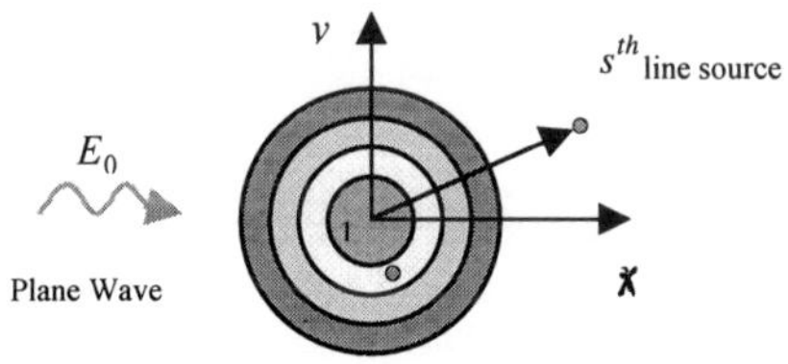

Fig. 3. *Concentrically layered dielectric cylinder.*

The incident field $\mathbf{E}^i$ due to a line source s residing in the mth region, or a plane wave in region $M+1$ can be expressed in terms of a cylindrical wave function as:

For a line source:

$$E_{zm}^{l_s} = \frac{-k_m \eta_m I_s}{4} \left| \begin{array}{ll} \sum_{n=-\infty}^{\infty} H_n^{(2)}(k_m \rho_s') J_n(k_m \rho) e^{jn(\phi-\phi')} & \rho < \rho_s' \\ \sum_{n=-\infty}^{\infty} J_n(k_m \rho_s') H_n^{(2)}(k_m \rho) e^{jn(\phi-\phi')} & \rho > \rho_s' \end{array} \right. \tag{17}$$

For a plane wave:

$$E_{z(M+1)}^{p} = E_0 \sum_{n=-\infty}^{\infty} j^{-n} J_n(k_{M+1}\rho) e^{jn\phi} . \tag{18}$$

The scattered electric field $\mathbf{E}^s$ in the mth region is expressed as

$$E_{zm}^{s} = \sum_{n=-\infty}^{\infty} \left[a_{mn} J_n(k_m \rho) + b_{mn} H_n^{(2)}(k_m \rho) \right] e^{jn\phi} \tag{19}$$

where J_n and H_n are the cylindrical Bessel function and Hankel function of nth order, k_m is the complex propagation constant in the mth region, and η_m is the intrinsic impedance of the mth region. The expansion coefficients a_{mn} and b_{mn} are determined by applying the boundary conditions at the dielectric interfaces.

The design goal is to determine the permittivity and thickness of each layer for a five-shell-30λ diameter 2-D Luneburg lens antenna as shown in Fig. 4, such that maximum directivity is achieved.

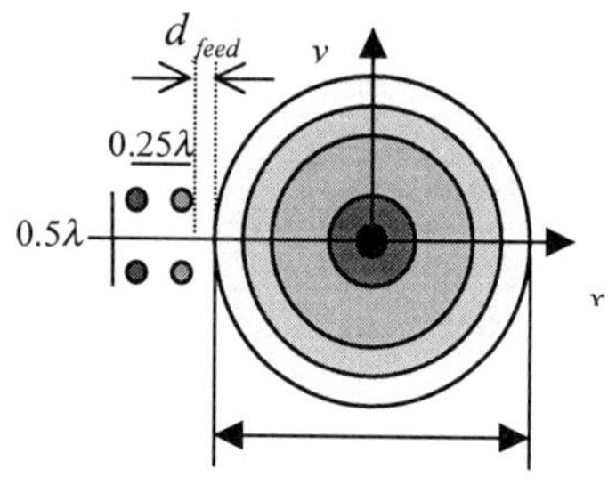

30λ

Fig. 4. *Five-shell-30λ diameter 2-D Luneburg lens antenna Four line sources are used to create horn-type illumination pattern, separation between the feed and the lens is indicated by* d_{feed}.

In the context of present problem, the fitness function f is the design's maximum directivity. It depends upon the various parameters as

$$f(\varepsilon_{r1}, t_1, \varepsilon_{r2}, t_2, \ldots, \varepsilon_{r5}, t_5) = D_0 \tag{20}$$

where ε_{ri} and t_i represent the material choice and thickness of shell i, respectively. To apply a GA to the synthesis of this structure, the material and thickness of each shell i are coded as genes $2i-1$ and $2i$, respectively. Thus, the chromsome's genes may be expressed as

$$\begin{aligned} g_{2i-1} &= a_{(i-1)(N_m+N_t)+1} a_{(i-1)(N_m+N_t)+2} \cdots a_{(i-1)(N_m+N_t)+N_m} \\ g_{2i} &= a_{(i-1)(N_m+N_t)+N_m+1} a_{(i-1)(N_m+N_t)+N_m+2} \cdots a_{i(N_m+N_t)} \end{aligned}, \tag{21}$$

where N_m bits are used to code the material choices, N_t bits are used to code the thicknesses, and i ranges from 1 to 5 (the number of shells). However, because the total lens diameter is fixed, the thickness of layer five is not encoded as it is can be computed once the thicknesses of the other layers are known. Therefore, the whole structure is represented by the sequence $\mathbf{p} = g_1g_2g_3g_4g_5g_6g_7g_8g_9$. In this case $N_{mb} = N_{tb} = 10$ and a population with size $N_{pop} = 100$ is selected. In the GA implementation $p_{cross} = 0.7$ and $p_{mut} = 0.01$ were used [*Holland, 1975; Goldberg, 1989; Haupt, 1995; Johnson and Rahmat-Samii, 1997; Weile and Michielssen, 1997*]. Tournament selection and elitism were also employed [*Goldberg, 1989*]. Several cases were studied as outlined below.

Table 1. *Optimum material for a five-shell-30λ diameter 2-D Luneburg lens antenna. Feed is on the surface of lens, F=D₀.*

Shell	ε_r	t/λ
1	1.933	4.320
2	1.824	2.577
3	1.670	2.523
4	1.476	2.129
5	1.217	3.451

Table 2. *Optimum material for a five-shell-30λ diameter 2-D Luneburg lens antenna. Uniform design.*

Shell	ε_r	t/λ
1	1.980	3
2	1.900	3
3	1.740	3
4	1.500	3
5	1.180	3

Table 1 lists the optimal materials and thicknesses for a five-shell-30λ diameter 2-D Luneburg lens. The directivity pattern of the lens is plotted in Fig. 5 and compared with that of a uniform design (cylindrical shells are uniform and their permittivities are those based on the average radius of each shell, as shown in Table 2). As is seen in Fig. 5, a maximum directivity of $D_0 = 21.34\,dB$ is obtained for the optimum design and the level of the grating lobe is decreased with respect to that of the uniform design. Note that, because of the limited number of shells and their effective periodic behavior, sets of grating lobes occur at relatively high levels. To further improve the characteristics of 2-D Luneburg lens, the fitness function *f* was modified to represent the difference between maximum directivity and maximum grating lobe. Table 3 lists optimal lens parameters, and Fig. 6 shows that improved performance is achieved.

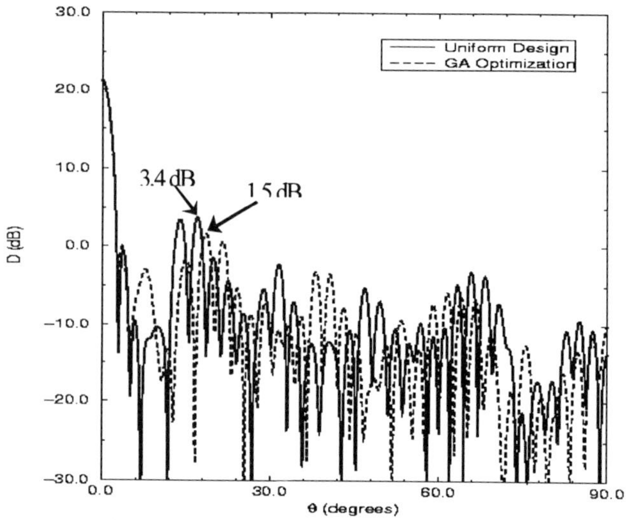

Fig. 5. *Radiation pattern of a Five-shell-*30λ *diameter 2-D Luneburg lens antenna, the feed is on the surface of the lens,* $F = D_0$.

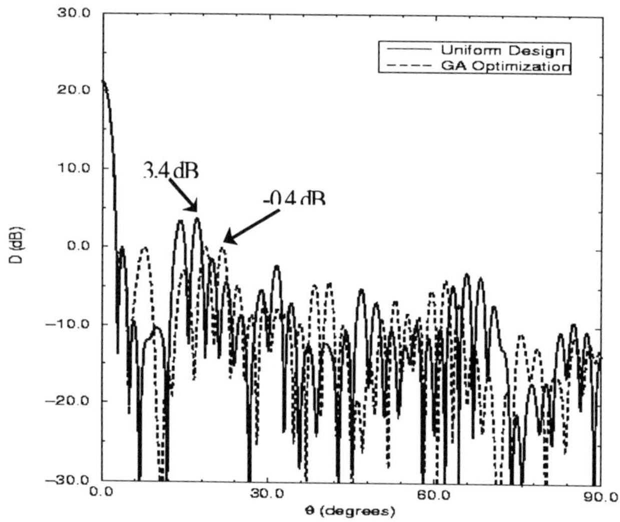

Fig. 6. *Radiation pattern of a five-shell-*30λ *diameter 2-D Luneburg lens antenna. Feed is on the surface of the lens,* $F = D_0 - Max[grading\ lobe]$.

Table 3. *Optimum material for A five-shell-30λ diameter 2-D Luneburg lens antenna. Feed is on the surface of lens, F=D_o –Max[grading lobe].*

Shell	ε_r	t/λ
1	1.925	4.410
2	1.810	2.455
3	1.665	2.396
4	1.494	2.230
5	1.209	3.509

Table 4. *Optimum material for a five-shell-30λ diameter 2-D Luneburg lens antenna. Feed is 1λ away from surface of lens F=D_o –Max[grading lobe].*

Shell	ε_r	t/λ
1	1.874	4.718
2	1.750	2.930
3	1.601	2.187
4	1.451	2.206
5	1.190	2.959

Another interesting design is obtained when the phase center of the feed is moved 1λ away the surface of lens. The optimized material choices and thicknesses, and the resulting patterns for this geometry, are shown in Table 4 and Fig. 7. It is clearly observed that the optimal design considerably improves the performance of the lens for the displaced feed.

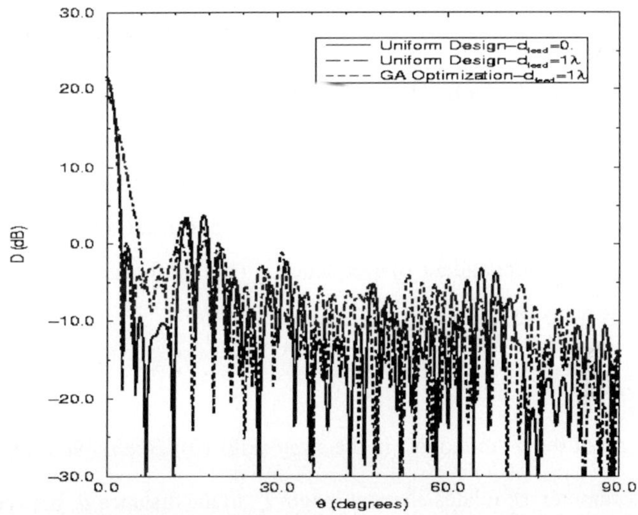

Fig. 7. *Radiation pattern of a five-shell-30λ diameter 2-D Luneburg lens antenna, the feed is 1λ away the surface of the lens,* $F = D_0 - Max[grading\ lobe]$.

The pattern of a 3-D Luneburg lens constructed using the material choices and shell thicknesses listed in Table 4 is shown in Fig. 8. Radiation pattern is computed using a 3-D code

[*Kim, 1998*] and again much improved performance is observed. This result clearly shows that use of 2-D lens data provides a viable approach for designing 3-D configurations.

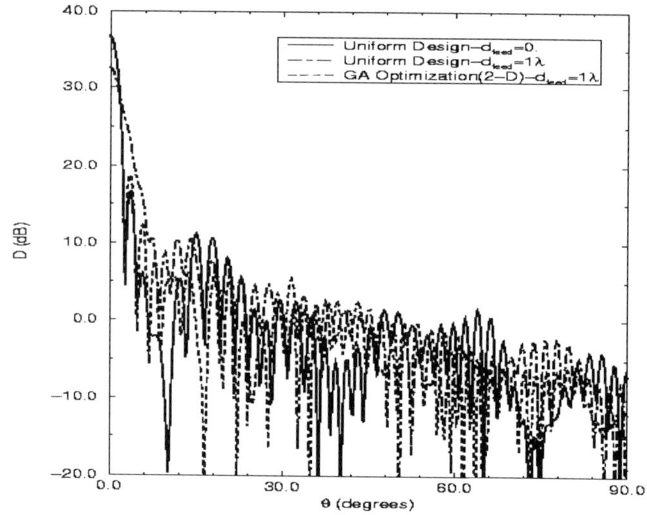

Fig. 8. *Radiation pattern of a five-shell-30λ diameter 3-D Luneburg lens antenna, the feed is 1λ away the surface of the lens,* $F = D_0 - Max[grading\ lobe]$.

3.3 Pareto Design of Log-Periodic Monopole Array Mounted on a Realistic Platform

In this section, the design of a log-periodic monopole array (LPMA) mounted on the wing of an airplane is considered (Fig. 9) [*Mayes et al., 1961; Ingerson and Mayes, 1968*]. LPMAs are often used as broadband radiators because they are easy to construct and because their radiation characteristics are nearly independent of frequency. This performance is achieved by scaling successive cells of the array by a constant

$$\tau = \frac{l_{n+1}^e}{l_n^e} \tag{22}$$

where l_n^e is the length of the nth monopole in the array (Fig. 10) [*Isbell, 1960; Ingerson and Mayes, 1968*]. A second parameter, σ, relates element length l_n^e to the distance d_n between elements n and n+1 as:

$$\sigma = \frac{d_n}{4l_n^e}. \tag{23}$$

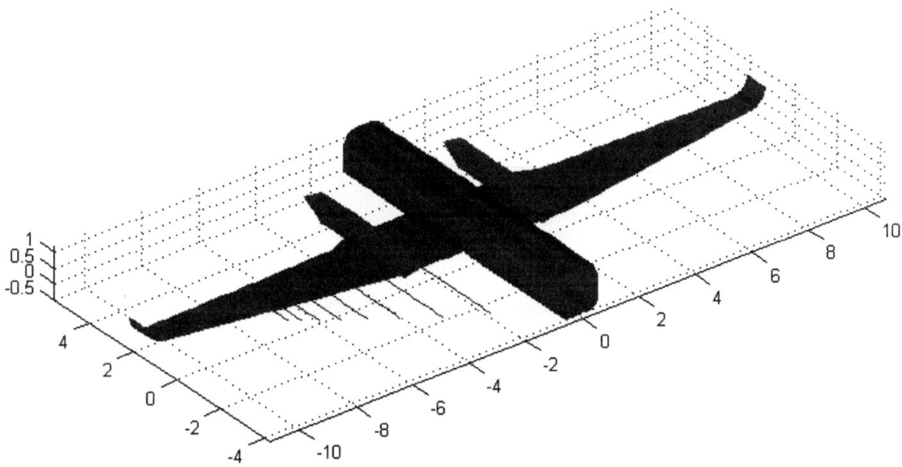

Fig. 9. *Airplane with LPMA attached to one wing.*

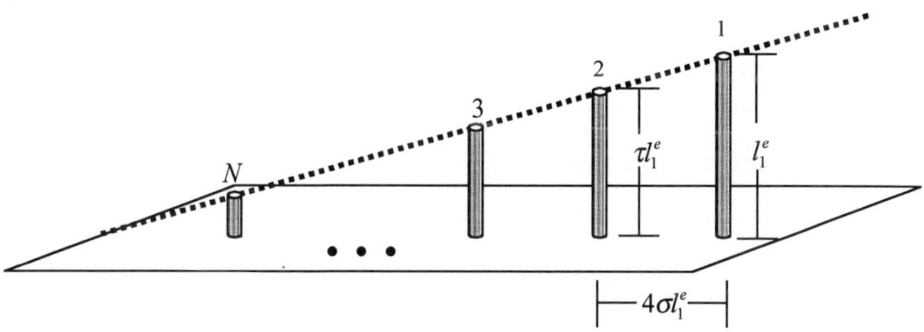

Fig. 10. *Classical LPMA residing on an infinite ground plane.*

The standard design techniques for LPMAs assume that the array is mounted on a large ground plane so that the extensive design theory for log-periodic dipole arrays (LPDAs) is applicable [*Isbell, 1960*]. Because proper phasing between array elements is achieved in LPDA design by feeder transposition (a method unavailable to the LPMA designer) standard LPMAs incorporate an impedance-modulated feed [*Ingerson and Mayes, 1968*]. In other words, the feed contains transmission lines adjacent to each monopole with characteristic impedance different from the main line impedance (Fig. 11). Although the standard design equations (discussed in [*Ingerson and Mayes, 1968*]) for the impedance modulated feed do not apply to the design of an LPMA mounted on a complicated structure such as an airplane wing, this section will demonstrate that a suitable redesign of the modulated feed can result in a practical antenna.

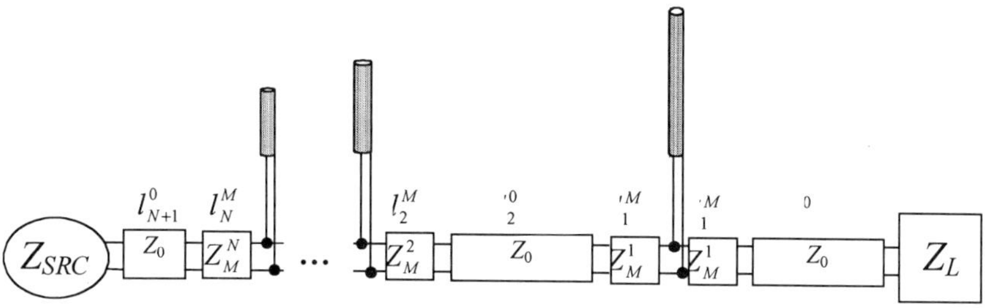

Fig. 11. *Modulated feed for an LPMA.*

To design an LPMA mounted on an arbitrary platform, the GA presented here will minimize

- the field degradation Δ, defined as the maximum weighted mean square deviation of the field pattern of the antenna from a desired pattern over the frequency band of operation of the antenna,
- the feed reflection Γ, defined as the maximum absolute reflection coefficient of the antenna over the operating band, and
- the feed inefficiency Ξ, defined as the maximum percentage of energy not reflected by the feed network that is absorbed in its terminating load (i.e., not radiated).

To accomplish these goals, the GA will design the modulated feed of Fig. 11, by optimizing

- l_i^0 for $i = 1,2\ldots(N+1)$: the nonmodulated length of feed line,
- l_j^M for $j = 1,2\ldots N$: the modulated length of feed line,
- Z_{SRC}: The source impedance,

- Z_0: The characteristic impedance of all nonmodulated lines,
- Z_M^j for $j = 1,2\ldots N$: the characteristic impedance of the modulated lines, and
- Z_L: the load impedance.

The optimizer treats l_i^0 and l_j^M as continuous values and selects Z_{SRC}, Z_0, and Z_M from a discrete set of readily available impedance values. While at first glance this seems a problem well suited for the application of a simple GA, note that the design requires the consideration of three goals simultaneously, or alternatively an optimization of the goal vector $f(\mathbf{x}) = [\Delta \quad \Gamma \quad \Xi]$. Thus, a multiobjective optimization criterion needs to be defined.

Specifically, consider two designs with goal vectors $f(\mathbf{x}_1) = [\Delta_1 \quad \Gamma_1 \quad \Xi_1]$ and $f(\mathbf{x}_2) = [\Delta_2 \quad \Gamma_2 \quad \Xi_2]$. Design 1 is said to *dominate* design 2 if $\Delta_1 \le \Delta_2$, $\Gamma_1 \le \Gamma_2$, $\Xi_1 \le \Xi_2$, and $\mathbf{x}_1 \ne \mathbf{x}_2$. The *Pareto front* for a given multiobjective optimization problem is the set of all nondominated designs in the search space [*Cohon and Marks, 1975; Cohon, 1978; Goldberg, 1989*]. The Pareto front thus represents the set of all optimal trade-offs between the various design goals, because any design not on the front is dominated by a design on the front that is superior with repect to all goals. Given the Pareto front for a specific design problem, an engineer can choose which trade-off is appropriate for the application at hand.

The construction of a GA that accomplishes such a Pareto optimization requires that the SGA (or, more specifically, the selection operator) be altered in two ways:

- The selection operator must be based on a measure of Pareto optimality instead of the typical scalar comparison.
- Selection must ensure that the population converges to the entire Pareto front, instead of to a single design as in the SGA.

The evaluation of designs on the basis of their relative Pareto optimality is fairly straightforward. Each design in the population is assigned a *Pareto rank* given by the number of designs that dominate it, plus one [*Fonseca and Fleming, 1993; Fonseca and Fleming, May 1993*]. Thus, all designs in a given population that are currently nondominated have rank one and all others have higher rank.

To ensure that the GA constructs a broad Pareto front, a method based on the competition for limited resources in the environment known as *fitness sharing* is used [*Goldberg and Richardson, 1987; Goldberg, 1989*]. Under fitness sharing, each chromosome i in the population is assigned a *niche count* n_i related to the population density surrounding it. The niche count is used to penalize chromosomes for their banality so that the population remains diverse throughout the entire GA run. Specifically, the niche count for chromosome *i* is calculated as

$$n_i = \sum_{j=1}^{N_{pop}} Sh\left(\sqrt{\left(\Delta_i - \Delta_j\right)^2 + \left(\Gamma_i - \Gamma_j\right)^2 + \left(\Xi_i - \Xi_j\right)^2}\right) \tag{24}$$

where $\left[\Delta_k \quad \Gamma_k \quad \Xi_k\right]$ is the goal vector for design *k*, and the function *S*(*d*) is given by

$$Sh(d) = \begin{cases} 1 - \left(\dfrac{d}{\sigma_{share}}\right)^E & \text{for } d \le \sigma_{share} \\ 0 & \text{for } d > \sigma_{share} \end{cases} \tag{25}$$

where σ_{share} is a *niche radius* controlling the area of the space considered "near" a given chromosome, and *E* is an exponent controlling the shape of the sharing function.

Pareto ranking and sharing can be used together to assign a single, scalar fitness value to each chromosome in the population as follows. First, all chromosomes in the first Pareto rank are given an arbitrary "unshared" fitness value. Then, for each chromosome of the first rank, this value is divided by the niche count for that chromosome, resulting in its "shared" fitness value. Then, the second rank chromosomes are given an unshared fitness equal to the lowest "shared" fitness of the first rank, and this value is then divided by the niche count for each chromosome. Continuing in this fashion, each chromosome is assigned a fitness that is used for the GA selection operator [*Fonseca and Fleming*, *1993; Srinivas and Deb*, *1995; Fonseca and Fleming*, *1993*]. (Unfortunately, this selection operator cannot be tournament-based due to well-known instabilities caused by using tournament selection with sharing [*Oei et al.*, *1991*]. Very often, roulette-wheel selection is used in Pareto GAs, see [*Weile and Michielssen*, *1996; Weile et al.*, *1996*].) Crossover and mutation are as in the SGA.

The LPMA designed in this section to demonstrate the above design technique is mounted on the trailing edge of an airfoil. This configuration has potential for use in synthetic aperture radar systems and is shown in Fig. 12.

The LPMA is designed to operate from 25.0 MHz to 88.0 MHz, and consists of eight monopoles, with $\tau = 0.850$, $\sigma = 0.096$, and $l_1^e = 3.125$ m. The antenna is required to exhibit a low voltage standing wave ratio (VSWR), low power consumption in the terminating load, and its pattern must remain a certain shape over the entire band of operation. The Pareto GA will find tradeoffs between these three goals. Although the first two goals are self-explanatory, the last requires some discussion. The *E-plane* is defined as a conical cut with its vertex at the origin and a base that intersects the unit sphere at –30° elevation. The *H-plane* pattern is measured around a circle that is concentric with the fuselage and aligned with the trailing edge of the wing. Fig. 12 illustrates the H- and E-planes as defined above, and illustrates the patterns supplied as ideal. Each pattern cut was evaluated at 180 equally spaced points, and weight vectors were constructed that permit the evaluation of the field degradation measure Δ.

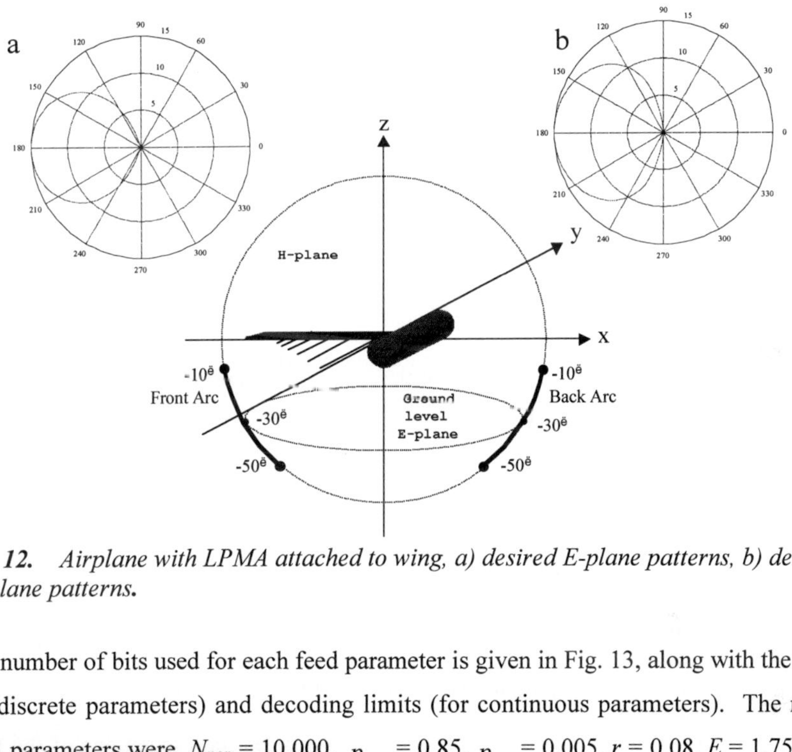

Fig. 12. *Airplane with LPMA attached to wing, a) desired E-plane patterns, b) desired H-plane patterns.*

The number of bits used for each feed parameter is given in Fig. 13, along with the databases used (for discrete parameters) and decoding limits (for continuous parameters). The remaining Pareto-GA parameters were N_{pop} = 10,000, p_{cross} = 0.85, p_{mut} = 0.005, r = 0.08, E = 1.75, and N_{freq} = 8. The N_{freq} = 8 frequencies used for optimization were 25.0, 30.0, 35.0, 45.0, 55.0, 65.0, 75.0, and 88.0 MHz. Note that the allowable ranges of the feed lengths scale by τ, which encourages log-periodicity while not strictly enforcing it. Figure 14 shows the front returned by the Pareto GA, as well as a front generated by over 140 runs of the simple GA, optimizing different linear combinations of the three goals. The combined population of all of the simple GA runs was

39,050, almost four times the population allotted to the Pareto GA, and all the GAs (including the Pareto GA) were run for 65 generations to ensure convergence. Figure 14 shows that all of the simple GA results are dominated by the Pareto GA results.

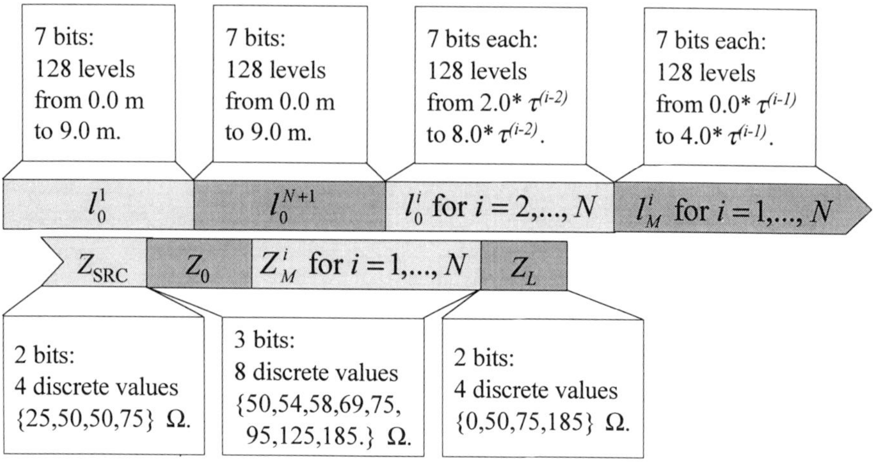

Fig. 13. *Chromosome structure used for optimizing LPMA design.*

To further demonstrate the superiority of the Pareto GA, representative designs were chosen from the two fronts shown in Fig. 14, and are compared to each other and to a design obtained with a purely log-periodic feed with the parameters l_0^n, K, Z_M, Z_0, Z_L, and Z_{SRC} optimized by the simplex method. In the case of the Pareto GA, since small improvements in Γ for $\Gamma < 0.2$ were associated with large increases in Δ, a design with desirable Ξ and $\Gamma \approx 0.2$ was selected for comparison. Because neither the simple GA nor the simplex method designs could produce such a low reflection coefficient for any design, the designs exhibiting the lowest Γ were chosen for comparison with the Pareto GA result. Figure 15 shows polar representations of the E- and H-plane patterns for representative frequencies for all three designs. Figure 16 shows the corresponding VSWR characteristics.

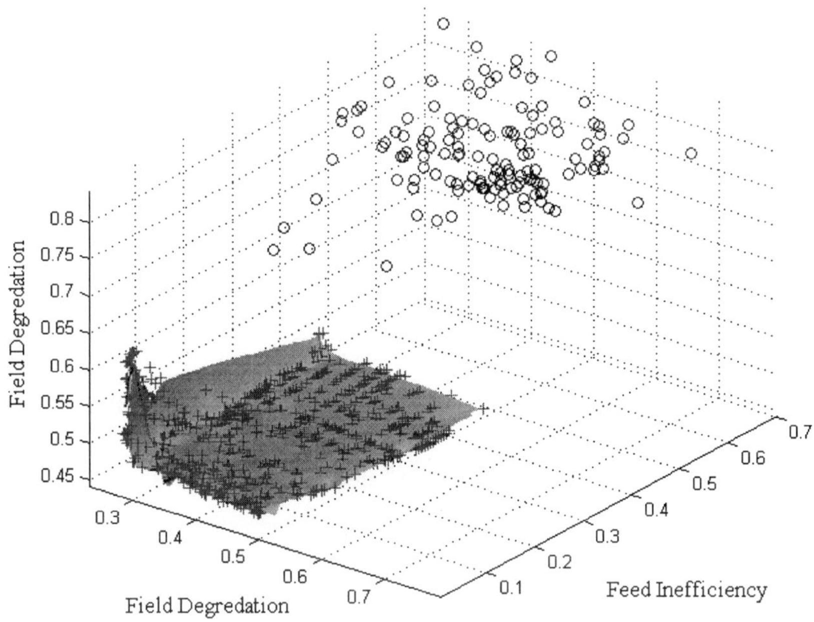

Fig. 14. Pareto fronts obtained using Pareto GA (+) and repeated application of single-goal SGA with varying weights for each goal (o).

These figures unambiguously demonstrate the power of the Pareto GA relative to the other two methods. Figure 14 shows that the entire simple GA front is dominated by the front returned by the Pareto GA. Moreover, the simplex front is so much worse than the other two it cannot even be shown on the same graph. More importantly, however, Figure 16 shows that the design returned by the Pareto GA is quantitatively superior to the other two: only the Pareto GA design exhibits frequency independent patterns similar to the ideal patterns supplied in Fig. 12. (The Pareto GA patterns are shown with a solid line, the simple GA patterns are shown with a dashed line, and the simplex patterns are shown with a dotted line.) Perhaps even more importantly, Fig. 17 shows that neither the simple GA nor the simplex method could generate a design with a VSWR characteristic as good as that generated by the Pareto GA, even when allowing for inferior patterns.

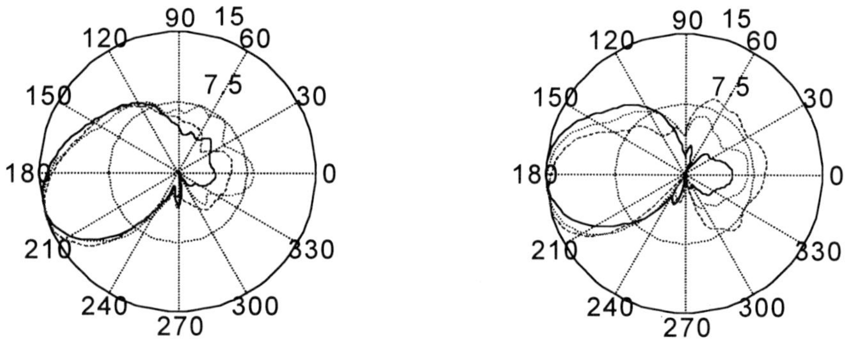

(E-plane at 4th and 5th frequency

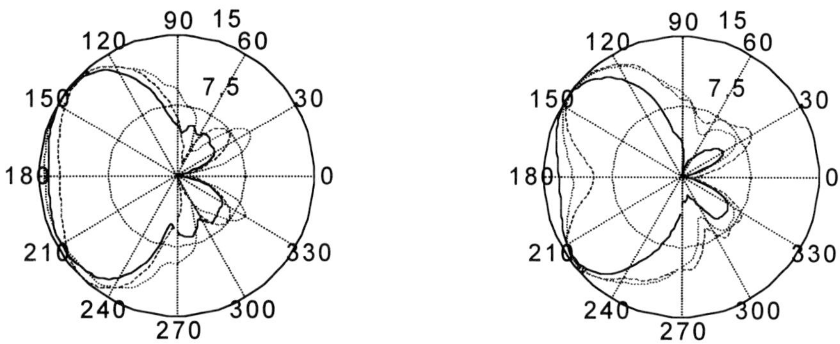

(H-plane, same frequencies)

***Fig. 15**. E and H-plane patterns obtained using Pareto GA, SGA, and simplex method.*

The fact that the Pareto GA returns designs that dominate the simple GA designs is very telling. The superiority of the Pareto technique may be largely due to the sharing operator, which disallows stagnation and forces the GA to search novel regions of the space [*Goldberg, 1995; Weile and Michielssen, 1996; Weile et al., 1996*]. For the same reason, the sharing operator has even been found useful in single parameter problems. This behavior is characteristic of advanced

GA techniques: Because the simple GA is such a broadly applicable algorithm, a suitably modified advanced GA will usually return better results for a specific problem.

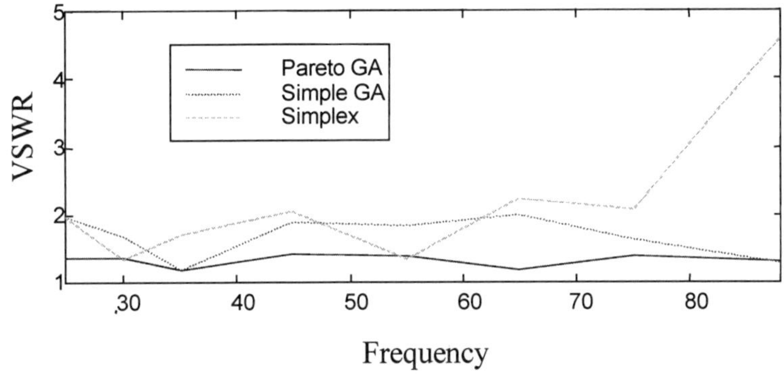

Fig. 16. *VSWR of designs resulting from various optimizers.*

4. Conclusions

This chapter introduced simple genetic algorithms, some of the basic theory underlying their operation, and advanced operator extensions. In addition, the history of the application of GA-based optimizers in electromagnetic system design was reviewed, and two recent applications, viz. the design of Luneburg lenses and log-periodic monopole arrays mounted on airborne vehicles, were highlighted. To date, many electromagnetic design problems have succumbed to the application GA-driven design paradigms. The authors expect the continued development of GA theory and advanced operator schemes to sustain the future application of GAs to ever more complex electromagnetic synthesis tasks.

Acknowledgments

This work was supported in part by a grant from the National Science Foundation (grant ECS 95-02138). The authors would like to thank Mr. H. Mosallaei of UCLA for his assistance in designing the Luneburg lenses, and Mr. Stephen Fisher of UIUC for his help in synthesizing LMPA feeds.

References

Altschuler, E.E. and D.S. Linden [1997], "Design of a loaded monopole having hemispherical coverage using a genetic algorithm," *IEEE Trans. Antennas Propag.*, 45, 1, pp. 1-4.

Altschuler, E.E. and D.S. Linden [1997], "Wire-antenna design using genetic algorithms," *IEEE Antennas Propag. Mag.*, 39, 2, pp. 33-43.

Ares, F., E. Villanueva, J.A. Rodríguez , and S. R. Rengarajan [1997], "Application of genetic algorithms in the design and optimization of array patterns," *IEEE Antennas Propag. Soc. Symp.*, pp. 1684-1687.

Aygün, K., D.S. Weile , and E. Michielssen [1997], "Design of multilayered strip gratings by genetic algorithms," *Microwave and Optical Techn. Letters*, 42.

Bäck, T. and H.-P. Schwefel [1995], "Evolution Strategies I: variants and their computational implementation," in G. Winter, J. Périaux, M. Galán and P. Cuesta (ed.), Genetic Algorithms in Engineering and Computer Science, New York, John Wiley and Sons.

Cohen, N. [1997], "Fractal coding in genetic algorithm (GA) antenna optimization," *IEEE Antennas Propag. Soc. Digest,* pp. 1692-1695.

Cohon, J.L. [1978], *Multiobjective Programming and Planning,* New York, Academic Press.

Cohon, J.L. and D.H. Marks [1975], "A review and evaluation of multiobjective programming techniques," *Water Resources Research,* 11, 2, pp. 208-219.

Darwin, C. [1985], *The Origin of Species*, London, Penguin Books.

Delabie, C., M. Villegas, and O. Picon [1997], "Creation of new shapes for resonant microstrip structures by means of a genetic algorithm," *Electronics Letters*, 30, 18, pp. 1509-1510.

Eclercy, D., A. Reineix, and B. Jecko [1997], "FDTD genetic algorithm for antenna optimization," *Microwave and Optical Techn. Letters*, 16, pp. 72-74.

Edwards, R.M., G.G. Cook, S.K. Khamas, R.J. Aidley, and B. Chambers [1998], "Design of circularly polarized printed spiral antenna using dual objective genetic algorithm," *Electronics Letters*, 34, 7, pp. 608-609.

Fonseca, C.M. and P.J. Fleming [1993], "Genetic algorithms for multiobjective optimization: Formulation, discussion and generalization," *Proc. Fifth Internl. Conf. Genetic Algorithms*, pp. 416-423.

Fonseca, C.M. and P.J. Fleming [1993], "Multiobjective genetic algorithms," *Proc. IEE Conf. Genetic Algorithms for Control Engineering*, pp. 6/1-5.

Goldberg, D.E. [1989], *Genetic Algorithms in Search, Optimization and Machine Learning, Reading*, MA, Addison-Wesley.

Goldberg, D.E. [1991], "Real-coded genetic algorithms, virtual alphabets and blocking," *Complex Systems*, 5, 2, pp. 129-167.

Goldberg, D.E. [1995]. Personal communication.

Goldberg, D.E. [1998], "The design of innovation: Lessons from genetic algorithms, lessons for the real world," University of Illinois Genetic Algorithms Laboratory, IlliGAL Report No. 98004.

Goldberg, D.E. and K. Deb [1991], "A comparative analysis of selection schemes used in genetic algorithms," in G. J. E. Rawlins (ed.), *Foundations of Genetic Algorithms*, San Mateo, CA, Morgan Kaufmann.

Goldberg, D.E., K. Deb, and J.H. Clark [1991], "Genetic algorithms, noise, and the sizing of populations," *Complex Systems*, 6, 3, pp. 333-362.

Goldberg, D.E. and J.J. Richardson [1987], "Genetic algorithms with sharing for multimodal function optimization," Genetic Algorithms and Their Applications: *Proc. Second Internl. Conf. Genetic Algorithms*, pp. 41-49.

Harik, G., E. Cantu-Paz, D.E. Goldberg, and B.L. Miller [1996], "*The gambler's ruin problem, genetic algorithms, and the sizing of populations*," University of Illinois at Urbana Champaign, IlliGAL Report No. 96004.

Harik, G., F. Lobo , and D.E. Goldberg [1997], "*The compact genetic algorithm*," Illinois Genetic Algorithms Laboratory, University of Illinois, 97006.

Haupt, R.L. [1994], "Thinned arrays using genetic algorithms," *IEEE Trans. Antennas Propag.*, 42, 7, pp. 993-999.

Haupt, R.L. [1995], "An introduction to genetic algorithms for electromagnetics," *IEEE Antennas Propag. Mag.*, 37, 2, pp. 7-14.

Haupt, R.L. [1997], "Phase-only adaptive nulling with a genetic algorithm," *IEEE Trans. Antennas Propag.*, 45, 6, pp. 1009-1015.

Himdi, M. and J.P. Daniel [1997], "Synthesis of slot coupled loaded patch antennas using a genetic algorithm through various examples," *IEEE Antennas Propag. Soc. Internl. Symp.*, pp. 1700-1703.

Holland, J.H. [1975], *Adaptation in Natural and Artificial Systems*, Ann Arbor, MI, University of Michigan.

Horn, J. [1993], "Finite Markov chain analysis of genetic algorithms," *Fifth Internl. Conf. Genetic Algorithms*, pp. 10-17.

Horrell, J.M. and L.J. duToit [1996], "Array pattern synthesis using PBIL," *IEEE AFRICON*, pp. 276-281.

Ingerson, P.G. and P.E. Mayes [1968], "Log-periodic antennas with modulated impedance," *IEEE Trans. Antennas Propag.*, 16, 11, pp. 633-642.

Isbell, D.E. [1960], "Log-periodic dipole arrays," *IRE Trans.*, AP-8, pp. 260-267.

Johnson, J.M. and Y. Rahmat-Samii [1997], "Genetic algorithms in engineering electromagnetics," *IEEE Antennas Propag. Mag.*, 39, 4, pp. 7-21.

Johnson, J.M. and Y. Rahmat-Samii [1997], "Genetic algorithms and method of moments (GA/MoM): a novel integration for antenna design," *IEEE Antennas Prop. Soc. Symp.*, pp. 1664-1667.

Johnson, J.M. and Y. Rahmat-Samii [1994], "Genetic algorithm optimization and its application to antenna design," *Proc. IEEE Antennas Propag. Soc. Internl. Symp.*, pp. 326-329.

Jones, E.A. and W.T. Joines [1997], "Design of Yagi-Uda antennas using genetic algorithms," *IEEE Trans. Antennas Propag.*, 45, 9, pp. 1386-1392.

Kim, K.W. [1998], "Spherical Luneberg lens antennas: Engineering characterizations including air gap effects," *IEEE-APS/URSI Meeting*, pp. 2062-2065.

Linden, D.S. [1997]. Automated Design and Optimization of Wire Antennas Using Genetic Algorithms, Ph. D. Thesis, Cambridge, MA: Massachussetts Institute of Technology.

Luneberg, R.K. [1944], *Theory of Optics,* Providence, RI, Brown University.

Marcano, D. [1997], "Synthesis of linear and planar antenna arrays using genetic algorithms," *IEEE Antennas Propag. Soc. Internl. Symp.,* pp. 1688-1691.

Mayes, P.E., G.A. Deschamps, and W.T. Patton [1961], "Backward wave radiation from periodic structures and application to the design of frequency independent antennas," *Proc. IRE*, 49, 5, pp. 962-963.

Michielssen, E., S. Ranjithan, and R. Mittra [1992], "Optimal multilayer filter design using real coded genetic algorithms," *IEE Proc.-J*, 139, 12, pp. 413-420.

Mittra, R. [1997], "Genetic algorithms: the last word for solving all of your design problems?," *IEEE Antennas Propag. Soc. Internl. Symp.*, pp. 1672-1675.

Mohammed, O.A. [1997], "GA optimization in electrical devices," *IEEE Antennas Propag. Soc. Internl. Symp.*, pp. 1696-1699.

Mohammed, O.A., and G.F. Üler [1997], "A hybrid technique for the optimal design of electromagnetic devices using direct search and genetic algorithms," *IEEE Trans. Magnetics*, 33, 2, pp. 1931-1934.

Oei, C.K., D.E. Goldberg , and S.J. Chang [1991], "*Tournament selection, niching, and the preservation of diversity,*" University of Illinois at Urbana-Champaign, IlliGAL Report Number 91011.

Reineix, A., D. Eclercy , and B. Jecko [1997], "FDTD/genetic algorithm coupling for antennas optimization," *Annals of Telecommunication,* 52, 9-10, pp. 503-508.

Rong, R., D.A. Lowther, Z. Malik, H. Su, J. Nelder, and R. Spense [1997], "Applying response surface methodology in the design and optimization of electromagnetic devices," *IEEE Trans. Magnetics*, 33, 2, pp. 1916-1919.

Sarabandi, K. [1997], "Characterization of optimum polarization for multiple target discrimination using genetic algorithms," *IEEE Trans. Antennas Propag.*, 45, 12, pp. 1810-1817.

Sarabandi, K. and E.S. Li [1997], "Characterization of optimum polarization for multiple target discrimination using genetic algorithms," *IEEE Antennas Propag. Soc. Internl. Symp.*, pp. 502-505.

Shimizu, M. [1994], "Determining the excitation coefficients of an array using genetic algorithms," *Proceedings of the IEEE Antennas Propag. Soc. Internl. Symp.*, pp. 530-533.

Sim, D-J., H-K. Jung, S-Y. Hahn, and J-S. Won [1997], "Application of vector optimization employing modified genetic algorithm to permanent magnet motor design," *IEEE Trans. Magnetics*, 33, 2, pp. 1888-1891.

Srinivas, N. and K. Deb [1995], "Multiobjective Optimization Using Nondominated Sorting in Genetic Algortihms," *Evolutionary Computation*, 2, 3, pp. 221-248.

Thierens, D. and D.E. Goldberg [1993], "Mixing in genetic algorithms," *Proc. Fifth Internl. Conf. Genetic Algorithms*.

Treyer, D., D.S. Weile, and E. Michielssen [1997], "The application of novel genetic algorithms to electromagnetic problems," *13th Ann. Rev. Prog. Applied Comp. Electromagnetics*, pp. 1382-1386.

Vasconcelos, V.A., R.R. Saldanha, and I. Krähenbühl [1997], "Genetic algorithm coupled with a deterministic method for optimization in electromagnetics," *IEEE Trans. Magnetics,* 35, 2, pp. 1860-1863.

Weile, D.S. and E. Michielssen [1996], "Integer coded Pareto genetic algorithm design of antenna arrays," *Electronics Letters,* 32, 19, pp. 1744-1745.

Weile, D.S., and E. Michielssen [1997], "Genetic algorithm optimization applied to electromagnetics: A review," *IEEE Trans. Antennas Propag.*, 45, 3, pp. 343-353.

Weile, D.S. and E. Michielssen [1999], "Design of doubly periodic filter and polarizer structures using a hybridized genetic algorithm," *Radio Science*, to be published.

Weile, D.S. and E. Michielssen [1999], "The design of E-plane microwave filters using a community genetic algorithm," *Microwave and Optical Techn. Letters*, to be published.

Weile, D.S. and E. Michielssen [1999], "The use of domain decomposition genetic algorithms exploiting model reduction for the design of frequency selective surfaces," *Computer Methods in Applied Mechanics Eng.*, to be published.

Weile, D.S. and E. Michielssen [1997], "Evolutionary optimization of electromagnetic devices using advanced operators and population structures," *IEEE Antennas Propag. Internl. Symp.*, pp. 1668-1671.

Weile, D.S., E. Michielssen, and A. Boag [1996], "Community-based evolutionary optimization of frequency selective surfaces," *Proc. USNC/URSI Radio Science Mtg.*, p. 345.

Weile, D.S., E. Michielssen, and A. Boag [1996], "Optimization of broad-band loaded wire antennas in real environments using genetic algorithms," *Proc. 12th Annual Rev. Prog. Applied Comput. Electrom.*, pp. 726-733.

Weile, D.S., E. Michielssen, and D.E. Goldberg [1996], "Genetic algorithm design of Pareto optimal broad band microwave absorbers," *IEEE Trans. Electromagnetic Compatibility*, 38, 4, pp. 518-524.

Wurtz, F., M. Richomme, J. Bigeon, and J.C. Sabonnadiere [1997], "A few results for using genetic algorithms in the design of electrical machines," *IEEE Trans. Magnetics*, 33, 2, pp. 1892-1895.

Yan, K-K., and Y. Lu [1997], "Sidelobe reduction in array-pattern synthesis using genetic algorithm," *IEEE Trans. Antennas Propag.*, 45, 7, pp. 1117-1122.

Yegin, K. and A.Q. Martin [1997], "Very broadband loaded monopole antennas," *IEEE Antennas Propag. Internl. Symp.*, pp. 232-235.

Zuffada, C., T. Cwik , and C. Ditchman [1997], "Synthesis of novel all-dielectric grating filters using genetic algorithms," *IEEE Antennas Propag. Society Internl. Symp.*, pp. 1676-1679.

INTELLIGENT ANTENNAS FOR FUTURE WIRELESS COMMUNICATIONS

Gilles Delisle, Khelifa Hettak and Godfrey Lucas

1. Introduction

Design engineers are continually looking for solutions that increase coverage and capacity within a cellular or PCS system without the need for more cell sites. Re-evaluating antenna technology is one approach [*Litva and Lo, 1996; Okamoto and Xu, 1997; Swales, et al., 1990; Winters, 1993; Naguib and Paulraj;, Tsoulos, et al., 1994; Andersson, et al., 1991.*]. To date, have been the most neglected of all the components in personal communications systems. Yet, the manner in which radio frequency energy is distributed into and collected from space has a profound influence upon the efficient use of spectrum, the cost of establishing new personal communications networks, and the service quality provided by these networks. The combination of the enormous increase in low-cost computing capacity and the development of new algorithms for processing signals from arrays of simple antennas make possible "intelligent antennas" for cellular-like communications systems, providing greater coverage area for each cell site, the rejection of interference and substantial capacity improvements [*Naguib, et al., 1994; Zetterberg and Ottersten, 1995; Naguib and Paulraj, 1995; Ogawa and Ohgane, 1996; Ward, et al., 1996; Lindmark, et al., 1998; Paulraj, 1997; Jeng, et al., 1998; Naguib, 1996*].

In light of this background, the first part of this tutorial will present an overview of the impact of intelligent antennas for next generation radio including a review of major challenges for the realization of such wireless systems, followed by the key issues and the state of the art of marked potential and expected impact and an assessment of real world performance gains. The second part of the tutorial will be a description of intelligent antenna systems. This section emphasizes clearly that the integration of basic antenna characteristics with fast, highly efficient signal processing solutions implemented in a Digital Signal Processor (DSP) provides a flexible system whose characteristics can respond to changes in the radio environment to provide efficient communications. This dynamic response includes a combination of techniques such as switched or adaptive beamforming, distributed antennas, leading to the overarching Intelligent antenna concept. In addition, the fundamentals of the technology will be examined: radiating element and arrays, summary of relative complexities and benefits of intelligent antenna system design over conventional omnidirectional approaches. The third part of this tutorial will address the RF aspects of intelligent antennas. Major factors relevant to radio frequency (RF) integrated circuit design will be discussed, including the main directions in the antenna modeling and CAD development, antenna technologies and microwave techniques. The fourth part of this tutorial will be dedicated to practical applications of smart antennas to broadband wireless communications. This will lead into a discussion of the potential gain of deploying intelligent antennas at a base station. Emphasis will be placed particularly on the integration of array antennas for wireless broad band multimedia communications systems which play the central role in the deployment of intelligent antenna systems. A concluding summary, perspectives and the extensive bibliography provide the reader with a starting point to further explore the topics covered in this tutorial in greater depth.

2. Intelligent Antennas for Next Radio Generation

Personal communications services are of great importance for the third-generation mobile radio communications systems such as UMTS. The objectives of third-generation systems are to go far beyond that what is known from second generation systems such as IS-54, GSM, or IS-95, especially with respect to the wide range of multimedia services (speech, audio, image, video, data) with bit rates up to 2 Mbit/s, the high quality of service requirements (speech/image quality, bit error rate (BER) performance, and user capacity), and the operation in mixed cell scenarios (macro, micro, pico) and different environments (indoor/outdoor, business/domestic, cellular/cordless) [*Ogawa and Ohgane, 1996; Ward, et al., 1996; Lindmark, et al., 1998; Paulraj, 1997*]. Present second generation systems are primarily designed as low-rate voice and data systems (usually single-environment such as indoor or outdoor) and fall short of meeting important requirements of the third generation systems *[Roy, 1997; Weinmann, et al., 1998; Roy, 1996*].

2.1 Major Challenges

There are many problems that exist with current second generation systems. The biggest problem is that the current systems do not have enough capacity to deal with a rapid increase in communication volume. With the current AMPS cellular system, the conventional three-sector antenna at each cell site can handle a maximum of only about 40 mobile phones which is far short of the predicted future demand. The growth in the number of customers is fast and there is a need to rapidly expand cellular phone facilities to handle additional call traffic. The only solution seems to be the addition of more antenna sites. However, sites and obtaining zoning approval to put up new antenna towers can take a very long period of time. The bottom line is also that antenna systems are very expensive and have associated political risks. Another big problem is the quality of the service. In today's installations, the antenna systems used in wireless networks send and receive radio waves in all directions. They cannot separate the desired signal from interfering signals on the same channel. Existing antennas also have difficulty dealing with multipath signals that scatter off building and hills. When a cellular phone is in a fast-moving vehicle, the strength of multipath signals can fluctuate rapidly. Since cellular utilizes radio channels to transmit voice and data, cellular transmissions can be subject to interruptions due to geographic and atmospheric conditions. Also, problems surface as the number of users increase. The frequencies in a given cell site become crowded, increasing interference and cross-talk as cell-phones inadvertently intercept signals on nearby frequencies.

2.2 Key Issues

Design and development of sustainable transmission systems, with very high capacity is a major challenge for the realization of the Universal Personal Telecommunication Systems (UPTS). Intelligent antennas are a new technology that have the potential to p trial equipment and conducting the tests, authoritative input will be obtained to contribute to the standardization process of third generation systems.

Essentially, the intelligent antenna is an evolutionary step from the existing cellular technology. Unlike traditional cellular networks in which the cell is typically divided in three 120 degree sectors to provide directionality, it is provided through an array of antennas and a powerful DSP that handles the signal switching [*Tsoulos, et al., 1997; 1998; Roy, 1997; Weinmann, et al., 1998; Roy, 1996*]. So instead of using sectors, intelligent antennas typically split the 360 degree field into 12 distinct beams as illustrated by Figure 1. This is accomplished by performing a detailed analysis of the antenna signals across the entire frequency band occupied by the network, identifying individual channels and then applying appropriate processing. The array setup makes each of the beams extremely focused which helps to reduce interference between the beams. Since there is less interference, each of the 12 sectors could use the full spectrum of transmission instead of dividing the spectrum between 3 sectors.

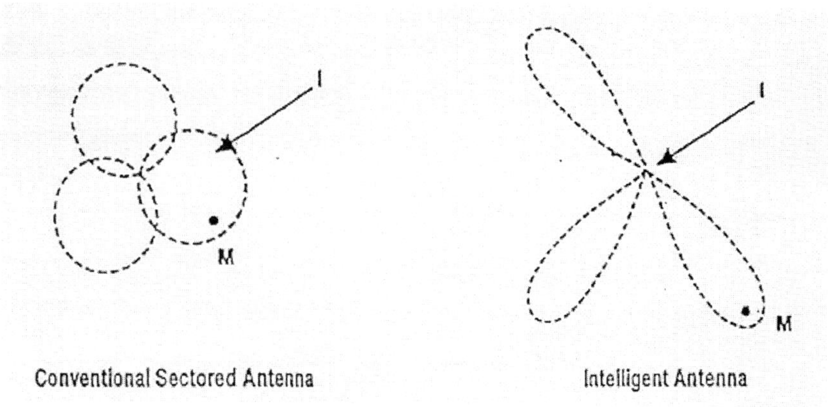

Fig. 1 *Base station (a) conventional sectored antenna (b) Intelligent antenna configurations.*

2.3 Expected Impact

Intelligent antennas offer greatly improved spectral efficiency by making full use of the spatial domain. When translating this operational need to a technical requirement for an antenna system, the need is not for a steerable "higher gain" array, but instead an array which can increase the carrier to interference radio signal (C/I) at both the base station and the mobile terminal. If this can be achieved, then more RF channels can be added to the base stations in the area without driving the C/I below an acceptable limit.

The first and most obvious step is to create gain of the array in the intended direction using antenna aperture. This is done by using physically directive elements, or by combining the outputs of more elements, or by combining the outputs of more elements to create aperture, and thus gain. It is important to note that is not the gain but the directivity which has value in trying to maximize C/I. The lack of gain in the direction of interference, ideally a null, may be more effective than gain in the direction of the desired signal.

The second approach is the mitigation of multipath fading. C/I budgets are generally designed with significant margin for slow and fast fading. Preventing the significant losses in C/I induced by multipath propagation is one very effective use of an intelligent antenna array. The final approach used to maximize C/I is the identification and nulling of interferers. When designing an antenna element combiner, it is generally much easier to form sharp nulls than to form narrow beams for a given number of elements in the array. Thus, smart antennas have an integral part to play in helping third generation systems reach their declared capacity objectives.

2.4 Market Potential

With the current growth rate of cellular use at 30-40% each year, the demand for intelligent antennas can be expected to remain high. Domestically, most cellular providers will be upgrading their existing networks to keep up with demand caused by the new cellular phone user, PCS services, and new cellular services that are being rapidly developed. Analysts are estimating that cellular and PCS networks will provide a market of $4 billion [*Feuerstein, 1998*]. Internationally, many countries that are becoming industrialized are looking at wireless as the medium for their networking infrastructure. At this pace, there will be more demand internationally than there is domestically in North America.

3. Description of the Intelligent Antenna System

3.1 Definition

The antennas are not really "smart". Antenna systems are generally co-located with a base station, a smart antenna system combines the outputs of multiple antenna elements with signal processing capabilities to transmit and/or receive RF signals in an adaptive, spatially sensitive manner. In other words, a smart or adaptive antenna is one which can automatically change its radiation pattern in response to its signal environment. Smart antennas are based on " phased arrays" or " a beamformer " which combine the signals from an array of antennas. A composite antenna pattern is produced which can be controlled by adjusting the amplitude and phase with which the individual antenna signals are combined. This enables the array to act as "spatial filter" which can enhance or reject signals based on their direction of arrival *[Litva and Lo, 1996*].

3.2 Types of Intelligent Antenna Systems

The categories of smart antenna systems embraces such terminology as intelligent, phased array, spatial division multiple access (SDMA), spatial processing, digital beamforming, adaptive antenna systems. But, in the most general sense, intelligent antenna systems are categorized as either switched beam or adaptive array systems *[Roy, 1996]*.

The following are distinctions between the two major categories of intelligent antennas regarding the choices in transmit strategy:

* Switched Beam: A finite number of fixed, pre-defined patterns or combining strategies (sectors).
* Adaptive array: A very large number of patterns that are adjusted in real time.

Adaptive antennas are antenna systems whose properties-polarization, directivity, radiation pattern and input impedance can be changed adaptively to optimize the antenna performance. These adaptive changes are accomplished by varying the excitations of the individual antenna elements of the antenna system. To this end, the antenna must be combined with a unit which measures the antenna performance and determines, on the basis of some algorithm, the changes in element excitations needed for the optimization (Figure 2).

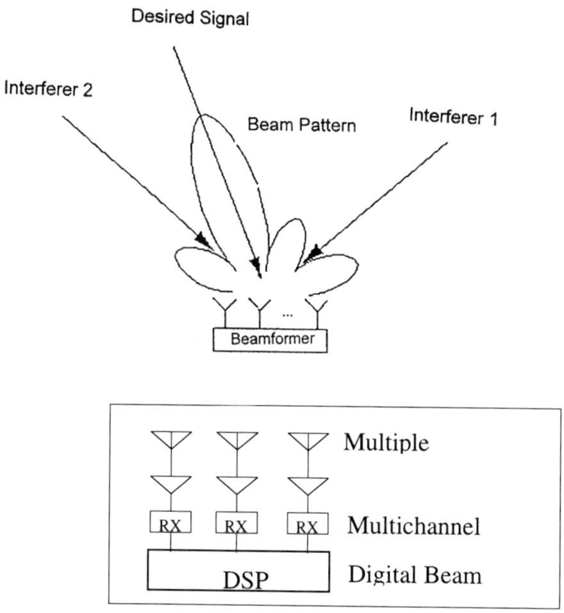

Fig. 2 *Intelligent antenna implementation with switched beam antenna.*

In other words, adaptive antenna technology represents the most advanced intelligent antennas approach to date. Using a variety of new signal processing algorithms, the adaptive system is able to effectively locate and track various types of signals so as to dynamically minimize interference and maximize intended signal reception.

Switched beam antenna systems form multiple fixed beams with heightened sensitivity in particular directions. These antenna systems detect signal strength, choose from one of several predetermined, fixed beams, and switch from one beam to another as the mobile moves throughout the sector (Figure 3).

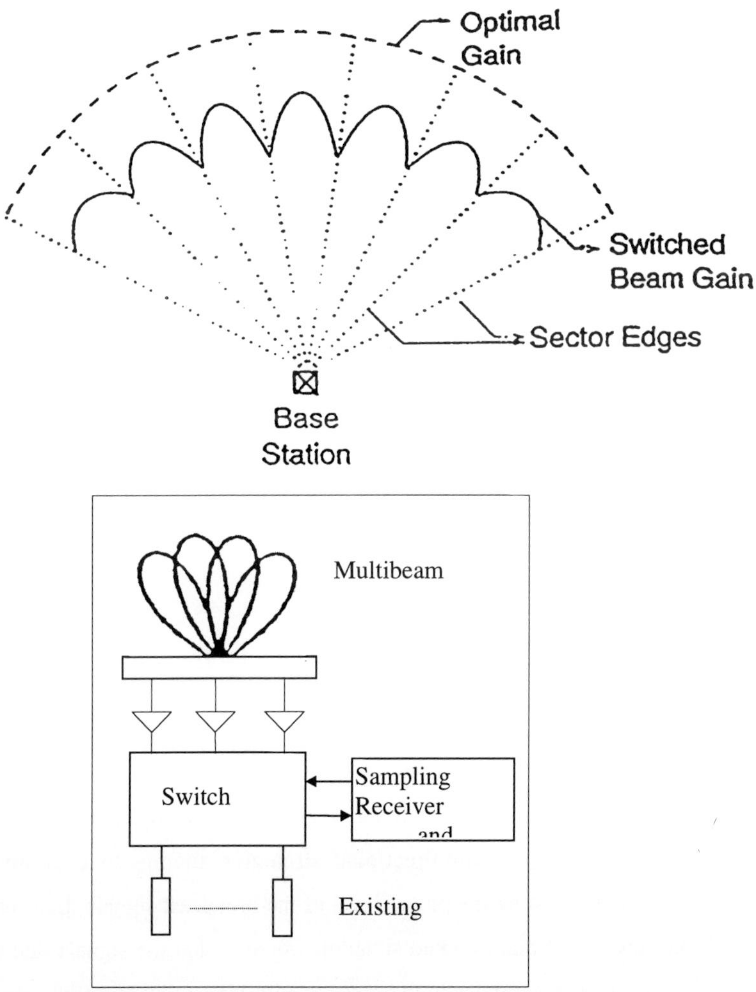

Fig. 3 *Intelligent antenna implementation with adaptive antenna.*

Instead of shaping the directional antenna pattern with the metallic properties and physical design of a single element like a sectorized antenna, switched beam systems combine the outputs of multiple antennas so as to form finely sectorized beams with a more spatial selectivity than can be achieved with conventional, single element approaches.

One of the challenges in the development of intelligent antennas is the availability of efficient and affordable radiating elements. Since the early days of wireless communications, there has been the " simple " dipole antenna, which radiates and receives equally well in all directions (Figure 4) [*Balanis, et al., 1997*]. To find its users, this single element design broadcasts omnidirectionally in a pattern resembling ripples radiating outward in a pool of water. While adequate for simple RF environments where no specific knowledge of the users' where abouts is available, this unfocused approach scatters signals, reaching desired users with only a small percentage of the overall energy sent out into the environment.

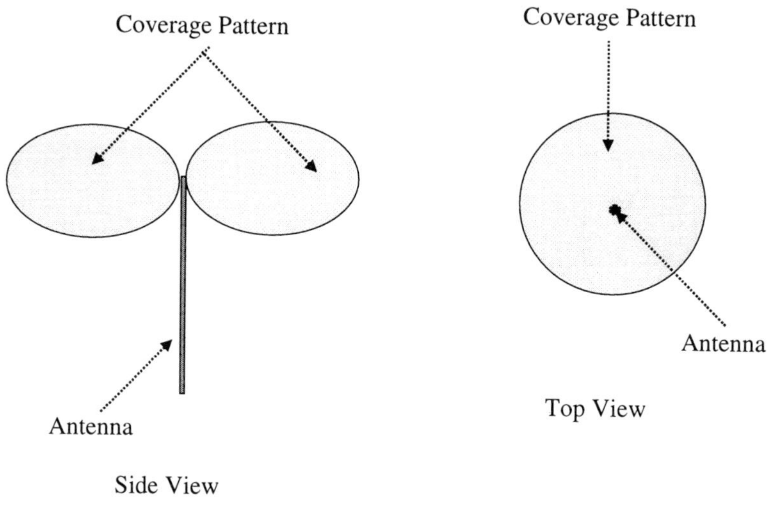

Fig. 4 *Omnidirectional antenna and coverage patterns.*

Given this limitation, omnidirectional strategies attempt to overcome environmental challenges by simply boosting the power level of the broadcast signals. In a setting of numerous users and interferers, this makes a bad situation worse in that the signals that miss the intended user become interference for those in the same or adjoining cells. In uplink applications e.g. user to base station, omnidirectional antennas offer no preferential gain for the signals of served users.

In addition, the single element approach cannot selectively reject an interfering signal from those of served users and it has no spatial multipath mitigation or equalization capabilities.

Omnidirectional strategies directly and adversely impact spectral efficiency, limiting frequency reuse. These limitations force system designers and network planners to devise increasingly sophisticated and costly remedies. In recent years, the limitations of broadcast antenna technology on the quality, capacity, and coverage of wireless systems have prompted an evolution in the fundamental design and role of the antenna in a wireless system.

A single antenna can also be built to have certain fixed preferential transmission and reception directions (Figure 5) [*Balanis, 1997*]. As an alternative to the brute force method of adding new transmitter sites, many conventional antenna towers today split or sectorize cells. A 360° area is often split into 120° subdivisions, each of which is covered by a slightly less broadcast method of transmission. Everything else being equal, a sectored antenna provides increased gain over a restricted range of azimuths as compared to an omnidirectional antenna. This is commonly referred to as antenna element gain and should not be confused with the processing gains associated with smart antenna systems. While sectorized antennas multiply the use of channels, they do not overcome the major disadvantages of a standard omnidirectional antenna broadcast, such as co-channel interference.

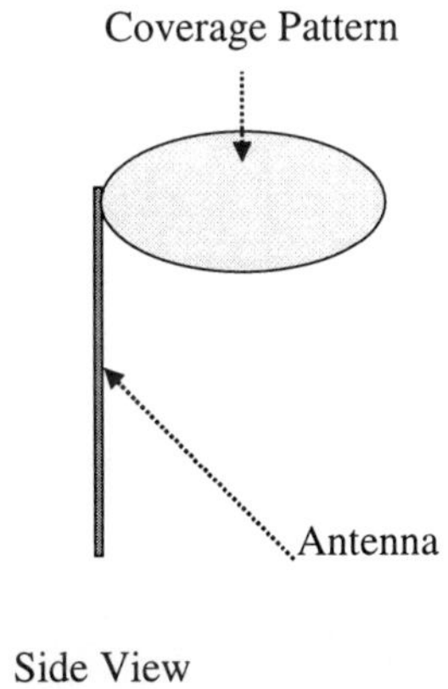

Fig. 5 *Directional antenna and coverage patterns.*

By combining the outputs of the individual antennas in array, a single effective antenna can be created with gain and directional characteristics that are very different from those of the individual elements comprising the array. For example, consider a row of m, simple and identical antenna elements (a linear array) with elements spaced one-half wavelength apart. Suppose that one simply adds the outputs of all the elements. Signals arriving at the array from broadside or perpendicular to the axis of the array, arrive simultaneously at each element of the array and their sum will therefore be m times as large as the signal received by a single antenna. Simply adding the outputs of the array element results in an amplitude gain for signals arriving from broadside [*Balanis, 1997*].

Figure 6 depicts the effective antenna pattern resulting from the above strategy for linear arrays consisting of a single antenna, two antennas, and six antennas (omnidirectional element patterns and half-wavelength spacing are assumed in all cases). The radius of the pattern in Figure 6 is proportional to the gain or strength of the signal at the output of the array. The increased gain in the broadside direction and the increases in gain with the number of array elements are evident in Figure 6. It is also evident that there are certain directions in which the effective antenna has reduced sensitivity, or nulls.

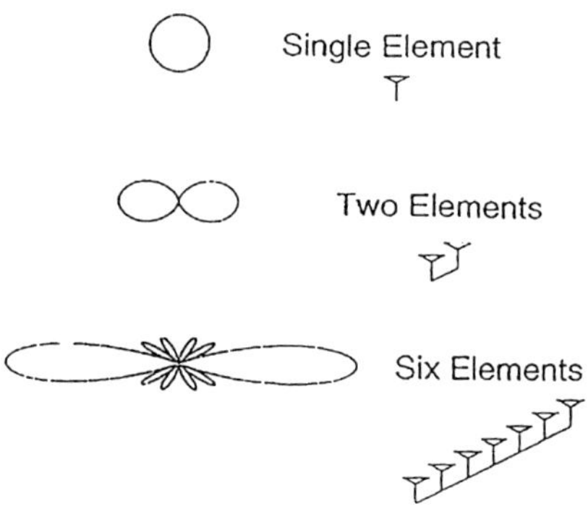

Fig. 6 *Broadside antenna patterns for arrays with difference numbers of elements.*

Classical phased-array antenna systems have gain patterns or beam patterns that look very much like the one in this figure, except that the direction of maximum gain might not be the broadside direction. More advanced systems have patterns that are optimized to enhance a particular user's signal while simultaneously rejecting interferers. Another differentiator of antenna array systems is whether or not the beam patterns can change with time. Antenna arrays can be built with combining strategies that are fixed. From an operational standpoint, this sort of array is no different from a conventional antenna with the same directional sensitivity.

Alternatively, arrays can be equipped with combining hardware and software that make it possible for the pattern to be changed over time and adapted to the current operational scenario. Because radio reception and transmission are reciprocal, directive transmission with gain is also possible from an array. Any directivity pattern achievable for reception is also achievable for transmission.

4. Comparative Study Between Intelligent and Omnidirectional Antennas

4.1 Antenna Arrays to Intelligent Antenna System Transition

Omnidirectional antennas are obviously distinguished from their intelligent counterparts by the number of antennas employed. Switched beam and adaptive array systems, however, share many hardware characteristics and are distinguished primarily by their adaptive intelligence.

The process information that is directionally sensitive requires an array of antenna elements, the inputs from which are combined to adaptively control signal transmission. Antenna elements can be arranged in linear, circular, or planar configurations and are most often installed at the base station.

Moreover, a simple antenna works for a simple environment. Intelligent antenna solutions are required as the number of users, interference, and propagation complexity grows. Their intelligence resides in their digital signal processing facilities [*Paulraj, 1997*].

Like most modern advances in electronics today, the digital format for manipulating the RF data offers numerous advantages in term of accuracy and flexibility of operation. Speech starts and ends as analog information. Along the way however, digital communication systems capture, convert, and modulate analog signals for transmission as digital signals and reconvert them to analog information at the other end. In adaptive antenna systems, this fundamental signal processing capability is augmented by advanced techniques e.g. algorithms that are applied to control operation in the presence of complicated combinations of operating conditions.

Finally, the concept of using multiple antennas and innovative signal processing to more intelligently serve cells has existed for many years. In fact, varying degrees of relatively costly smart antenna systems have already been applied in defense systems. Until recent years, cost barriers have prevented their use in commercial systems. The advent of powerful low-cost digital signal processors, general-purpose processors, as well as innovative software-based signal processing technology, make it now possible to implement this type of intelligent antenna solution.

4.2 The Architecture of Intelligent Antenna Systems

The conceptual situation illustrating the deployment of intelligent antennas in wireless communication system is depicted in Figure 7. In this figure, the potential ability of intelligent antenna technology to handle multiple cochannel signals are clearly manifest [*Tsoulos, et al., 1998*].

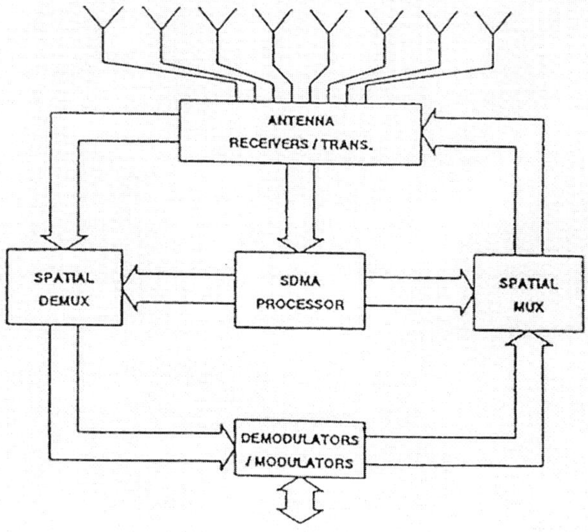

Fig. 7. *Typical intelligent antenna system configuration.*

The following sequence of events occurs many times each second in an intelligent antenna system:

* A snapshot, or sample, is taken of the signals coming from all of the antenna elements, converted into digital form, and stored in memory.

* A computer, the SDMA processor, then analyzes the sample to obtain an estimate of the radio environment identifying users and interferers and their locations.

* The processor calculates the combining strategy for the antenna signals that optimally recovers the users signals. With this strategy, each user's signal is received with as much gain as possible and with the other user's and interferers' signals rejected as much as possible.

* An analogous calculation is done to allow spatially selective transmission from the array. Each user's signal is now effectively delivered through a separate communications channel.

* The system now has the ability to both transmit and receive information on each of the spatial channels making them two-way channels.

4.3 Intelligent Antenna Purposes

The dual purpose of intelligent antenna system is to augment the signal quality of the radio based system through more focused transmission of radio signals while enhancing capacity through increased reuse. More specifically, the benefits derived from intelligent antenna system are as follows:

* Range extension - the coverage area of the array is greater than that of any single element as a result of the gain provided by the array. When a system is built using a intelligent antenna, the number of cells required to cover a given area can be substantially reduced.

* Interference from other systems and from users in other cells is significantly reduced. In noisy areas where range is limited by interference, spatially selective transmission and reception result in range extension.

* The destructive effects of multipath signals - copies of the desired signal that have arrived at the antenna after bouncing from objects between the signal source and the antenna can often be mitigated. In certain cases, the multipath can actually be used to reinforce the desired signal.

* The capacity of the system is increased in two distinct ways:

** Channel reuse patterns in cellular systems can be significantly tighter because the average interference resulting from co-channel signals in other cells is markedly reduced

** Separate spatial channels can be created in each cell on the same conventional channel.

* Because the intelligent antenna employs spatially selective transmission, a base station radiates much less total power than a conventional base station. One immediate result is a reduction in network-wide RF pollution. Another is a reduction in power amplifier size. First, the power is divided among the elements, and then the power to each element is reduced because the energy is being delivered directionally.

* The direction of each spatial channel is known and can be used to accurately establish the position of the signal source, a prerequisite for location based services.

* Intelligent antenna is compatible with almost any modulation method. bandwidth, frequency band or standards including AMPS, GSM, PHP, DECT, IS-54, IS-95 [22], and other formats. Intelligent antenna can be implemented with a broad range of array geometries and antenna types.

Finally, designers of a communications system will avail themselves of the benefits described above in different ways and in different mixtures as each system matures. When a communications system is initially constructed, costs can be minimized and roll-out can occur more rapidly by using the range extensions available from intelligent antennas. Because the cells are larger, better channel reuse is possible than would be the case with small cells. It is unnecessary to add sites and reduce cell size until required by subscriber demand and then only after the benefits of spatial channels have been absorbed. The major benefit of intelligent antenna is in the flexibility offered to the system implementers.

5. RF Aspects of Intelligent Antennas

The area of intelligent antennas is interdisciplinary, encompassing electromagnetics software, microwave, antenna design, circuit engineering and signal processing. In other words, the electromagnetics is required to develop communications signals and the digital signal processing is required to make the systems intelligent.

The work involves a mix of theoretical and technology issues that are best studied in close cooperation between users, industry research institutes and universities. One of the challenges in the development of intelligent antennas is the availability of efficient and affordable radiating elements with interfacing to the beamformer. Therefore, a considerable amount of the research effort is concentrated in two areas: novel antenna architectures and MMIC components and circuits with the objective to demonstrate that it is feasible and cost effective to deploy intelligent antennas within the infrastructure of third generation mobile systems [*Hettak, 1996*]. This activity comprises two basic tasks :

* Theoretical modeling, development, validation and integration into CAD tools for specific software modules for intelligent antennas.
* Development and demonstration of intelligent antenna concepts, radiators, beam formers and critical components, to meet expected smart antenna requirements.

5.1 Antenna Modeling and CAD Tool Development

Low cost requirements calls for the use of multilayer printed antenna technologies for the easy mass production of radiators and printed on substrates. Electromagnetic modeling of smart multi-element, multi-polarization and/or multi-frequency antennas is required to develop accurate CAD tools and avoid costly breadboard iterations [*Hettak, 1996*]. Frequency re-use, with multiple interference cancellation and adaptive polarization matching in a multipath environment, requires accurate modeling and characterization of the intelligent antenna. The research efforts are focused on the following tasks:

* Improvement of multi-layer antenna analysis software to cover finite array effects.
* Fast multi-layer antenna design/optimization including sub-array circuits.
* Modeling and design software for arrays with three-dimensional features.
* Modeling, analysis and synthesis of conformal arrays.
* Computer modeling of complete element chains, including circuits.
* Study and modeling of adaptive beam forming and polarization matching algorithms.

5.2 Radiating Element Technologies and Components

The development of appropriate antennas and associated technology will be important to the success of intelligent antenna systems. Research in these areas have been carried out with particular emphasis on integrating antennas with multi-function GaAs MMICs. Furthermore, MMIC based circuitry is useful for intelligent antennas and will be consolidated to reach a higher level of integration and reliability [*Hettak, et al., 1998*]. As is well known, planar antennas are extremely useful due to their low cost, light weight, and flexibility of design. In spite of their ease and low cost of manufacturing, printed antennas gave specific problems of parasitic radiation from feeding circuitry, frequency sensitivity and interfacing with phase control and amplification modules for which original solutions are required. Research efforts are devoted to the following tasks:

* Development of a new element excitation and matching techniques by use of electromagnetic coupling.

* Finding new ways to: decouple transmission from reception; to increase bandwidth; to improve repeatibility; and to lower manufacturing costs.

6. Application to Smart Base Station Antenna

Unfortunately, the geometrical size of the antenna array and the complexity of both radio frequency and baseband signal processing prevent the use of intelligent antennas at the user terminal. A very considerable amount of development and miniaturization will need to be done before this possibility could be considered as viable. Therefore, capacity enhancement has to be achieved by spatial array processing only at the base station (BS), at the moment. This means adaptive direction selective reception at the uplink and adaptive directional transmission at the downlink [*Swales, et al., 1990; Fuhl and Bonek, 1996*].

Six future focus topics clearly identified:

1. Third generation mobile handset antennas
2. Third generation mobile base station antennas
3. Multimedia Ku- and Ka-band LEO satellite home terminal antennas
4. Future Wireless LAN antennas
5. Future synthetic aperture radar antennas
6. Array antennas for radar and communications

6.1 The Potential Gain of Deploying Intelligent Antennas at Base Station

For the operating carrier, the higher gain and narrower beam base station antennas can provide increased signal-to-interference levels and range extension. The intelligent antenna system features can [*Ogawa and Ohgane, 1996; Paulraj, 1997*]:

a) extend the range of base station by up to 40%

b) extend the service area for portables in a system designed for mobiles

c) increase coverage in shadowed areas or within buildings

d) decrease co-channel interference at the base station

e) decrease battery drain on portables

f) improve overlapping coverage between adjacent cells

g) allow the use of fewer base stations. In an optimal situation, the cost of infrastructure can be decreased by up to 25%

h) decrease cochannel interference at the mobile

i) increase system channel capacity

6.2 Access System with Intelligent Antenna and Radio on Fiber

In recent years, the increasing world wide demand of wireless services has been the catalyst for the deployment of sophisticated networks such as digital cellular, local area networks (LAN) or satellite [*Talbi and Delisle, 1996*]. The wireless systems are based on different infrastructures and RF/baseband standards and require specialized user terminals for access. Naturally, the development of a small, cheap, universal mobile communication device would be highly desirable. In light of this, access networks from mixed technologies, optic and copper (ADSL-Asymetric Digital Subscriber Line), or coaxial (HFC - Hybrid Fiber Coax), will allow in the short term the introduction of multimedia services up to the subscriber. This will prepare for the arrival in the next decade of a high rate services which will be sufficient effective to really complete with optical fiber (FTTH - Fiber to the home). Radio systems have offered until now, for mainly technological reasons, rates limited to 1 Mbits/s. LMDS Systems type (Local Multi point Distribution Systems) are undergoing considerable development in North America in the 27.5 - 29.5 GHz band and are driving the market to produce components in these frequency bands. In Europe, LMDS systems appear to be directed towards the 40 GHz band. These systems, originally designed for digital image distribution, will be able soon to support a return path, and therefore be able to offer interactive services.

Future systems will integrate optical fiber communications and millimeter wave radio communications technologies. The interest of such integration lie on capability to ensure continuum high rate wireless system in transparency manner. The development of the nomadism and the necessity to improve the ergonomy of services militate in favor of wireless system in indoor environment.

However, the integration of such systems is conditioned by the development of low cost radio components, antennas, packaging as well as the optical/radio transductors. Furthermore, some problems still need to be addressed. Due to large propagation losses, in this MMW/optical network system, large transmission power is required to cover the wide area of cell zones. In addition, the effect of multipath becomes serious when the signal bandwidth is wide. To cope with these problems, intelligent antennas are expected to solve the problems of large propagation losses and multipath fading in MMW. Therefore, an intelligent antenna that enables beam control is one of the key components in the millimeter wave fiber wireless systems with options such as range extension, improved immunity to flat fading, interference suppression, and Spatial Division Multiple Access [*Uehara and Kagoshim, 1997; Hettak and Delisle, 1998*].

6.3 System Concept of Intelligent Antenna for Millimeter-Wave over Fiber

For any wireless network based on mm-wave radio systems to be viable, the cost of the infrastructure must be much lower than is currently the case. One approach that promises this cost reduction is radio over fiber, where radio signals at the carrier frequency are delivered over an optical network to the radio access point. The benefits of this centralized approach are that expensive and delicate pieces of equipment can be located in an central located environment and their costs can be shared between a number of radio access points. Figure 8 shows the configuration of the system in which adaptive circuits for beamforming (attenuators and phase shifters) modulator and demodulator are centralized in the control station (CS). In this case, a wavelength-division multiplexing (WDM) technique is used to decrease the number of fibers between the CS and the BS. In other words, the system is composed of a CS, several radio BS's, and many personal mobile stations (PS's). Feeder links between the CS and BS's are connected by optical fibers. Intelligent antennas installed at the BS's have access to a number of PS's by multiple spotbeams. The zones of the BS's overlap each other, and this causes BS diversity as a contermeasure for shadowing and maintains efficient utilization of the frequency under traffic localization [*Ohtsuka, et al., 1996*].

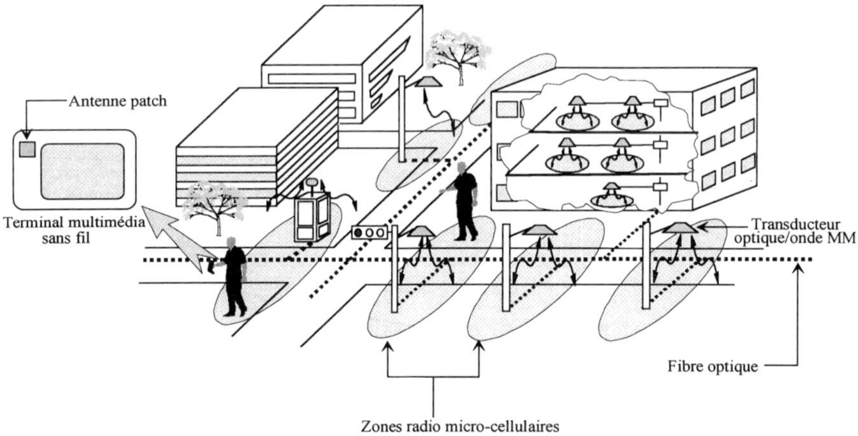

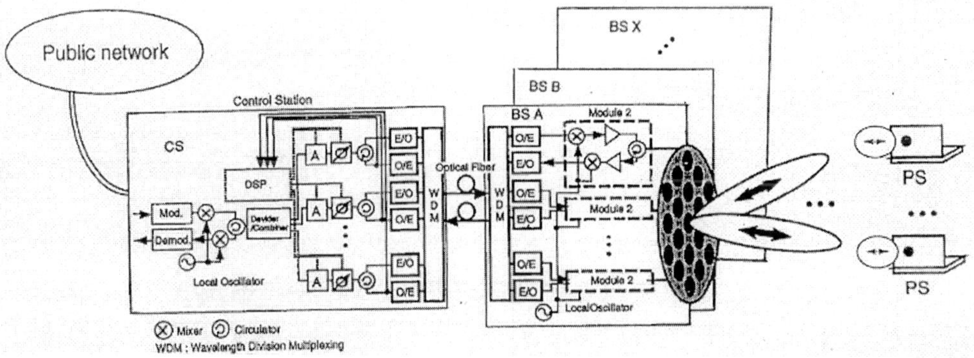

Fig. 8 *Millimeter Wave Fiber Wireless System using a Smart Antenna*
(a) A view of millimeter wave personal communication system concept
(b) Configuration of smart antenna for millimeter wave over optical fiber.

In this system, the BS should have an intelligent high gain antenna with a narrow beamwidth capability and the PS should have a simple low-gain antenna with a wide beamwidth to enable it to see the BS. The actual gain of a single patch antenna at 38 GHz has been estimated to be around 5 dBi. This necessitates a gain of 25 dBi for a BS antenna with at least 100 patch elements (array-factor gain of 20 dB). It would only be a few centimeters square in a two-dimensional (2-D) array configuration in the MMW band. If dual polarized sensors are used so that to orthogonal polarizations are received, this may enhance the array signal processing in two different ways. The first is that the number of signals available is increased which may further improve C/I after signal processing. The second alternative is to use a post detection maximum ratio combination of the signals from the two sensors. This method is similar to the one frequently used in today's polarization diversity systems for mobile communication base stations. Because of the nonoptimal combining used in these commercial systems, it is often preferable to receive equal mean power from the two sensors.

A problem associated with antenna arrays is mutual coupling, which changes the element patterns and, thus, degrades the array signal processing. The coupling can be compensated by matrix multiplication in the digital processing of the signals. An alternative approach is to minimize the mutual coupling when designing the antenna. Indeed, considering that the antennas play the central role in the deployment of such systems, the challenges arising from the development of this EHF-band antenna array include the minimization of the array's feed network loss, coupling between antennas and between feedlines and antennas, and the overall architecture that the system requires. It worth noting that the small wavelengths allow miniaturization of the size of components such as antennas which are very important for portable sets as well as base station (BS) facilities, which need to be easily installed. Thereby, the objective of a truly monolithic array with integrated planar antennas and phase shifters was the driving force behind the development of the aperture coupled patch element which is especially relevant for these systems [*Yamazaki, et al., 1994; Zurcher and Gardiol, 1995; Yamazaki, et al., 1994; Menzel and Grabher, 1991; Hettak and Delisle, 1998; 1998*].

6.4 Intelligent Antennas for Broadband Unlicenced U-NII Bands

In January 1997 [41], the Federal Communications Commission (FCC) allows the use in U.S.A 300 MHz of spectrum at 5.15-5.35 GHz and 5.725-5.825 GHz for use by a new category of unlicensed equipment, called Unlicensed National Information Infrastructure ("U-NII") devices.

These devices will provide short-range, high speed wireless digital communications on an unlicensed basis. It is anticipated that U-NII devices will support the creation of a new wireless local area networks ("LAN") and will facilitate wireless access to the National Information Infrastructure ("NII"). The FCC's allocation of the U-NII band offers computer users in schools, libraries, businesses, and hospitals the ability to connect wirelessly to one another and to the Internet, for the first time at the data rates required to communicate today's multimedia content. It also permits users to establish, for the first time, their own longer-reach wideband "community networks" in municipalities and throughout many rural areas. None of these U-NII applications requires a radio license, and the U-NII band is available to everyone. One portion (5150-5350) of the NII band includes radio frequencies already earmarked in Europe for local area wireless networks ("wireless LANs") adhering to a network standard called HIPERLAN, developed for Europe. Upon FCC reallocation to create the NII band, the band rules will permit, but not require, HIPERLAN-compliant products as well as many other networks technologies. U.S and European HIPERLAN technologies can advance the development of the Global Information Infrastructure ("GII"). The other portion of the NII band (5725-5875 MHz) is already available for low-power, unlicensed spread spectrum devices.

Furthermore, the best regulatory framework to facilitate the introduction of U-NII devices is one that provides the maximum technical flexibility in their design and operation by imposing only the minimum technical rules necessary to prevent harmful interference to primary operations and to provide for basic spectrum sharing among unlicensed devices. The minimum technical regulations which will most facilitate the introduction of U-NII devices, will adequately protect primary services, and will promote sharing among U-NII devices. These rules specify power limits (in terms of peak power and power spectral density), emission limits, radio frequency requirements, and other basic technical rules appropriate for unlicensed operations.

The 100mW power limit proposed in the NPRM (Notice of Proposed Rule Making) appears insufficient to accommodate the range and scope of communications envisioned for U-NII devices. The increasing of U-NII device power limits will enable these devices to provide for a variety of operations including local areas networks, campus-type settings, or as part of community networks. At the same time, the need to ensure that primary operations are adequately protected from harmful interference is recognized. In this regard, it is noted that the primary users and the considerations that relate to interference with their operations vary in different parts of the spectrum. Specifically, the 5.15-5.25 GHz band will be shared with MSS feeder links, the 5.25-5.35 GHz band will be shared with Government radiolocation operations, and the 5.725-5.825 GHz band will be shared with Government radiolocation, Amateur, ISM, and other operations. Therefore, the sharing environment for U-NII devices will be different for each of these three 100 MHz segments. This approach will provide the needed flexibility to allow U-NII proponents to design and manufacture equipment to meet a variety of communications needs while ensuring a successful spectrum sharing environment with other spectrum users.

To permit manufacturers flexibility in designing U-NII equipment, the use of higher directional antenna gain provided there is a corresponding reduction in transmitter output power and increase the carrier to interference radio signal (C/I). In the other words, limits on emission levels outside the bands of operation and frequency stability requirements are necessary to protect adjacent spectrum occupants and sensitive operations that may operate on harmonic frequencies. Therefore, as already specified, intelligent antennas is one key issue providing rules which permit maximum technical flexibility in the design and development of U-NII devices capable of providing high data rate communications for a variety of multimedia applications in a shared spectrum environment. Such devices have not yet been designed, built, or tested.

7. Intelligent Antennas: Summary and Perspectives

Increased user capacity and expanded geographic coverage from existing installations are the two strong economic motivations voiced by service providers as the impetus for these developments. Furthermore, the need for mobile radio systems with increased spectrum efficiency is paramount in the drive towards third generation systems. Currently favoured solutions in today's systems, which are generally applicable to any modulation scheme or access technique, include the deployment of smaller cells as well as fixed sector, or multi-beam antennas at the base station site. In terms of modulation schemes and access techniques the Code Division Multiple Access (CDMA) and especially Direct Sequence (DS) CDMA is one of the currently favoured approaches. Thus, intelligent antenna systems are currently receiving considerable attention as a means of increasing the spectrum efficiency of wireless networks. Here the application of this technology is considered at the base station site to support spatial separation of the spectrally and temporally overlapping mobile users. Beside, the flexibility of intelligent antenna technology allows for the creation of new value added products and services that give operators a significant competitive advantage. Intelligent antennas are not restricted to any particular modulation format or radio interface protocol. They are compatible with all current radio interface modulation schemes. Several challenges remain in the development of intelligent antennas. In the near future, a new class of intelligent antennas which have potential applications throughout the whole microwave and millimeter wave spectrum into the next millenium is expected. This will require the spread of intelligence to all of the system components and the coordination of the antenna and microwave units of subsystems with signal and data processing functions. Finally, key issues which remain to be solved before these expectations become reality, include: propagation modeling; antenna structures; estimation and detection; channel allocation; and finally mobility management.

References

Andersson, S., M. Millnert, M. Viberg and B. Wahlberg, "An adaptive array for mobile communication systems", *IEEE Trans. Vehicular Techn.*, 40 (1), Jan. 1991, pp. 230-236.

Balanis, C.A., *Antenna Theory: Analysis and Design*, 2nd ed., John Wiley & Sons, Inc., 1997, pp.941.

FCC: Digital Wireless Spectrum Allocated: January 9, 1997, Technical report, ET Docket N° 96-102

Feuerstein, M., "Smart antennas know how to optimize CDMA networks" *America Network*, Feb. 15, 1998, pp.26-30.

Fuhl, J. and E. Bonek, "Optimum antenna topologies and adaptation strategies for SDMA", in *Proc. IEEE GLOBECOM*, London, U.K., Nov. 1996, pp.575-580.

Goldburg, M. and R.H. Roy , "The impacts of SDMA on PCS system Design", *1996 Ann. Rev. Communications*, 944-947.

Hettak, K., G. Delisle and M. Boulmalf, "A novel integrated antenna for millimeter wave personal communication systems", Accepted for publication *IEEE Trans Antennas Propag*. June 1998.

Hettak, K. and G. Delisle, "A 38 GHz integrated phased-array antennas for PCS", *Symp. Antenna Techn. Applied Electromagnetics* (ANTEM), Aug. 1998, pp.75-78.

Hettak, K. and G. Delisle, "Millimeter wave mobile access system with smart antenna and radio on fiber", *Symp. Antenna Techn. Applied Electromagnetics* (ANTEM), Aug. 1998, pp. 87-90.

Hettak K., "Uniplanar technology of RF subsytems for broadband wireless communications at 38 GHz", *Ph.D dissertation*, Ecole Nationale Supérieure des Télécommunications (ENST), Brest, France, 1996.

Jeng, S., T. Okamoto, G. XU, H. Lin and W. J. Vogel, "Experimental evaluation of smart antenna system performance for wireless communications", *IEEE Trans. Antennas Propag.*, 46 (6), June 1998, pp. 749-757.

Lindmark, B., S. Lundgren, J. R. Sanford and C. Beckman, "Dual polarized array for signal processing applications in wireless communications", *IEEE Trans. Antennas Propag.*, 46 (6), June 1998, pp. 758-763.

Litva, J and T. Lo, *Digital beamforming in wireless communications*, Artech House, Boston, 1996.

Menzel, W. and W. Grabher, "A microstrip patch antenna with coplanar feed line", *IEEE Microwave Guided Lett.*, pp. 340-342, Nov. 1991.

Naguib, A.F. and A. Paulraj, "Performance of CDMA cellular networks with base station antenna arrays", *Lecture Notes in Computer Science*, No. 783, pp. 87-100, Springer-Verlag.

Naguib, A.F., A. Paulraj and T. Kailath, "Capacity improvement with base station antenna arrays in cellular CDMA", *IEEE Trans. Vehicular Techn.*, 43 (3), 1994, pp. 691-698.

Naguib, A.F. and A. Paulraj, "Performance enhancement and trade-offs of smart antennas in CDMAcellular networks", *45th Vehicular Techn. Conf.*, Chicago, July 25, pp. 40-44, 1995.

Naguib, A.F., *Adaptive antenna for CDMA wireless network.* PhD thesis, Stanford University, August 1996.

Ogawa, Y. and T. Ohgane, "Adaptive antennas for future mobile radio", *IEICE Trans.*, E79-A (7), 1996, pp.961-967.

Ohtsuka, H., R. Ohmoto, T. Shimizu and H. Arai, "Potential SEFA technology in fiber-oriented wireless acess systems", *IEEE Symp. Personal Indoor and Mobile Radio Communications*, Tapei, Taiwan, Oct. 15, pp. 1629-1633, 1996.

Okamoto, G. and G. Xu, "The smart wireless LAN system: physical layer design and results" *47th Vehicular Techn. Conf.*, pp. 47-52, 1997.

Paulraj, A., "Smart antennas in mobile communications-where is it useful", *1997 Ann. Rev. Communications,* pp.949-954

Pozar, D.M., "Microstrip antennas: The analysis and design of microstrip antennas and arrays", *IEEE PRESS*, 1995.

Rheinschmitt, R. and Tangemann, "Performance of sectorized spatial multiplex systems", *IEEE Vehicular Techn. Conf.*, 1, pp. 426-430, Atlanta, GA, April 28, 1996.

Roy, R.H., "Spatial Division Multiple Access Technology and its application to wireless communication systems", *IEEE Vehicular Techn. Conf.*, 2, pp. 730-734, Phoenix, 1997.

Roy, R.H., "Application of smart antenna technology in wireless communication systems", *1996 Ann. Rev. Communications*, pp. 940-943.

Swales, S.C., M.A. Beach and J.P. McGeehan, "The performance enhancement of multi-beam adaptive base station antennas for cellular land mobile radio systems", *IEEE Trans, Vehicular Techn.,* 39 (1), Feb. 1990, pp. 56-67.

Talbi, L. and G. Delisle, "Experimental characterization of EHF multipath indoor radio channels", *IEEE J. Selected Area in Communications*, 14, pp. 431-440, April 1996.

Tangemann, T., U. Bigalk, C. Hoek, M. Hother, "Sensitivity enhancements of GSM/DCS 1800 with smart antennas" *Proc. 2nd European Personal Mobile Communications Conf.* (EPMCC), Bonn, Germany, Oct.2, 1997, pp. 87-94.

Tsoulos, G.V., M.A. Beach, S. C. Swales, "Applications of adaptive antenna technology to Third Generation mixed cell radio architectures", *44th Vehicular Techn. Conf.*, Stockholm, Sweden, June 8-10, 1994, pp. 615-619.

Tsoulos, G.V., M. Beach, J.McGeehan, "Wireless personal communications for the 21st century: European technological advances in adaptive antennas", *IEEE Communications Mag.,* Sept. 1997, pp. 102-109.

Tsoulos, G.V., G.E.Athanasiadou, J.McGeehan, M. Beach, "Adaptive antennas for UMTS microcellular operational environments", *IEEE Vehicular Tech. Conf.,* V. 2, pp.805-808, Ottawa, 1998.

Uehara, K. and K. Kagoshima, "A novel beam combination method for indoor high-speed wireless communication systems", *IEEE Symp. Personal Indoor Mobile Radio Communications,* Helsinki, Finland, Sept. 5, pp. 1758-1762, 1997.

Ward, C., M. Smith, A. Jeffries, D. Adams and J. Hudson, "Characterising the radio propagation channel for smart antenna systems", *Electronics Comm. Eng. J.*, 1996, pp. 191-200.

Winters, J.S., "Signal aquisition and tracking with adaptive arrays in digital mobile radio system IS-54 with flat fading", *IEEE Trans. Vehicular Techn.*, 42 (4), Nov. 1993, pp. 377-384.

Wong, K.K., K.B. Letaief and R.D. Murch, "Investigating the performance of smart antenna systems at the mobile and base stations in the down and uplinks", *IEEE Vehicular Techn. Conf.,* 2, pp. 880-884, Ottawa, 1998.

Yamazaki, M., E.T. Rahardjo and M. Haneishi, "Construction of a slot-coupled planar antenna for dual polarisation", *Electron. Lett.*, 30, pp. 1814-1815, Oct. 1994.

Yamazaki, M., E.T. Rahardjo and M. Haneishi, "Construction of a slot-coupled planar antenna for dual polarisation", *Electron. Lett.,* 30, pp. 1814-1815, Oct. 1994.

Zetterberg, P. and B. Ottersten, "The spectrum efficiency of a base station antenna array system for spatially selective transmission", *IEEE Trans. Vehicular Techn.*, 44 (3), Aug. 1995, pp. 651-660.

Zurcher, J.F. and F.E. Gardiol, *Broadband Patch Antennas*, Artech House, 1995.

HIGH-IMPEDANCE ELECTROMAGNETIC SURFACES

Dan Sievenpiper and Eli Yablonovitch

1. Introduction

In some situations, the presence of electric conductors can have consequences that adversely affect the performance of electromagnetic devices, particularly those involving radio communication. A flat metal sheet is used in many antennas as a reflector, or ground plane. [*C. Balanis, 1997*] The presence of a ground plane redirects half of the radiation into the opposite direction, improving the antenna gain by 3dB, and partially shielding objects on the other side. Unfortunately, if the antenna is too close to the conductive surface, the phase of the impinging wave is reversed upon reflection, resulting in destructive interference with the wave emitted in the other direction. This is equivalent to saying that the image currents in the conductive sheet cancel the currents in the antenna, resulting in poor radiation efficiency.

Another property of metals is that they support surface waves [*R. Collin, 1991*]. These are propagating electromagnetic waves that are bound to the interface between metal and free space. They are called surface plasmons at optical frequencies, [*H. Raether, 1988*] but at microwave frequencies, they are nothing more than the normal AC currents that occur on any electric conductor. If the conductor is smooth and flat, the surface waves will not couple to external plane waves. However, they will radiate if scattered by bends, discontinuities, or surface texture.

Surface waves appear in many situations involving antennas. If an antenna is placed near a metal sheet, such as a reflector or ground plane, it will radiate plane waves into free space, but it will also generate currents that propagate along the sheet. On an infinitely large ground plane, the surface currents would be evident only as a slight reduction in radiation efficiency. In reality, the ground plane is always finite in size, and these currents propagate until they reach an edge or corner. Any break in the continuous translational symmetry of the smooth, flat surface allows the currents to radiate. The result is a kind of multipath interference or "speckle", which can be seen as ripples in the far field radiation pattern. Moreover, if multiple antennas share the same ground plane, surface currents can cause unwanted mutual coupling between them.

By applying a special texture to a conducting surface, it is possible to alter its surface properties [*D. Sievenpiper, 1998*]. Made of solid metal, the structure conducts DC currents, but over a particular frequency range, it does not conduct AC currents. This new surface is characterized by having high electromagnetic surface impedance. It does not reverse the phase of reflected waves, and the image currents appear in-phase, rather than out-of-phase as they are on normal conductors. Furthermore, it does not support propagating surface waves. Instead, radio-frequency currents on the surface radiate efficiently into free space. This high-impedance surface provides a useful new ground plane for novel low-profile antennas, and other electromagnetic structures.

In the limit where the period of the surface texture is much smaller than the wavelength, the structure can be described using an effective medium model, and its qualities can be summarized into a single parameter, the surface impedance. A smooth conducting sheet has low surface impedance, while with a specially designed geometry, the textured surface can have high surface impedance.

Fig. 1 *Cross-section of a high-impedance surface.*

A high-impedance surface, shown in cross section in Fig. 1, consists of an array of metal protrusions on a flat metal sheet. They are arranged in a two-dimensional lattice, and are usually formed as metal plates, connected to the continuous lower conductor by vertical posts. They can be visualized as mushrooms or thumbtacks protruding from the surface. An example of a top view is shown in Fig. 2. The hexagonal metal patches are raised above the surface, and the dots in the center are vertical connecting posts.

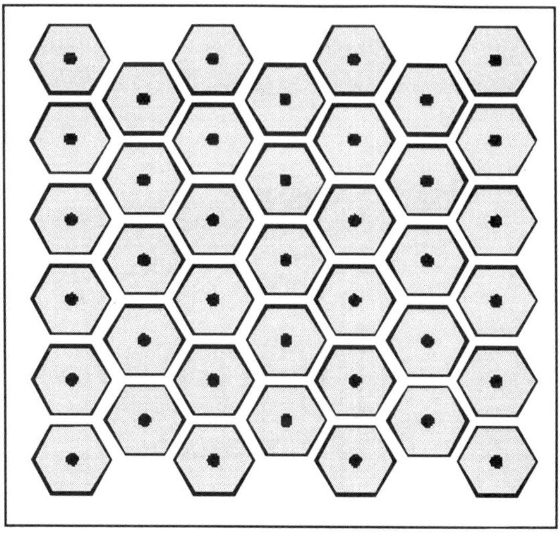

Fig. 2 *Top view of the high-impedance surface.*

If the protrusions are small compared to the wavelength, their electromagnetic properties can be described using lumped circuit elements – capacitors and inductors. The proximity of the neighboring metal elements provides the capacitance, and the long conducting path linking them together provides the inductance. They behave as parallel resonant LC circuits, which act as electric filters to block the flow of currents along the sheet.

In the frequency range where the surface impedance is very high, the tangential magnetic field is small, even with a large electric field. Such a structure is sometimes described as a "magnetic conductor". This is a mathematical idea that is used in certain electromagnetics problems, but does not exist in reality. Having high impedance and being nearly lossless, this new surface can be regarded as a kind of magnetic conductor over a certain frequency range.

2. Corrugated Surfaces

The high-impedance surface can be understood by examining a similar structure, the corrugated surface, which is discussed in various textbooks, [*S. Ramo, et. Al., 1984*] and review articles. [*A. Harvey, 1959*] Numerous authors have also contributed general treatments of corrugated surfaces, [*L. Brillouin, 1948, R. Elliot, 1954*] and specific studies of important structures. [*P.-S. Kildal, 1990*] A corrugated surface is a metal slab, into which a series of vertical slots have been cut, as depicted in Fig. 3. The slots are narrow, so that many of them fit within one wavelength across the slab. Each slot can be regarded as a parallel-plate transmission line, running down into the slab, and shorted at the bottom.

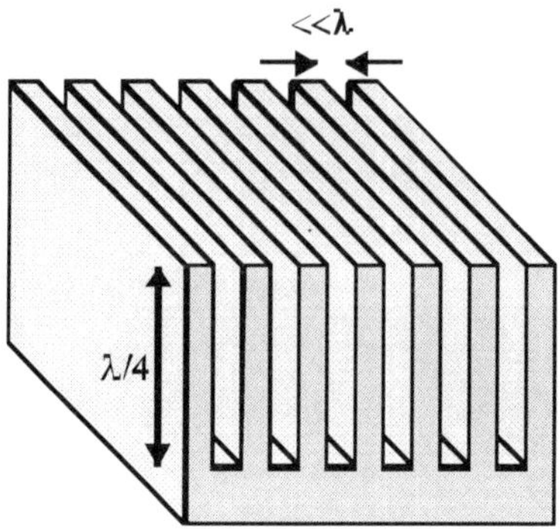

Fig. 3 *A corrugated metal slab.*

If the slots are one-quarter wavelength deep, then the short-circuit at the bottom end is transformed by the length of the slot into an open-circuit at the top end, and the impedance at the top surface is very high. In the high-impedance surface, the quarter-wavelength slots have simply been folded up into lumped elements, capacitors and inductors, and distributed in two dimensions. Periodic two or three-dimensional dielectric, [*E. Yablonovitch, 1987, K.-M. Ho, et. Al., 1990*] metallic, [*D. Sievenpiper, et. Al., 1996, J. Pendry, et. Al., 1996*] or metallodielectric [*E. Brown, et. Al., 1995, D. Sievenpiper, et. Al., 1998*] structures that prevent the propagation of plane waves are known as photonic crystals. [*J. Joannopoulos, et. Al., 1995*]. The high-impedance surface can be considered as a kind of two-dimensional photonic crystal that prevents the propagation of radio-frequency surface currents.

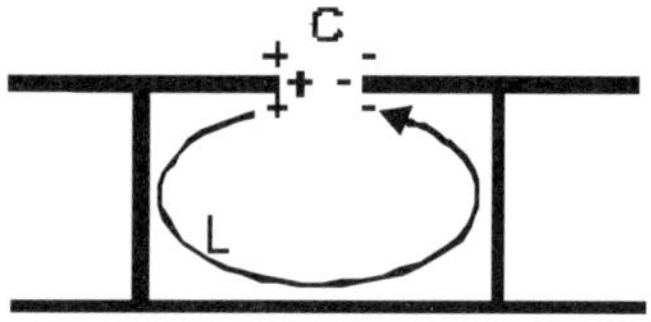

Fig. 4 *Origin of the equivalent circuit elements*

The two-dimensional array of resonant elements can be viewed as a kind of electric filter, and many of its properties can be explained using a simple circuit model. The capacitance is due to the proximity of the top metal patches, while the inductance originates from current loops within the structure, as shown in Fig. 4. The electromagnetic properties of the surface can be reduced to an equivalent LC circuit, shown in Fig. 5.

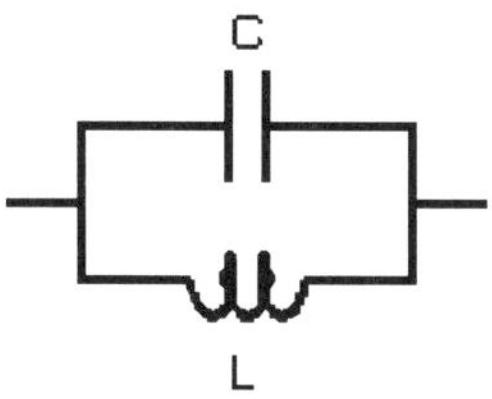

Fig. 5 *Equivalent circuit model for the high-impedance surface.*

In a narrow band around the LC resonance, the impedance is very high. In this frequency range, currents on the surface radiate very efficiently, and the structure supports neither type of surface wave. Having high surface impedance, it also reflects external electromagnetic waves without the phase reversal that occurs on a flat conductor. By using lumped elements, we retain the reflection phase and surface wave properties of the quarter-wave corrugated slab, while reducing the overall thickness to a small fraction of a wavelength.

3. Measuring Surface Properties

Techniques have been developed for measuring various properties of electromagnetic surfaces. The presence of surface wave modes is detected in the transmission between two antennas positioned near the surface. These antennas can be simple monopole probes, or they can be specifically designed to efficiently launch surface waves. By varying the polarization of the antennas, one can distinguish between TM and TE modes. The measurements of the textured surface indicate a frequency band in which there are no propagating surface waves.

3.1 TM Surface Waves

In TM surface waves, the electric field forms loops that extend vertically out of the surface. TM waves can be measured using a pair of small monopole antennas oriented normally with respect to the surface, as shown in Fig. 6. The vertical electric field of the probe couples to the vertical electric field of the TM surface waves.

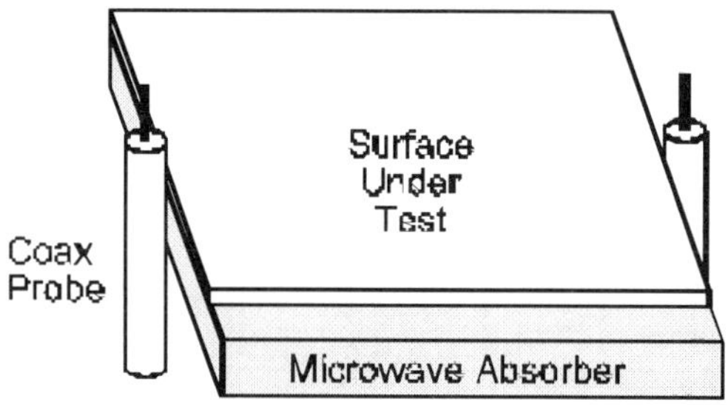

Fig. 6 *TM surface wave measurement using monopole probe antennas.*

On a flat metal sheet, a TM surface wave measurement produces the results shown in Fig. 7. The data has small variations of 10-15 dB, but remains relatively flat over a broad spectrum. The fluctuations are produced by multipath interference, or speckle, which occurs in coherent measurements whenever multiple signal paths are present. This effect can be distinguished from a surface wave band gap because it is characterized by narrow-band fading that depends on the antenna position. The transmission drops off at low frequencies because the small probes are inefficient at exciting long wavelengths.

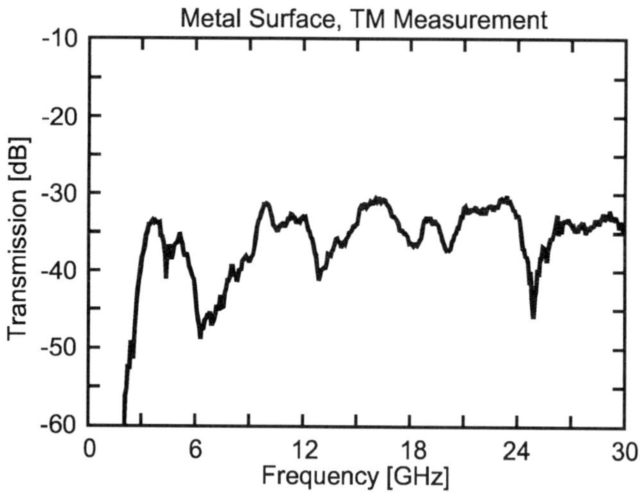

Fig. 7 *TM surface wave transmission data on a flat metal surface.*

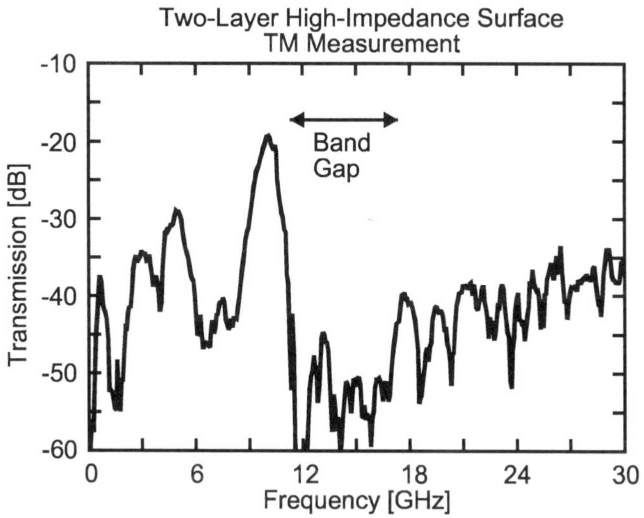

Fig. 8 *TM surface wave transmission data on a textured surface.*

A typical TM surface wave measurement on a textured surface is shown in Fig. 8*Fig. 8*. The transmission is strong at low frequencies, and exhibits the same multipath interference seen on the metal surface. At 11 GHz, the transmission drops by about 30 dB, indicating the edge of the TM surface wave band. Beyond this frequency, the transmission level remains low and flat, eventually sloping upward at much higher frequencies because of weak coupling to TE surface waves. The TE band edge is not apparent in this measurement, but the region corresponding to the surface wave band gap is indicated on the graph with an arrow.

3.2 TE Surface Waves

In TE surface waves, the electric field is parallel to the surface, and the magnetic field forms vertical loops that arc out of the surface. They can be measured with a pair of small monopole probes oriented parallel to the sheet, as shown in Fig. 9. The horizontal electric field of the antenna couples to the horizontal electric field of the TE waves. Since this configuration lacks the symmetry of the vertical monopole, there will be much greater cross-coupling to TM waves, that may complicate the measurement. On a metal surface, a TE surface wave measurement produces no significant signal, because any antenna that excites TE waves is shorted out on a conducting surface. It is only on the textured surface, with its unusual surface impedance, that significant signal levels can be obtained.

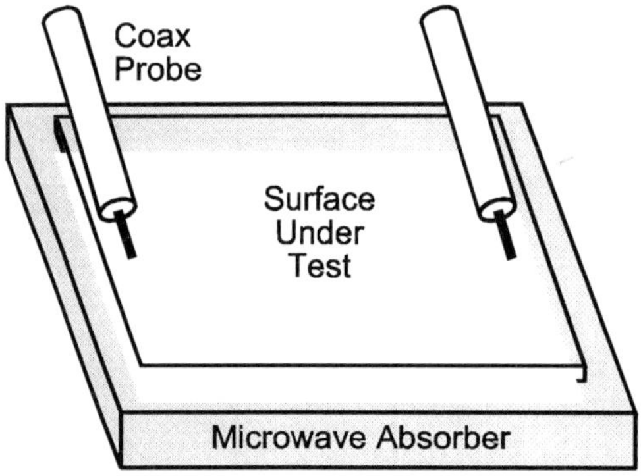

Fig. 9 *TE surface wave measurement using monopole probe antennas.*

The measurement shown in Fig. 10 was taken using a pair of small straight-wire coaxial probes, oriented parallel to the high-impedance surface. The transmission is weak at low frequencies, and strong at high frequencies, the reverse of the TM data. A sharp jump of 30 dB occurs at 17 GHz, indicating the TE band edge. Beyond this frequency, the transmission is flat, with only small fluctuations due to speckle. The TE probes also couple to TM surface waves, so there is an additional transmission peak at 11 GHz, at the TM band edge, where the density of TM states is high. Both TM and TE probes tend to couple slightly to both polarizations. However, the cross-coupling is weaker between the TM probe and the TE surface waves because of the symmetry of the vertical monopole.

A surface wave band gap exists between the TM band edge at 11 GHz and the TE band edge at 17 GHz. Within this range, neither type of measurement produces significant transmission. Currents cannot propagate across the surface, and any currents induced in the surface radiate rapidly into free space.

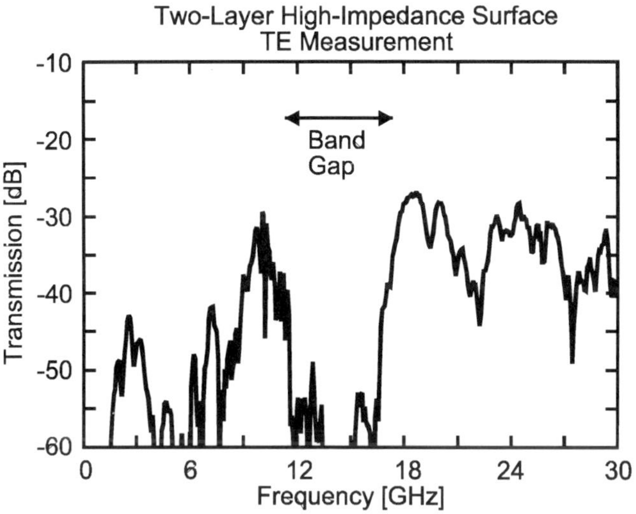

Fig. 10 TE surface wave transmission data on a textured surface.

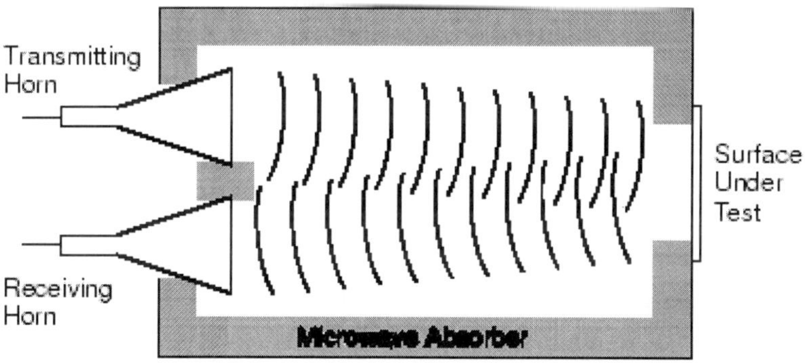

Fig. 11 Reflection phase measurement using a pair of horn antennas.

3.3 Reflection Phase

The reflection phase of the high-impedance surface can be measured using two microwave horn antennas, as shown in Fig. 11. A single antenna can also be used, but the data contains much more noise because of additional reflections that occur inside the horn itself. The measurement is done in an anechoic chamber lined with microwave absorbing foam. The two horns are placed next to each other, aimed at the surface. Two windows are cut in the chamber, one for the antennas, and one for the surface under test.

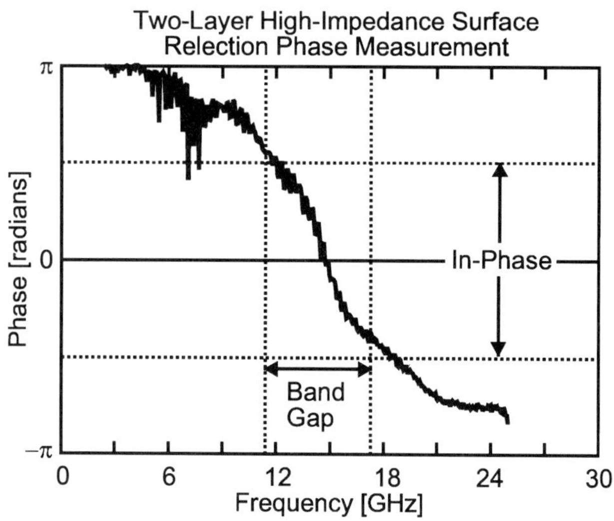

***Fig. 12** Reflection phase of a textured surface.*

The reflection phase of the high impedance surface is shown in Fig. 12. At low frequencies, it reflects with a π phase shift, just like a metal surface. As the frequency is increased, the phase slopes downward, and eventually crosses through zero at the resonance frequency of the structure. At higher frequencies, the phase continues to slope downward, and eventually approaches $-\pi$. Within the region between $+\pi/2$ and $-\pi/2$, indicated on the graph by arrows, plane waves are reflected in-phase, rather than out-of-phase as occurs on a smooth metal surface. This range also corresponds to the surface wave band gap, indicated on the graph by an arrow, with the TM and TE band edges falling approximately at the points where the phase crosses through $+\pi/2$ and $-\pi/2$, respectively.

4. Antennas

Antennas have been demonstrated that take advantage of both the suppression of surface waves, and the unusual reflection phase of the high-impedance surface. For example, because of the suppression of surface waves, an antenna on a high-impedance ground plane produces a smoother radiation profile than a similar antenna on a conventional metal ground plane, with less power wasted in the backward direction. Various workers have studied antennas on photonic crystals, [*E. Brown, et. al., 1993, M. Sigalas, et. al., 1997*] which are closely related to the high-impedance surface. The unique reflection phase properties of the high-impedance surface can allow for new, low-profile antenna designs, with radiating elements lying flat against the ground plane.

The high-impedance surface is particularly applicable to the field of portable communications, in which the interaction between the antenna and the user can have a significant impact on antenna performance. Using this new ground plane as a shield between the antenna and the user in portable communications equipment can lead to higher antenna efficiency, lower weight, and longer battery life.

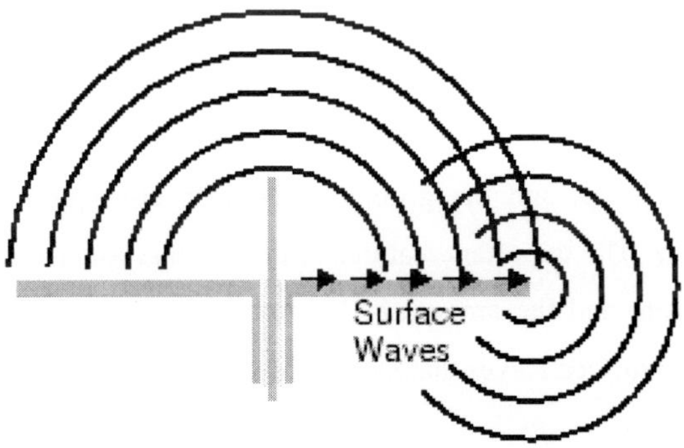

Fig. 13 *Surface waves radiating at ground plane edges.*

4.1 The Vertical Monopole Antenna

One of the simplest antennas is a vertical monopole, shown in Fig. 13. On an infinitely large ground plane, an antenna of this type would ideally have a smooth, half-doughnut shaped radiation pattern, with a null on the axis of the wire, and no radiation in the backward direction. In reality, the ground plane is always finite, and its edges contribute to the radiation pattern. In addition to space waves, the antenna generates surface waves in the ground plane, which then radiate from edges and corners. The combined radiation from the wire and the ground plane edges interfere to form a series of lobes and nulls at various angles. The edges radiate backwards as well as forwards, causing a significant amount of wasted power in the backward hemisphere.

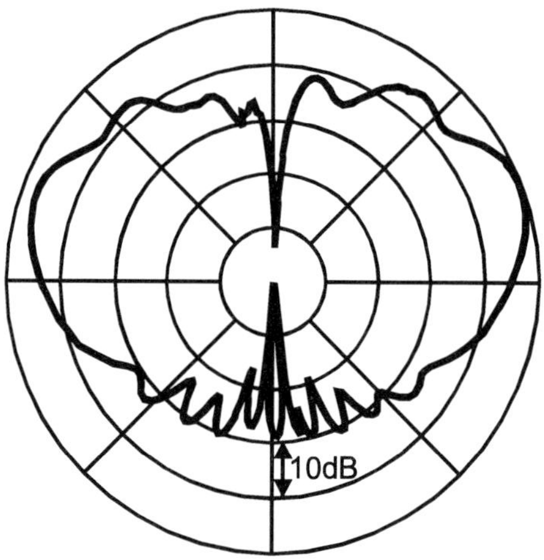

***Fig. 14** Radiation pattern of a monopole antenna on a metal ground plane.*

The radiation pattern of a monopole antenna on a metal ground plane is shown in Fig. 14. The antenna is 3 mm long, and the ground plane is 5 cm square. The frequency of the measurement is 35 GHz. The important features of the antenna pattern are the ripples that appear in the forward direction, and the amount of wasted power in the backward direction. These features are both due to surface waves that propagate away from the antenna and radiate from the ground plane edges.

If the metal ground plane is replaced with a high-impedance ground plane, as shown in Fig. 15, the surface waves are suppressed. While driven currents can exist on any reflective surface, there are no freely propagating surface waves. Any induced currents are restricted to a localized region around the antenna, and never reach the edges of the ground plane. The result is a smoother radiation pattern, with less wasted power in the backward hemisphere, as shown in Fig. 16.

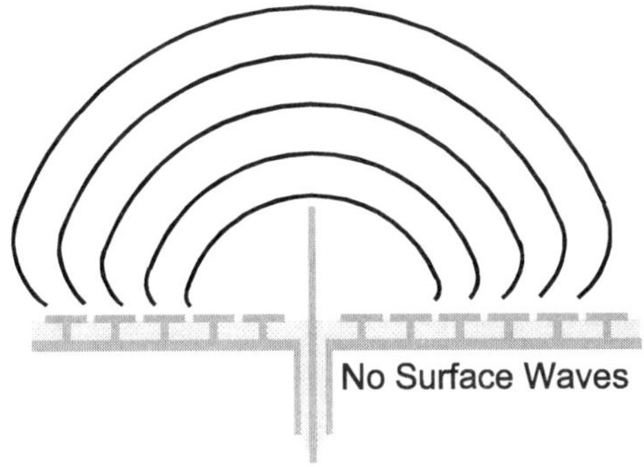

Fig. 15 *Surface wave suppression on a high-impedance ground plane.*

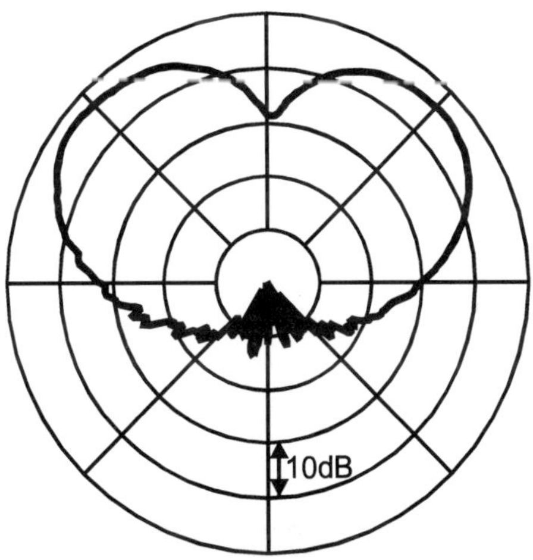

Fig. 16 *Radiation pattern of a monopole on a high-impedance ground plane.*

4.2 The Horizontal Wire Antenna

The high-impedance surface reflects in-phase, rather than out-of-phase. This is equivalent to a reversal of the direction of the image currents. This unusual property allows antennas to be constructed that are not possible on a flat, conducting ground plane. An example, shown in Fig. 17, is a wire antenna that has been bent over so that it lies parallel to a conducting surface.

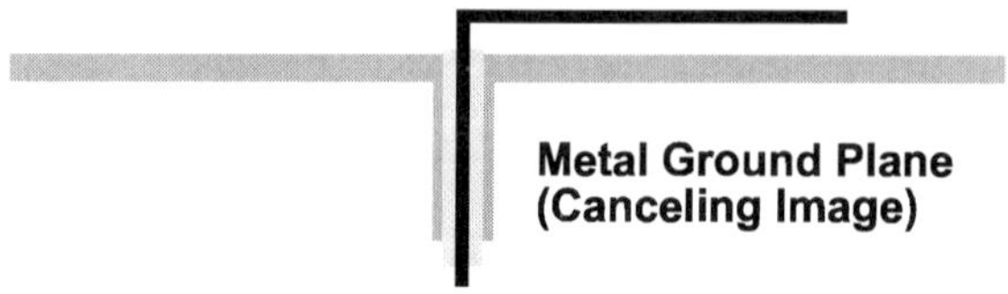

Fig. 17 *Horizontal wire antenna on a metal ground plane.*

A horizontal wire radiates very poorly on an ordinary, metal ground plane because the image currents cancel the currents in the antenna. The S11 measurement, or return loss of this antenna is shown in Fig. 17. Most of the power is reflected back toward the generator, so the radiation efficiency is poor. In this example, the antenna was 1 cm long, and separated by a distance of 1 mm from a 3 cm square ground plane.

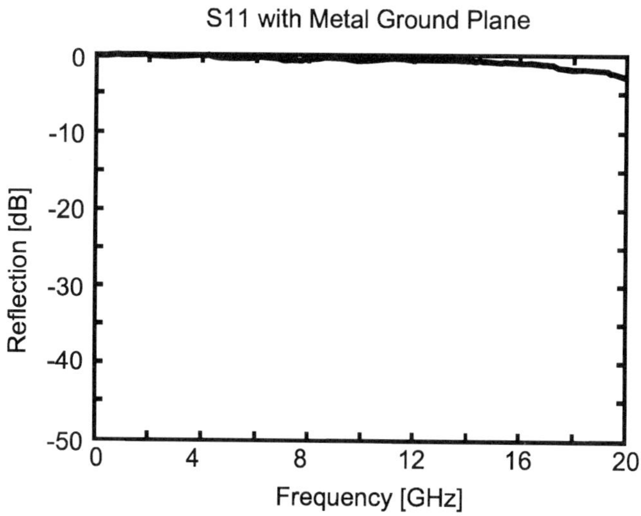

Fig. 18 *S11 of a horizontal wire antenna on a flat metal ground plane.*

A similar wire antenna on a high-impedance ground plane is shown in Fig. 19. Within the band gap, which spans from 11 GHz to 17 GHz, this antenna has a much lower return loss, as shown in Fig. 20. Below the TM band edge, the antenna performance is similar to the one on the metal ground plane. The surface impedance is low, and the image currents cancel the currents in the antenna. Within the band gap however, the return loss is about -10 dB, indicating that only 10% of the power is being reflected back to the generator. In this range, the image currents are reversed, and they reinforce the antenna currents. Above the TE band edge, the return loss also low because of coupling to TE surface waves. There is also strong coupling to TM waves near the TM band edge, where the density of states is very high.

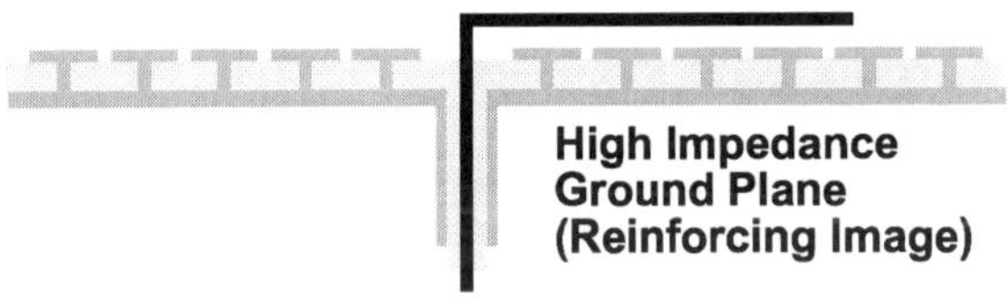

Fig. 19 *A horizontal wire antenna on a high-impedance ground plane.*

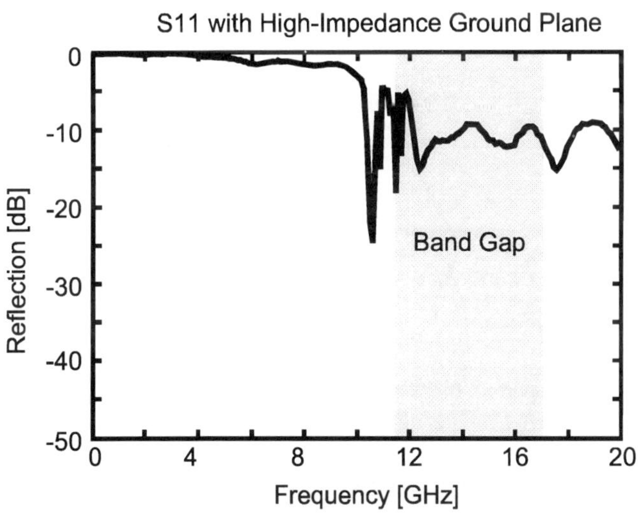

Fig. 20 *S11 of a horizontal wire antenna on a high-impedance ground plane.*

The radiation pattern of both antennas is shown in Fig. 21 and Fig. 22. The dotted line is for the metal ground plane, and the solid line is for the high-impedance ground plane. The data in the radiation patterns supports the return loss data, in that the signal level is about 10 dB higher on the high-impedance ground plane. The maximum gain of the horizontal wire is about 7 dBi, and the radiation pattern is rotationally symmetric about the axis.

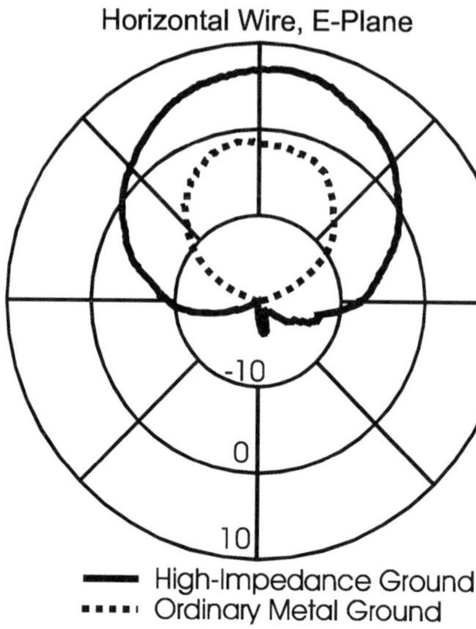

Fig. 21 E-Plane radiation pattern of two horizontal wire antennas.

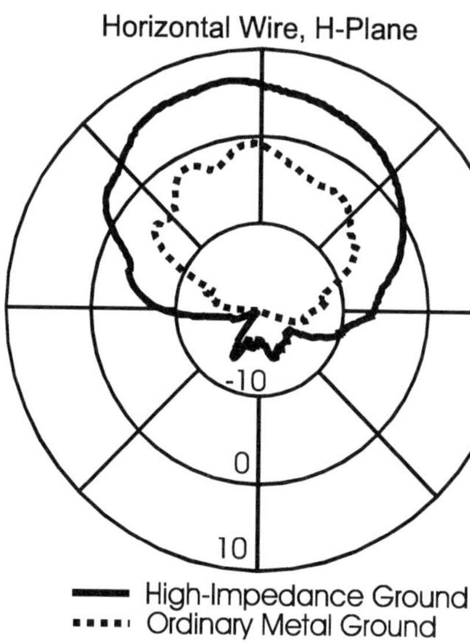

Fig. 22 H-plane radiation pattern of two horizontal wire antennas.

4.3 Portable Communications

One important application of the high-impedance ground plane is in the area of portable communications. At present, the standard antenna geometry is a monopole, which extends from a handset and lies in close proximity to the user's head. It has been shown that with this configuration, the user absorbs roughly 50% of the radiated power. [*Jensen and Rahmat-Samii, 1995*]. The effect of the radiation on the user has not been evaluated conclusively. However, the effect of the user on the antenna results in a significant reduction in battery life. The power lost to absorption can be reduced by using a directional antenna. Directional antennas might appear impractical for portable communications. However, in a real environment, scattering from buildings and other objects tends to randomize the signal direction

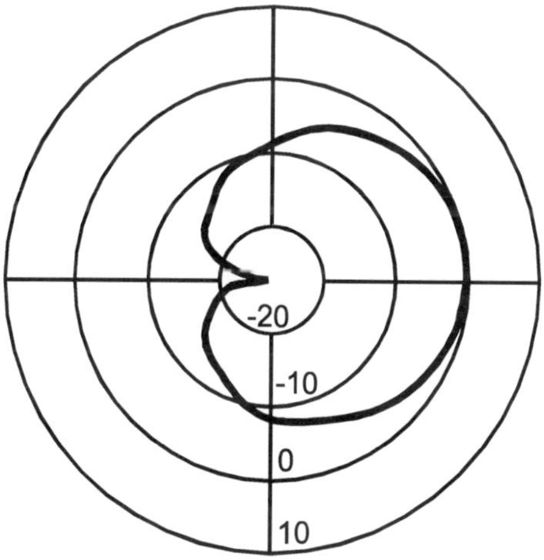

Fig. 23 *Radiation pattern of a dipole antenna near a jar of water.*

The plot shown in Fig. 23 is the radiation pattern of a dipole antenna placed near a jar of water. The jar of water simulates the presence of the user, as water has a similar dielectric constant to many human tissues. The radiation is blocked from propagating through the water, as indicated by the deep null in the direction of the jar. The overall signal level is also reduced in all directions due to a combination of absorption, and modification of the antenna input impedance. The maximum gain is about 0dBi, in the direction opposite to the jar of water.

A similar antenna on a high-impedance ground plane produced the radiation pattern shown in Fig. 24. The jar of water was placed behind the ground plane, on the opposite side from the antenna. The antenna has a maximum gain of about 6dBi, in the direction normal to the ground plane. The high-impedance surface acts as a shield between the radiating antenna and the absorbing water, blocking not only the direct radiation, but also the surface currents that would wrap around to the back side. The ground plane redirects the radiation that would otherwise be absorbed into the user, recovering it as useful power in the other direction. The combination of the antenna and the ground plane is compact and lightweight, allowing it to fit within the packaging of most portable communications equipment.

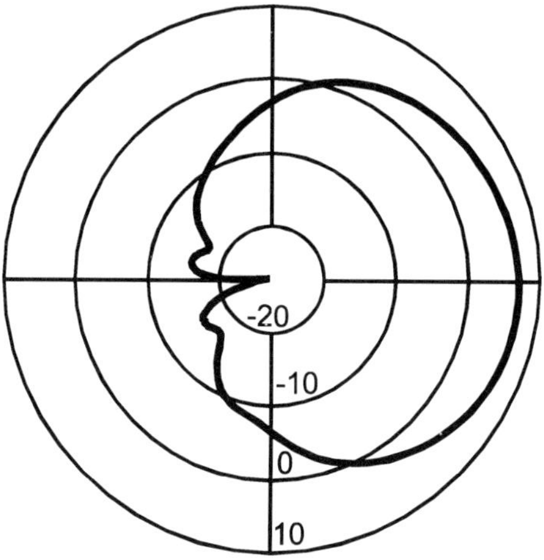

Fig. 24 *Wire antenna on high-impedance ground plane, near jar of water.*

5. Conclusion

A new type of metallic electromagnetic structure has been presented that is characterized by having high surface impedance. It is made of continuous metal, and conducts DC currents, but it does not conduct AC currents within a forbidden frequency band. Instead, currents that are induced in the surface radiate efficiently into surrounding space. This new surface also reflects electromagnetic waves with no phase reversal, behaving as a kind of magnetic conductor. The structure can be described using a lumped parameter circuit model, which accurately predicts many of its electromagnetic properties. This unique material is applicable to a variety of electromagnetic problems, including new kinds of low-profile antennas.

References

Balanis, C., [1997] *Antenna Theory, Analysis, and Design*, 2nd ed., John Wiley and Sons, New York.

Brillouin, L., [1948] "Wave Guides for Slow Waves", *J. App. Phys.* 19, 1023.

Brown, E., C. Parker, E. Yablonovitch, [1993] "Radiation properties of a planar antenna on a photonic-crystal substrate", *J. Opt. Soc. Am.* B 10, 404.

Brown, E., O. McMahon, [1995] "Large Electromagnetic Stop Bands in Metallodielectric Photonic Crystals", *Appl. Phys. Lett.* 67, 2138.

Collin, R., [1991] *Field Theory of Guided Waves*, 2nd ed., IEEE Press, New York.

Elliot, R. [1954] "On the Theory of Corrugated Plane Surfaces", *IRE Trans. Ant. Prop.* 2, 71.

Harvey, A., [1959] "Periodic and Guiding Structures at Microwave Frequencies", *IRE Trans.* 8, 30.

Ho, K-M., C. Chan, C. Soukoulis, [1990] "Existence of a Photonic Gap in Periodic Dielectric Structures", *Phys. Rev. Lett.* 65, 3152.

Jensen, M., Y. Rahmat-Samii, [1995] "EM Interaction of Handset Antennas and a Human in Personal Communications", *Proc. IEEE*, 83, 7.

Joannopoulos, J., R. Meade, J. Winn, [1995] *Photonic Crystals: Molding the Flow of Light*, Princeton University Press, Princeton.

Kildal, P-S. [1990] "Artificially Soft and Hard Surfaces in Electromagnetics", *IEEE Trans. Ant. Prop.* 38, 1537.

Pendry, J., A. Holden, W. Stewart, I. Youngs, [1996] "Extremely Low Frequency Plasmons in Metallic Mesostructures", *Phys. Rev. Lett.* 76, 4773.

Raether, H., [1988] *Surface Plasmons on Smooth and Rough Surfaces and on Gratings*, Springer-Verlag, New York.

Ramo, S., J. Whinnery, T. Van Duzer, [1984] *Fields and Waves in Communication Electronics*, 2nd ed., John Wiley and Sons, New York.

Sievenpiper, D., M. Sickmiller, E. Yablonovitch, [1996] "3D Wire Mesh Photonic Crystals", *Phys. Rev. Lett.* 76, 2480.

Sievenpiper, D., E. Yablonovitch, J. Winn, et al., [1998] "3D Metallo-Dielectric Photonic Crystals with Strong Capacitive Coupling Between Metallic Islands", *Phys. Rev. Lett.* 80, 2829.

Sievenpiper, D., [1999] Ph. D. Thesis, University of California, Los Angeles.

Sigalas, M., R. Biswas, Q. Li, et al., [1997] "Dipole antennas on photonic band-gap crystals: Experiment and simulation", *Microwave Optical Technol. Lett.* 15, 153.

Yablonovitch, E., [1987] "Inhibited Spontaneous Emission in Solid-State Physics and Electronics", *Phys. Rev. Lett.* 58, 2059.

ELF SFERICS AND LIGHTNING EFFECTS ON THE MIDDLE AND UPPER ATMOSPHERE

D. Llanwyn Jones

1. Introduction

This tutorial paper is concerned with ELF sferics and aspects of lightning effects on the atmosphere of interest to URSI Commission E whose terms of reference encompass 'electromagnetic noise and interference'. The effects considered are those which are involved in the generation and propagation of terrestrial radio noise. Such radio noise consists of 'sferics' or 'atmospherics' produced by, or associated with, lightning discharges. Attention is confined to the ELF (Extremely Low Frequency – 3 Hz to 3 kHz) band and particularly to the lower-ELF ('Schumann resonance') band – 5 Hz to 60 Hz. This reflects the upsurge in research publications in this area over the past decade and the recent comparisons of the occurrence of transient optical emissions in the mesosphere and lower ionosphere called sprites (or red sprites*) [Sentman et al., 1995; Lyons, 1996*], elves [*Fukunishi et al, 1996*] and blue jets [*Pasko et al., 1996*] with the characteristics of the causative lightning source as deduced from observations of ELF sferics propagated to great, even global, distances.

Within the past 5 years or so, it has been recognised that many of the lightning sources responsible for generating transient ELF electromagnetic signatures ('events') in the Schumann resonance band and at higher ELF frequencies (where the signal is referred as a 'slow tail' sferic), are the same sources which frequently, but not necessarily, produce the luminous phenomena in the upper atmosphere above some of the thunderclouds. It is now established that the essential connection between the electromagnetic and optical effects is positive cloud-to-ground (+CG) lightning discharges with high peak currents and large charge transfers. Mesoscale convective systems (MCSs) appear to play an important role in both the optical and sferic cases. However, sprites are observed over smaller convective systems [*Lyons, 1996*] and only a minority of ELF sferics are associated with the spectacular upper atmosphere occurrences. Some sprites and elves have an association with ELF events [*e.g. Boccippio et al., 1995; Füllekrug and*

Reising, 1998], and with ELF 'slow tail' sferics [*e.g. Marshall et al., 1998; Huang et al., 1999*]. To set the scene we begin with a brief description of the optical and electromagnetic phenomena.

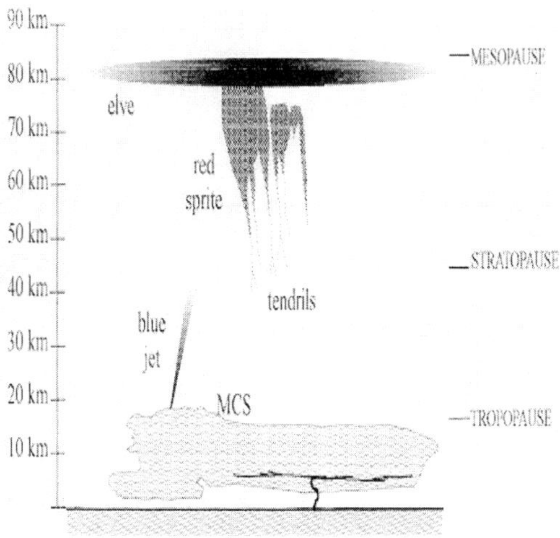

Fig. 1 *Locations of red sprites, elves and blue jets in the atmosphere.*

2. Background

2.1 The Optical Phenomena in the Middle and Upper Atmosphere

Figure 1, (adapted from *Rowland* [*1998*]), is a sketch illustrating the location in the atmosphere of the optical emission phenomena known as red sprites, elves and blue jets together with the parent convective system. Red sprites exist in the altitude range 80-55 km with dangling wispy tendrils below 55 km. Their lateral extent is around 25 km and their lifetime is a few tens of milliseconds. Elves (emissions of light and VLF perturbations due to EMP sources) appear between 90 and 70 km altitude and extend horizontally for a few hundred km. They have a much shorter existence than sprites, less than 1 ms. Blue jets spout from the top of thunderclouds upwards to some 40 km with a velocity of 100 km/s and a lifetime of a few hundred milliseconds.

There is an extensive literature concerned with mathematical models seeking to explain these optical emissions. These theories include the effects of the pre-stroke electrostatic field of the thundercloud charges, the electromagnetic pulse (EMP) generated from the lightning return stroke (the spectacular phase of the lightning flash) and the quasi-static fields due to the redistribution of cloud charges following strokes.

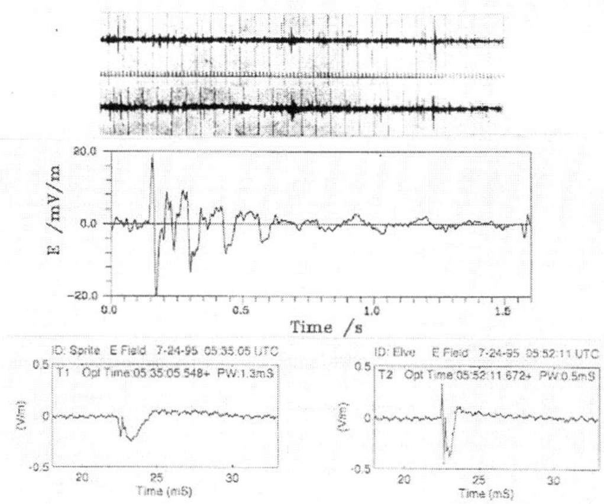

Fig. 2. *ELF waveform record, an ELF event and two slow tails.*

There have been several reviews concerned with the characteristics of sprites, elves and blue-jets and the theory of their formation – *e.g.* [*Rowland, 1998; Rodger, 1999*]. Special journal issues concerning these phenomena have also been published in the past few years – *e.g.* [*Holzworth, 1996; Sentman, 1998*]. These theories are not discussed in this tutorial but the electrodynamic treatment of *Ma et al.* [*1998*], which concerns the electromagnetic fields generated by high-atmosphere currents, appears to be particularly thorough. Additionally, non-sferic-material such as the important topic of VLF perturbations (including *Dowden's* 'RORDs') produced by sprites and elves, is excluded – see the review of *Strangeways,* [*1996*] and the recent review of *Rodger* [*1999*].

2.2 The ELF Electromagnetic Phenomena

In Fig. 2 (top) a time-domain record of ELF noise recorded in the band 3-50 Hz is shown. The data in the upper trace were recorded in the UK and those in the lower trace in Japan. The time markers here indicate seconds. The 'spikes' in the record are the ELF events referred to. Note the frequent simultaneity of spike occurrence in the UK and in Japan showing the global nature of the propagation. A record of the electric field of a single ELF transient event is shown on an expanded time scale (centre) [*Burke and Jones, 1996*]. The typical time-scale of these signals is a few hundred milliseconds. The lower two waveforms are slow tails associated with a sprite and an elve as observed in a bandwidth of 5 kHz [Marshall et al., 1998]. These have a time-scale of a few milliseconds.

2.3 Positive CG Lightning, Sferics, Sprites and MCSs

A positive cloud-ground (+CG) lightning stroke is defined as a stroke that lowers positive electrical charge to the ground. In the electromagnetic theory used for sferic analysis, the convention is to take the positive direction as *upwards*. This convention, adopted here, is opposite to that used in studies of atmospheric electricity where the positive direction is *downwards* – the direction of the 'fair weather' field. A positive stroke lowers positive charge to the ground by a 'conventional' current I flowing in a channel of length dl and has a current moment $I\,dl\,(t)$ (Amp metre) which is *negative* in the theory, *i.e.* directed downwards. This negative current moment generates a vertical electric field pulse that is positive (upwards). Confusingly, this positive field pulse generates a negative voltage pulse on the receiving antenna.

It is generally accepted that the vast majority (over 90%) of sferics observed in the upper ELF band (where they are called 'slow tails' [*Jones, 1970b*]) and in the VLF band have a negative initial polarity [*Uman, 1987*] – at least in mid-latitudes. These are generated by 'normal' thunderclouds in which the charge distribution is 'positive over negative' as indicated in Fig. 3 (left). The normal –CG stroke taps the main negative charge centre at the base of the cloud.

Positive CG strokes have to be initiated from the upper region of the cloud. This can be displaced laterally from the base region by wind shear, facilitating a +CG discharge as shown in the figure. This is the conventional mechanism for +CG stroke generation and is invoked in many of the theories of the atmospheric optical phenomena – see *Williams* [*1998*] for a (perhaps controversial) discussion. A lightning *flash* usually comprises several *strokes* (about 4) blazing essentially the same channel below the cloud charge center(s). The number of strokes per flash is the multiplicity of the flash and is typically four. In the tropics the situation may be different. For example, *Pinto et al.,* [*1996*] present statistics on over 300,000 flashes observed in Brazil. Their sample contained 35% positive flashes.

Boccippio et al. [*1995*], *Lyons* [*1996*], and *Williams* [*1998*], have described laterally extensive mesoscale convective systems (MCSs) in which the normal 'positive over negative' charge distribution is inverted. A sketch of an MCS is shown in Fig. 3 (right), adapted from *Lyons* [*1996*]. Here the *positive* charge reservoir is near the base and the 0°C isotherm, at some 4–6 km height. The total amount of positive charge available is such a system may be thousands of coulombs [*Marshall and Rust, 1993*].

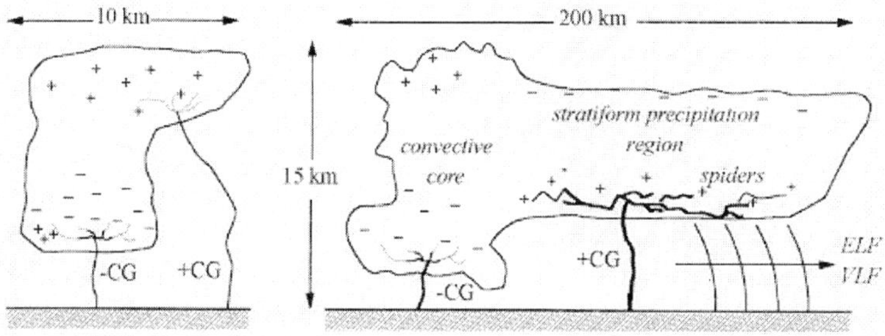

Fig. 3 *Charge distribution in a conventional thundercloud and in a mesoscale convective system (MCS).*

Positive CG lightning appears as an essential ingredient for generating sprites [*Sentman and Wescott, 1993; Boccippio et al., 1995; Lyons, 1996*]. The data of *Lyons* and that of *Boccippio et al.* support the hypothesis that sprites are *uniquely* associated with +CG flashes which have a large peak current. However, most +CG flashes (including some with a large peak current) do not produce sprites.

Much of the recent experimental work on sprites and sferics has used data from the US National Lightning Detection Network (NLDN). This locates sferic sources using a bandwidth of 5 kHz to 500 kHz and reports time-marked locations ('fixes') of individual sources and their peak amplitudes. The peak current in the discharge may be estimated from these data. However, it should be noted that the peak current occurs a few microseconds after leader touch-down when the return stroke channel is less than 1 km long. The peak current is not simply related to the total charge removed by the whole discharge that is an important parameter in ELF sferic characteristics.

3. ELF Sferics

By analysis of the ELF signature of the sferic received at a distant location it is possible to deduce both the propagation parameters of the Earth-ionosphere duct and certain characteristics of the source of the radiation. The ELF signatures are referred to as 'events' here but they have also been termed Q ('quiet')- or N ('noisy')-bursts, this appellation reflecting the complexity of the time-domain structure [*Ogawa et al., 1967*]. Data analysis procedures are well-established and have been developed over the past three decades – well before the high-altitude phenomena were discovered using low light level television (LLTV) cameras by *Franz, Nemzek* and *Winkler* in *1990*.

The sferic data analysis requires the application of the theory of propagation in the Earth-ionosphere duct. To apply the propagation theory it is necessary to determine the propagation constant of the duct. This can be computed theoretically if the variation with altitude of the ionospheric parameters is known. However, in view of the uncertainties of ionospheric models, it is much better to obtain the propagation constant from the experimental data themselves. We thus begin with a brief account of the propagation theory and describe how the propagation constant is determined from ELF event waveform data. While the main concern here is with research over the past few years, some discussion of the background to this work is included to form a coherent account.

3.1 ELF Propagation

3.1.1 Basic Concepts

Aspects of ELF propagation have been reviewed recently by *Jones* and *Burke* [*1995*]. The fundamental processes responsible for the propagation of ELF radio waves over the Earth's surface in the Earth-ionosphere spherical duct are well understood and are documented in the seminal texts of *Wait* [*1970*] and *Galejs* [*1972*]. The field propagates as a guided wave between the surface of the Earth (which behaves as an almost perfect conductor *i.e.* perfect reflector) and the ionosphere. The ionosphere is diffuse, extending from about 30 km altitude and reaching a maximum electron density at around 320 km by day.

The Earth-ionosphere system may be regarded as constituting a natural electromagnetic waveguide or a transmission line. It is the ionosphere which is almost wholly responsible for the absorption of ELF energy leading to propagational losses. Unlike the situation found at higher frequencies, ELF waves are not reflected from a specific height in the ionospheric plasma but are partially reflected over a relatively large range of altitudes. During daytime the average reflection height is around 50 km rising to 90 km at night. Owing to the low frequency, the low conductivity 'C-region' at the base of the ionosphere, consisting of heavy ions produced by galactic cosmic rays, plays a major role in determining the propagational attenuation of the duct. The Earth's magnetic field permeates the ionospheric plasma rendering it anisotropic. This anisotropy modifies the wave attenuation, the phase velocity and, particularly, the wave polarisation. This last factor affects the accuracy of goniometric location systems.

ELF sferics propagate in the duct as a 'slow' wave--the phase velocity v being less than that of light c, so that $c / v > 1$. The attenuation of the duct varies from less than 0.1 dB/Mm at a few Hertz to about 60 dB/Mm at 2 kHz (1 Mm = 1,000 km). In the lower ELF band the propagational attenuation is thus extraordinarily small, smaller than that at any other frequency. This fact enables lower ELF band lightning radiation to be detected at global distances.

3.1.2 The Mode Theory

The equations for the electromagnetic fields in a spherical Earth-ionosphere duct may be written as a summation of a set of waveguide modes as detailed in [*Wait, 1970 and Galejs, 1972*]. The significant feature of the theory for ELF is that only the transmission-line mode, *i.e.* the zero-order transverse magnetic mode (TM_0) mode, propagates. Higher order modes and TE (transverse electric) modes can exist at frequencies which exceed the cut-off frequency of the waveguide which is $f_n = nc/2h$ for a mode of order n and a waveguide of height $h \cong 50$ km (day) or 90 km (night). Thus for $f < 2$ kHz only the $n = 0$ mode is significant for propagation ranges greater than a few waveguide heights ([*Wait, 1970*], see Fig 1, p. 294).

The fields constituting the mode can be decomposed into upgoing and downgoing waves whose wave normals make a complex angle ϑ_n (the mode eigenangle) with the normal to the boundary. In mode theory the central problem is to determine $S = \sin \vartheta_n$ for specified boundaries because the propagational attenuation α and the phase velocity ratio c/v are determined by S through

$$S = (c / v) - i\,(5.49\alpha / f) \qquad (1)$$

where α expressed in dB/Mm and f is the wave frequency in Hertz. $|S|$ is somewhat larger than unity in the ELF band because $c / v > 1$.

An $e^{+i\omega t}$ time dependence is assumed. We use a spherical polar co-ordinate system (r, Θ, ϕ) with the pole located at the Earth's centre and a radiation source placed at $(a, 0, 0)$, a being the Earth's radius. A lightning stroke can be modelled as a vertical electric dipole source of length dl

and current I and hence with a *current moment* spectrum $I\,dl(i\omega)$ in the Fourier (frequency) domain. Although a stroke may have a substantial horizontal component, this component is a very ineffective ELF radiator [*Wait, 1970, eq.(9.39)*]. Nevertheless, and contrary to popular belief, horizontal currents *do* excite the *TM* waveguide modes. The radial (vertical) electric field E_r and azimuthal magnetic field H_ϕ on the Earth's surface, due to a vertical electric dipole (in this case a lightning flash) with a current I and a length dl at the point (α, Θ, ϕ), are given by the mode theory as– see [*Jones, 1970a*]:

$$E_r = i\frac{Idl(i\omega)\,\nu(\nu+1)P_\nu^0(-\cos\theta)}{4a^2\varepsilon_0\omega h\sin(\nu\pi)} \tag{2}$$

$$H_\phi = -\frac{Idl(i\omega)\,P_\nu^1(-\cos\theta)}{4ah\,\sin(\nu\pi)} \tag{3}$$

In these equations h is the effective ionospheric height, $P_\nu^{0,1}$ are associated Legendre functions of (complex) degree ν and order 0,1; ω is the angular frequency, and h is the effective ionospheric height. The parameter ν is the modal eigenvalue. It is related to S in eq.(1) by

$$\nu(\nu+1) = (kaS)^2 \tag{4}$$

with $k = \omega / c = 2\pi / \lambda$.

In the lower part of the ELF band $ka\,|\,S\,| \approx 1$. In this region the wavelength is of the same order as the Earth's circumference. It is this spectral region which is of most interest for the deduction of lightning source characteristics from fields measured at global ranges. Procedures for computing the Legendre functions in (2) and (3) (for a given value of ν) have been presented by a number of workers (see [*Jones and Burke, 1990*] and the references therein). For wave frequencies greater than 30 Hz (*i.e.* ν sufficiently large) a simple asymptotic formula may be used [*Wait, 1970; Jones, 1970a*].

To use the mode theory for data analysis it is necessary to determine the value of ν (or S) *i.e.* the propagation constant of the waveguide. If this is not to be done from the data themselves, but for an assumed ionospheric model, the ionospheric reflection coefficient has to be computed over the frequency band of interest. If the Earth's magnetic field $B_0 = 0$, the ionosphere is isotropic. [*Jones, 1967*] and [*Barr, 1970*] computed the modal propagation constant for several inhomogeneous isotropic ionosphere models to explain experimental Schumann resonance and sferics measurements respectively. A calculation technique for an anisotropic ionosphere in the ELF and VLF bands has been detailed by [*Jones, 1984*]. Such calculations indicate that goniometric bearing errors of a few degrees ($\approx 2-5^{\circ}$) can be produced by the anisotropy. An example of computed data, together with illustrative experimental measurements of the attenuation constant of the Earth-ionosphere duct is shown in Fig. 4. The measurements shown were obtained by analysis of ELF sferics and a series of experimental cw transmissions.

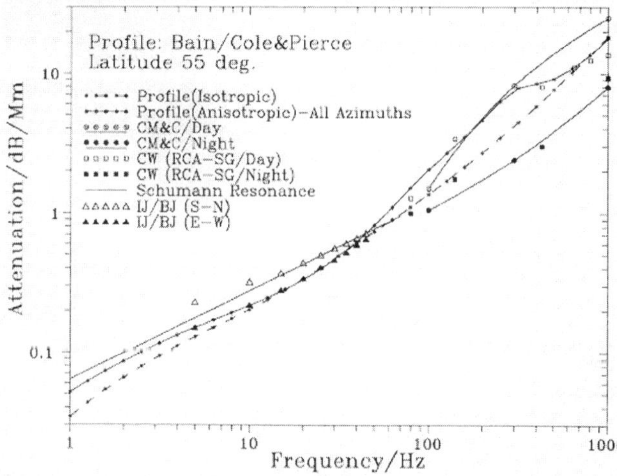

Fig. 4 *Experimental and theoretical ELF attenuation rates [Jones, 1984].*

3.2 The Schumann Resonances

The Schumann resonances are simply the electromagnetic resonances of the global earth-ionosphere (quasi) spherical-shell cavity. For lossless (perfectly reflecting) boundaries W. O. Schumann showed that the resonance frequencies are given by an equation of the form $f_n = c\sqrt{n(n+1)}\,/\,(2\pi a)$. This predicts a fundamental mode frequency (n = 1) of f_1 = 10.6 Hz and overtones at 18.4, 26.0, 33.5 and 41.1 Hz. Further developments in the theory may be found in the highly recommended texts of *Bliokh et al.*, [*1980*] together with those of *Wait* [*1970*] and *Galejs* [*1972*]. Recent reviews of both experimental data and theory have been given by *Sentman* [*1995*] and *Nickolaenko* [*1997*]. The celebrity of the phenomenon is such that it is now included a new edition of the classic undergraduate textbook by *Jackson* [*1998*].

The resonances are excited by global lightning activity and are evident in experimental atmospheric noise spectra in the band 5 Hz to 60 Hz. The noise spectra have maximums near 7.8, 14.2, 19.6, 25.9 and 32 Hz corresponding to the modes $n = 1$ to $n = 5$ in Schumann's formula. The measured frequencies are lower than predicted by the formula because of ionospheric losses and are thus diagnostic of these losses. The amplitude and frequency of the Schumann modes is determined by the temporal and spatial distribution of global lightning – this being most intense in the tropics.

3.3 ELF Transients in the Schumann Resonance Band

Unusually intense lightning discharges generate a large ELF transient signature or 'event' in the Schumann resonance band (5–60 Hz). Distinguishable transients can be detected at global ranges. They have a duration of order 1/2 sec and exceed the 'Schumann resonance background' (produced by global lightning activity) by a factor of three or more (Fig. 2). The resonance background amplitude in the frequency band 5 – 50 Hz is fairly constant with an amplitude of about 1 mV/m or 3 pT. Analysable events have typical amplitudes of 3 – 30 mV/m.

More information on the character of the source lightning can be obtained from the analysis of the events than from the background because they are produced by a single source at a fixed location on the Earth's surface. The aim of the data analysis is to compute the ELF propagation constant (α and c/v), the location of the source lightning on the Earth's surface, the time-scale of the source current moment change and the change in charge moment of the thundercloud produced by the lightning stroke (expressed in C km).

4. Technique for ELF Event Data Analysis

In this section the data processing techniques used for ELF event analysis are outlined.

4.1 The Propagation Constant

The most important ingredient in the data analysis procedure is the propagation constant of the duct. The desirable procedure is to deduce the propagation constant of the source – receiver path directly from the experimental data. This can be done by calculating the wave impedance of the duct from the sferic data *[Kemp and Jones, 1971; Ishaq and Jones, 1977*]. This wave impedance also enables the source receiver propagation distance *d* to be found.

The wave impedance *Z* is simply the (complex) ratio $E_r \,/\, H_\phi$. From (2) and (3)

$$Z = \frac{E_r}{H_\phi} = -i\frac{\nu(\nu+1)P_\nu^0(-\cos\theta)}{a\varepsilon_0\omega P_\nu^1(-\cos\theta)} \tag{5}$$

The huge advantage of this method is that *the wave impedance is independent of the source spectrum* ($I\,dl(i\omega)$ in equations (2) and (3)) and independent of the effective ionospheric height *h*. It depends only on the propagation parameter ν and the range of the source θ = *d/a*.

Burke and Jones, [*1992; 1996*] describe a technique for ELF event data processing. This enables the propagation constant and the source range to be determined by using an optimisation algorithm to find the value of $\nu(i\omega)$ which fits the experimentally determined wave impedance frequency spectrum to a theoretically computed impedance (equation (5)). The propagation constant appropriate to the propagation path of each individual event is then obtained. However, the optimisation is a lengthy procedure. *Burke and Jones* [*1992*] used this procedure to show that the ELF attenuation factor α varies with path latitude, increasing towards the Earth's poles.

A simpler analysis technique is to assume a value for the propagation constant. Various experimental measurements are available in the literature for the ELF band – examples are shown in Fig. 4. The most convenient procedure is to fit published data for the dependence of α and *c/v* on frequency to an analytic function which is then assumed to apply to all of the events analysed. This is the method adopted by *Huang* [*1997*] and *Huang et al.* [*1999*]. These workers computed ν from (1) and (4) using the analytic formulas:

$$c/v = 1.725 - 0.275\ln f + 0.052(\ln f)^2 - 0.0036(\ln f)^3 \tag{6a}$$

$$\alpha = 0.063 f^{0.64} \text{ dB/Mm}. \tag{6b}$$

4.2 Determination of the Source Location

If field measurements are available from two stations the source location can be found by triangulation. This has been done recently by *Füllekrug and Reising* [*1998*] using receiving sites in Germany and California.

The wave impedance method can be used to determine the range *d* of the source on the Earth's surface if field measurements are only available at a single recording station, [*Kemp and Jones, 1971; Nickolaenko and Kudintseva; 1994; Burke and Jones, 1996; Huang, 1997; Huang et al., 1999*]. The frequency-dependence of the modulus or argument of the wave impedance for any assumed propagation model is of the form of a damped oscillation. The 'period' τ_m of this oscillation (the 'qfrency') is a monotonic function of the range d. From a series of numerical simulations *Burke and Jones* [*1995*] suggest that the range may be determined with good accuracy from the formula $d = -125.5\ \tau_m + 20.5$, where τ_m is in seconds and d is expressed in Mm. This formula fails near the source and the source antipode but is quite accurate for 2 Mm $< d <$ 17 Mm. The numerical optimisation technique, however, is valid for all values of d.

From the two magnetic field components the bearing of the source at the receiving station is obtained – either from the usual goniometric technique (the 'hodograph') or by computing the direction of the Poynting vector using the measured electric and magnetic field components. The source range d and its bearing 'fix' the source location on the Earth's surface.

4.3 Determination of Source Characteristics

The 'characteristics' of the source are represented by its polarity and its current moment spectrum $Idl(i\omega)$ (refer to eq. (2) and (3)). As ν and $\theta = d/a$ are now known, the source current moment frequency spectrum can be computed from the experimentally measured electric field spectrum (by means of (2)) and/or from the magnetic field spectrum (by (3)). There should, of course, be no evidence of a Schumann resonance structure in the deduced current moment spectrum.

4.3.1 Source Polarity

It has been know for nearly three decades that that the sources of ELF events are predominantly positive. This contrasts with what is found for atmospherics in the upper ELF band (slow tails) and the VLF portion of atmospheric waveforms for which the sources are nearly all (90%) negative in mid-latitudes [*e.g. Lyons, 1996*]. The source polarity may be found from the polarity of the initial pulse of the E-field waveform (which is positive for a positive CG stroke). More reliably it can be determined from the phase of $Idl(i\omega)$ for $\omega \to 0$. This is zero for a positive stroke and π for a negative stroke.

4.3.2 Source Charge Moment Variation with Time

The data processing of each ELF event enables the current moment spectrum $Idl(i\omega)$ to be obtained within the bandwidth of the receiver. This is the Fourier transform of the time-domain current moment $I\,dl(t)$ as given by the Campbell-Foster Fourier pair,

$$I\,dl(t) = \int_{-\infty}^{\infty} I\,dl(\omega)\,e^{i\omega t}df\,;\;\; I\,dl(\omega) = \int_{-\infty}^{\infty} I\,dl(t)\,e^{-i\omega t}dt \tag{7}$$

where $\omega = 2\pi f$. Thus $I\,dl(t)$ can be computed from the experimental data. If it is assumed [*Burke and Jones, 1996; Sentman, 1996, Huang et al., 1999*] that $I\,dl(t)$ can be modelled by a simple exponential variation of the form $I\,dl(t) = A\,e^{-t/\tau}$, then, the Fourier integral gives

$$I\,dl\,(i\omega) = A\tau\,/\,(1 + i\omega\tau) \tag{8}$$

4.3.3 Source Charge Moment Change

The lightning stroke produces a change in the charge moment (charge times height of the charge-centre) of the thundercloud. With $\omega = 0$, the second integral in (7) gives $I\,dl(\omega = 0) = \int_{-\infty}^{\infty} I\,dl(t)\,dt$, which is the charge moment change ΔM_Q. So the zero-frequency value of the current moment spectrum is the charge moment change of the thundercloud. If the model in eq. (8) is used then $\Delta M_Q = A\tau$. Alternately, the change in charge moment produced by the flash can be deduced by integration of *I dl(t)* [*Huang et al., 1999*]. It should be noted that workers in atmospheric electricity who are concerned with the near-field electrostatic field change produced by a flash, normally include the image of the charges in the Earth and report the *dipole* moment value $2\Delta M_Q$. The factor of two here has led to some confusion!

5. Experimental Results

5.1 Source Locations

Burke and Jones [*1995*], *Huang* [*1997*], and *Huang et al.* [*1999*], present maps showing the location of ELF event sources. Burke and Jones used the laborious optimisation method described above and show source locations for 261 events. Huang has produced monthly maps [E. R. Williams, private communication] showing several thousand source locations for each month – an example is published in *Huang et al.,* [*1999*]. These locations are in accord with what is known of global thunderstorm activity with the storm centres mainly located in low latitudes and moving southwards in the northern hemisphere summer-winter transition.

An important feature of *Huang's* work is that the source locations found were compared with those for sferics located by the National Lightning Detection Network (NLDN). There was a systematic periodic bearing error (which was allowed for). It was noted that the ELF location method, using their propagation constant model (equations 6), overestimated the range by about 100 km.

Füllekrug et al. [*1996*] computed the arrival azimuth of 19 sprite-associated ELF events which were simultaneously located by the NLDN. The system bandwidth was 0.1-20 Hz. They found a bearing error of $\pm 5^0$ for events which were >3 times the rms Schumann background field strength but considerably larger errors for smaller signals. *Füllekrug and Reising* [*1998*] recorded 569 ELF events in Germany and California (9.1 Mm apart) and obtained the source locations by triangulation. Comparison with data provided by the NLDN showed that the location accuracy (comprised of bearing and range) was about 0.8 Mm. *Boccippio et al.* [*1998*] used satellite optical transient detector data to show that the accuracy of their range determination was 1 – 2 Mm for 40 events.

5.2 Source Polarity

In *Burke and Jones* [*1996*] rather small data set, 212 (76%) events had positive sources and 62 had negative. In *Huang et al.'s* [*1999*] larger data set of 400 events, 66% were of positive polarity. *Füllekrug and Reising* [*1998*] present the much larger data-set shown in Fig. 5. This shows the distribution of first return stroke peak currents of lightning flashes recorded by the NLDN on August 1, 1996. They note that the flashes which produce ELF events are overwhelmingly of positive polarity (confirming the two results just cited) and are associated with NLDN-deduced peak cloud-to-ground currents of 20–70 kA.

5.3 Charge Moment Amplitude and Time Constant

Burke and Jones [*1996*] data set is a selection of large amplitude events propagated over global paths. The system bandwidth was 5 – 50 Hz. They present statistical distributions of the two parameters A and τ in (8) with mean values of $A = 5.1\times10^7$ Am for positive strokes and $A = 3.2\times10^7$ Am for negative strokes. For τ the mean values were 32 ms and 26 ms for positive and negative strokes respectively. These give values of the charge moment change $\Delta M_Q = A\tau$ (which excludes the ground image) of 1630 C km and 830 C km for positive and negative strokes respectively. In this work it was not known if the event sources were associated with sprites or elves. It should be noted that the A and τ values obtained in this recent study are typical of those obtained over some three decades at King's College and refer to the largest sources which occur globally at a rate of about 1 per minute. Only about 1 in 10,000 strokes produce the ELF event signatures which correspond to the above parameter values.

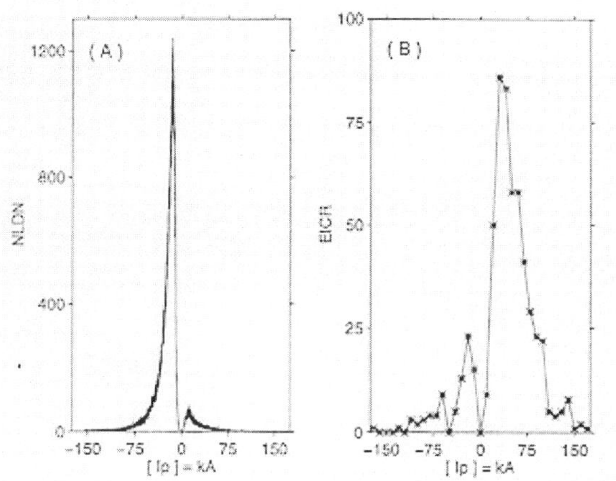

Fig. 5 *(A) First return stroke peak currents of all flashes, and (B) of those strokes which excite ELF events. [Füllekrug and Reising, 1998].*

5.4 ELF Transients Associated with Sprites and Elves

Boccippio et al. [*1995*] report the results of a comprehensive study in which NLDN and ELF sferic electromagnetic data were obtained in conjunction with simultaneous low-light monochrome LLTV imaging of sprites with 17 ms resolution at Yucca Ridge, N. Colorado. The ELF receiving system bandwidth was 3 – 120 Hz and was located in Rhode Island, 2300 km from the LLTV station. 56% of the +CG flashes produced sprites and these flashes fell within the upper 3 – 15% of the NLDN positive peak current distribution (currents of 100 – 200 kA). About 80% of the sprites were coincident with +CG strokes *and* ELF transients.

In *Huang et al.* [*1999*], simultaneous NLDN, ELF and LLTV data were obtained and elves as well as sprites were observed. The ELF recording system was the same as that used for the study of the preceding paragraph. The ELF data showed that the 40 sprites observed produced a 'red' ELF spectrum with an average value of τ (eq. 8) of 5.4 ms and ΔM_Q in the range 200 – 1500 C km. In contrast the 15 elves recorded produced a 'whiter' spectrum in the lower-ELF band with an average τ of 3.6 ms and smaller values of ΔM_Q. The τ values here, derived from ELF sferics produced by one MCS, are much less than those reported above and would not be atypical of a normal (–CG) stroke.

Marshall et al. [*1998*] used a recording system with a 5 Hz – 5 kHz bandwidth and recorded slow tail signatures at 2,200 km range. All of the recorded slow-tail sferics which were related to sprites and elves exhibited mainly unipolar signatures of about 1 – 2 ms in duration with a polarity corresponding to positive lightning sources. In a data-set of 356 sferics, 325 showed a slow tail with a polarity corresponding to +CG lightning and only 10 corresponded to a –CG source. According to the calculations of *Jones* [*1970b*], the duration is typical of 'normal' lightning at this range but the amplitudes (300 mV/m) are about ten times larger. The millisecond-scale of the electromagnetic signal does not correspond to the duration of the associated elves (which are shorter) or to sprites (which are longer and delayed in time).

Cummer et al. [*1998*] observed sferic signatures in a 5 Hz – 22 kHz bandwidth at Stanford, CA, 2,150 km from the parent storm system. Simultaneous photometer measurements of the high altitude optical sprites were made at Yucca Ridge, CO. Here the first evidence of electromagnetic signatures generated by in-sprite electrical currents is presented. The slow tail sferic waveforms had a characteristic 'double-hump' waveform with durations of 20 ms and a 4 ms separation between the two peaks. The authors infer *in-sprite* current moments of around 200 MAm, leading to a charge moment change of around 2,000 C km on a time-scale of around 10 ms. The order of magnitude here is similar to that obtained experimentally by *Burke and Jones* [*1996*] and *Huang et al.* [*1999*] using ELF event sferics and considered to be due to continuing currents in the parent discharge in the first reference. Earlier, *Cummer and Inan* [*1997*] had deduced that 25 to 325 C of charge were transferred during the first 5 ms of the parent +GC discharge (ΔM_Q = 250 – 3,250 C km for an assumed 10 km channel length).

6. Conclusions

The existence of slow tail sferics has been known for over 6 decades. The majority of these are well accounted for by accepted 'normal' lightning-stroke characteristics. The lower ELF band transient events have been the subject of study for 3 decades, with greatly increased activity in the past 10 years. There continues to be speculation as to the nature of the source currents which produce these signals which are so large that, for a time interval of the order of 100 ms, the electromagnetic signature dwarfs that of the whole global lightning distribution, even when the source is 10,000 or more kilometers distant.

The 1990s have seen the recognition of the various high-altitude luminous phenomena known as sprites, elves and blue jets. It appears that these are produced exclusively by +CG lightning and especially (but not exclusively) above mesoscale convective storm systems. Recent work shows that both in-sprite currents and continuing currents in the +CG discharge can produce ΔM_Q values of the order measured experimentally. It should be recalled, however, that a significant proportion of ELF events (and most slow tails) are generated by –CG discharges. The negative events involve somewhat lower values of ΔM_Q than positive events, but in the –CG case no sprites are generated.

Acknowledgements

Consultations with W. Lyons and the assistance of R. L. Dowden, U. S. Inan, A. P. Nickolaenko, K. Papadoupoulos, C. J. Rodger, H. L. Rowland and E. R. Williams in promptly supplying reprints of their groups' papers is gratefully acknowledged.

References

Barr, R. [1970], "The VLF and ELF Amplitude Spectrum of Amospherics," *J. Atmos. Terr. Phys.*, 32, pp. 977-990.

Boccippio, D. J., E. R. Williams, S. J. Heckman, W. A. Lyons, I. T. Baker and R. Boldi [1995], "Sprites, ELF Transients, and Positive Ground Strokes," *Science*, 269, pp. 1088-1091.

Boccippio, D. J., C. Wong, E. R. Williams, R. Boldi, H. J. Christian and S. J. Goodman [1998], "Global Validation of Single-Station Schumann Resonance Lightning Location," *J. Atmos. Sol-Terr. Phys.*, 60, pp. 701-712.

Burke, C. P. and D. Ll. Jones [1992], "An Experimental Investigation of ELF Attenuation Rates in the Earth-Ionosphere Duct," *J. Atmos. Terr. Phys.* 54, pp. 243-250.

Burke, C. P. and D. Ll. Jones [1995], "Global Radiolocation in the Lower ELF Frequency Band," *J. Geophys. Res.*, 100, (D12), pp. 26,263-26,272.

Burke, C. P. and D. Ll. Jones [1996], "On the Polarity and Continuing Currents in Unusually Large Lightning Flashes Deduced from ELF Events," *J. Atmos. Terr. Phys.*, 58, pp. 531-540.

Cummer, S. A. and U. S. Inan [1997], "Measurement of Charge Transfer in Sprite-Producing Lightning using ELF Radio Atmospherics," *Geophys. Res. Lett.*, 24, pp. 1731-1734.

Cummer, S. A., U. S. Inan , T. F. Bell and C. P. Barrington-Leigh [1998], "ELF Radiation Produced by Electrical Currents in Sprites," *Geophys. Res. Lett.*, 25, pp. 1281-1284.

Fukunishi, H., Y. Takahashi, M. Kubota, K. Sakanoi, U. S. Inan and W. A. Lyons [1996] "Elves: Lightning-Induced Transient Luminous Events in the Lower Ionosphere," *Geophys. Res. Lett.*, 23, pp. 2157-2160.

Füllekrug, M., S. C. Reising and W. A. Lyons [1996], "On the Accuracy of Arrival Azimuth Determination of Sprite-Associated Lightning Flashes by Earth-Ionosphere Cavity Resonances," *Geophys. Res. Lett.*, 23, pp. 3691-3694.

Füllekrug, M. and S. C. Reising [1998], "Excitation of Earth-Ionosphere Cavity Resonances by Sprite-Associated Lightning Flashes," *Geophys. Res. Lett.*, 25, pp. 4145-4148.

Galejs, J. [1972], *Terrestrial Propagation of Long Electromagnetic Waves*, New York, Pergamon Press.

Holzworth, R. W. [1996], "Introduction to Special Section: Atmospheric Electricity: An IUGG Symposium," *J. Geophys. Res.*, 101, D23, pp. 29,607-29,608.

Huang, E. [1997], "Electromagnetic Transients, Elves and Sprites in the Earth-Ionosphere Waveguide," *M. Eng. Thesis*, Massachusetts Institute of Technology, Cambridge, USA.

Huang, E., E. Williams, R. Boldi, S. Heckman, W. Lyons, M. Taylor, T. Nelson and C. Wong [1999], "Criteria for Sprites and Elves Based on Schumann Resonance Observations," *J. Geophys. Res.* In press.

Ishaq, M. and D. Ll. Jones [1977], "Methods of Obtaining ELF Propagation Parameters for the Earth-Ionosphere Duct at ELF," *Electronics Lett.*. 13, pp. 254-255.

Jackson, J. D. [1998], *Classical Electrodynamics* (3rd Edition), New York, John Wiley, pp. 374-378.

Jones, D. Ll. [1967], "Schumann resonances and ELF Propagation for Inhomogeneous, Isotropic Ionosphere Profiles," *J. Atmos. Terr. Phys.*, 29, pp. 1037-1044.

Jones, D. Ll. [1970a], "Numerical Computation of Terrestrial ELF Electromagnetic Wave Fields in the Frequency Domain," *Radio Sci.*, 5, pp. 803-809.

Jones, D. Ll. [1970b], "Propagation of ELF Pulses in the Earth-Ionosphere Cavity and Application to 'Slow Tail' Atmospherics," *Radio Sci.*, 5, pp. 1153-1162.

Jones, D. Ll. [1984], "The Interpretation of ELF Propagation Data Using a Full-Wave Matrix Method," *Technical Report RP840823*, pp. 16. (Available from the author).

Jones, D. Ll. and C. P. Burke [1990], "Zonal Harmonic Series Expansion of Legendre Functions and Associated Legendre Functions," *J. Phys. A: Math. Gen.*, 23, pp. 3159-3168.

Jones, D. Ll. and C. P. Burke [1995], "ELF Radio," in *I.E.E. Conference Publication 411, "100 Years of Radio,"* I.E.E., Savoy Place, London, pp. 101-106.

Kemp, D. T. and D. Ll. Jones [1971], "A New Technique for the Analysis of Transient Electromagnetic Disturbances in the Earth-Ionosphere Cavity," *J. Atmos. Terr. Phys.*, 33, pp. 567-572.

Lyons, W. A. [1996] "Sprite Observations above the US High Plains in Relation to their Parent Thunderstorm Systems," *J. Geophys. Res.*, 101, pp. 29,641-29,652.

Ma, Z., C. L. Croskey and L. C. Hale [1998], "The Electrodynamic Responses of the Atmosphere and Ionosphere to the Lightning Discharge," *J. Atmos. Sol.-Terr. Phys.*, 60, pp. 845-862.

Marshall, T. C. and W. D. Rust [1993], "Two Types of Vertical Electrical Structures in Stratiform Precipitation Regions of Mesoscale Convective Systems," *Bull. Am. Met. Soc.*, 74, pp. 2159-2170.

Marshall, L. H., L. C. Hale, C. L. Croskey and W. A. Lyons [1998], "Electromagnetics of Sprite- and Elve-Associated Sferics," *J. Atmos. Sol.-Terr. Phys.*, 60, pp. 771-786.

Nickolaenko, A. P. and I. G. Kudintseva [1994], "A Modified Technique to Locate the Sources of ELF Transient Events," *J. Atmos. Terr. Phys.*, 56, pp. 1493-1498.

Nickolaenko, A. P. [1997], "Modern Aspects of Schumann Resonance Studies'" *J. Atmos. Sol.-Terr. Phys.*, 59, pp. 805-816.

Ogawa, T., Y. Tanaka, A. C. Fraser-Smith and R. Gendrin [1967], "Worldwide Simultaneity of Occurrence of a Q-type ELF Burst in the Schumann Resonance Frequency Range," *J. Geomag. Geoelec.*, 19, pp. 377-385.

Pasko, V. P., U. S. Inan and T. F. Bell [1996], "Blue Jets Produced by Quasi-Electrostatic Pre-Discharge Thundercloud Fields," *Geophys. Res. Lett.*, 23, pp. 301-304

Pinto, O. Jr., R. B. B. Gin, I. R. Pinto, O. Mendes Jr., J. H. Diniz and A. M. Carvalho [1996], "Cloud-to-Ground Lightning Flash Characteristics in Southeastern Brazil for the 1992-1993 Summer Season," *J. Geophys. Res.*, 101, pp. 29,627-29,635.

Rodger, C. J. [1999], "Red Sprites, Upward Lightning and VLF Perturbations," *Rev. Geophys.*, In press.

Rowland, H. L. [1998], "Theories and Simulations of Elves, Sprites and Blue Jets," *J. Atmos. Sol.-Terr. Phys.,* 60, pp. 831-844.

Sentman, D. D. and E. M. Westcott [1993], "Observations of Upper Atmospheric Optical Flashes Recorded from and Aircraft," *Geophys. Res. Lett.*, 20, pp. 2867-2860.

Sentman, D. D. [1995], "Schumann Resonances," in H. Volland (ed.), *Handbook of Atmospheric Electrodynamics*, Boca Raton, CRC Press, pp. 267-310.

Sentman, D. D., E. M. Westcott, D. L. Osborne, D. L. Hampton and M. J. Heavner [1995], "Preliminary Results from the Sprites94 Aircraft Campaign: 1. Red Sprites," Geophys Res. Lett., 22, pp. 1205-1208.

Sentman, D. D. [1996], "Schumann Resonance Spectra in a Two-Scale-Height Earth-Ionosphere Cavity," *J. Geophys. Res.*, 100, pp. 7097-7103.

Sentman, D. D. [1998], "Effects of Thunderstorm Activity on the Upper Atmosphere and Ionosphere: Special Issue of the J. of Atmospheric and Solar-Terrestrial Physics," *J. Atmos. Sol.-Terr. Phys.,* 60, p. 667.

Strangeways, H. J. [1996], "Lightning, Trimpis and Sprites," in W. R. Stone (ed.), *Rev. Radio Science 1993-1996*, OUP, pp. 741-780.

Uman, M. A., [1987], *The Lightning Discharge*, Orlando FL, Academic Press.

Wait, J. R. [1970] *Electromagnetic Waves in Stratified Media (2nd ed.)*, New York, Pergamon Press. (Alternatively see [1996] 3rd *ed.*, New York, IEEE Press & Oxford University Press.)

Williams, E. R. [1998], "The Positive Charge Reservoir for Sprite-Producing Lightning," *J. Atmos. Sol-Terr. Phys.*, 60, pp. 689-692.

REMOTE CHARACTERIZATION OF GEOPHYSICAL PHENOMENA USING EM WAVES

Dag T. Gjessing

1. Introduction

In recent years we have witnessed a very impressive progress in radio science directed towards radio communication challenges. New powerful on-line signal processing systems involving adaptive coding and decoding have increased the rate of information transfer by orders of magnitude. It is the purpose of this contribution to present a unified approach to the remote sensing challenge based on general coding and signal processing procedures. Specifically, based on a general multistatic EM system (forward scattering in general, back-scattering as a special case) we shall focus the attention on geophysical phenomena (air, sea, ground).

Air:	Wind, turbulence, waves and vortexes.
Sea:	Surface current, wind-driven gravity waves, internal waves.
Ground:	Geological structure, structure of glaciers.

We shall endeavor to characterize the wide spectrum of geophysical phenomena with a unified set of concepts based on the same fundamental EM equations thus hoping to convey a physical understanding of the basic principles on which the remote characterization of geophysical processes are based.

For the purpose of unifying the approach also with regard to the experimental set up, we shall base our geophysical surveillance system on components that are well established among communication scientists as shown in Fig. 1. Based on this, we shall characterize the surveillance arena in 3D space and in time measuring the spatial correlation properties of the scattered signal $R(\Delta L)$, the temporal correlation $R(\Delta\tau)$ and the correlation properties in the frequency domain $R(\Delta F)$, the bandwidth.

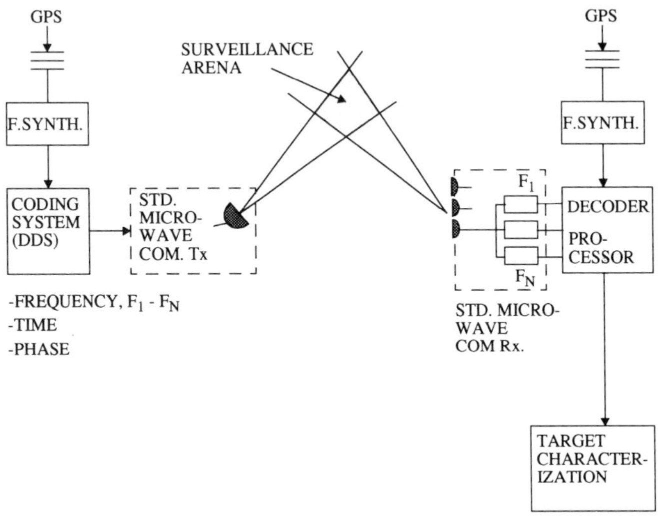

Fig. 1 *The surveillance system is based on well established radio communication techniques. Based on such a radio communication link we shall characterize (identify) the scattering phenomenon using frequency coding, time coding and space diversity reception in 3D space and in time.*

2. Brief Presentation of the Basic Theory

In this section we shall develop the basic equations from first principles so as to end up with a set of fundamental equations. Based on these together with the experimental set-up depicted in Fig. 1, we can characterize the geophysical phenomena of interest. For the purpose of confining this presentation to a minimum of space, let us focus on the non-ionized atmosphere when we develop the basic equations.

Consider a volume dv within which permittivity ε and/or field strength $\mathbf{E}_o$ vary. For a plane wave incident on the scattering volume, we have:

$$\mathbf{E}_0 = \mathbf{E}_1 e^{j(\omega t - \mathbf{k}_1 . \mathbf{x})} \tag{1}$$

where $\mathbf{k}_1$ is the wave number of the incident wave and the distance $\mathbf{x}$ from transmitter. If $\mathbf{k}_s$ is the wave number of the scattered wave and θ is the scattering angle (angle between $\mathbf{k}_1$ and $\mathbf{k}_s$), we can define a difference wave vector:

$$\mathbf{K} = \mathbf{k}_1 - \mathbf{k}_s \quad \left(|\mathbf{K}| = \frac{4\pi}{\lambda} \sin\theta/2 \right) \tag{2}$$

If the permittivity ε or the field strength $\mathbf{E}_o$ within the elementary scattering volume dv differs from the average values of ε and $\mathbf{E}$, a dipole moment is setup:

$$d\mathbf{P} = \Delta\varepsilon dv \mathbf{E}_o \;= \varepsilon \mathrm{dv} \mathbf{E}_1 e^{j(\omega t - \mathbf{k}_1 . \mathbf{x})} \tag{3}$$

At the distance $\mathbf{R}$ from this dipole moment we will have a polarization potential $\mathbf{\Pi}$, given by:

$$d\Pi = \frac{\Delta\varepsilon dv \mathbf{E}_1}{\varepsilon 4\pi R} e^{j(\omega t - \mathbf{k}_1 . \mathbf{x} - \mathbf{k}_s \mathbf{R})} \tag{4}$$

Here $\mathbf{R}$ is the distance along $\mathbf{k}_s$ from the scattering volume dv.

Integrating this over all contributions to the polarization potential within the total volume V illuminated by the transmitter and seen by the receiver (the scattering volume), we have:

$$\Pi = \frac{1}{4\pi R} \int f_\varepsilon(\mathbf{r}) \mathbf{E}_1(\mathbf{r}) e^{j(\omega t - \mathbf{K}\cdot\mathbf{r})} d^3\mathbf{r} \tag{5}$$

where $\mathbf{r}$ is a position vector, such that

$$\mathbf{k}_1 \cdot \mathbf{x} + \mathbf{k}_s \cdot \mathbf{R} = \mathbf{K} \cdot \mathbf{r} \tag{6}$$

where $\mathbf{K} = \mathbf{k}_1 - \mathbf{k}_s$.

Knowing the polarization potential $\mathbf{\Pi}$, we can calculate the field strength $\mathbf{E}_s$ from the well-known relationship:

$$\mathbf{E}_s = \nabla\nabla \cdot \mathbf{\Pi} + \mathbf{k}^2 \mathbf{\Pi} \tag{7}$$

Provided the scattering volume (spatial region illuminated by the radar) is small in comparison with the distance $\mathrm{R} (\mathrm{R} >> \mathrm{V}^{1/3})$:

$$\mathbf{K}^2 \mathbf{\Pi} >> \nabla\nabla \cdot \mathbf{\Pi} \tag{8}$$

Such that the scattered field $\mathbf{E}_s$ is given by:

$$\mathbf{E}_s \approx k^2 \mathbf{\Pi} \tag{9}$$

The scattered field resulting from an integral of elementary scattering elements is given by:

$$\mathbf{E}_s = \frac{k_s^{\,2}}{4\pi R} \int_V \mathbf{E}_1(\mathbf{r}) f_\varepsilon(\mathbf{r}) e^{-j\mathbf{K}\cdot\mathbf{r}} d^3\mathbf{r} \tag{10}$$

omitting the time factor e^{jwt}. The scattered/diffracted signal is the convolution of the permittivity spatial distribution $f_\varepsilon(\mathbf{r})$ and that of the electric field $\mathbf{E}_1(\mathbf{r})$.

Note that Equation 10, which is derived from Maxwell's equations, is perfectly general and does not consider the nature of the scattering object. The $\mathbf{E}_1(\mathbf{r})$ and $f_\varepsilon(\mathbf{r})$ functions may be stochastic, in which case statistical descriptions must be used (spatial auto correlation functions, spatial spectra), or we may be dealing with ordered variations, in which case well-behaved analytical functions may be used. This simple and intuitively obvious expression is based on the approximations of Born and Kirchhoff.

2.1 Correlation Properties of the Scattered EM Field

We define the function f(z) as the *delay function.* This has dimension field strength such that the scattering cross section as a function of distance z along the direction of propagation (z is measured along the direction of $\mathbf{K}$) is obtained by squaring the f(z) function. From Equation 10, we have:

$$\mathbf{E}_s(\mathbf{K}) \sim \int f(\mathbf{z}) e^{-j\mathbf{K}\cdot\mathbf{z}} d^3\mathbf{z} \tag{11}$$

where $\mathbf{K} = \omega/c$.

If we are dealing with a scattering medium that can only be determined in statistical terms such as the turbulence field, the delay function $f(\mathbf{z})$ should be replaced by the auto correlation function $\mathrm{R}(\Delta z)$. This relationship between the measured quantity $\mathbf{E}(\mathbf{K}) \sim \mathbf{E}(F)$ and the parameters describing the scattering arena is obtained by evaluating the auto correlation function also in the frequency domain:

$$R(\Delta K) = E^*(K)E(\mathbf{K}+\Delta\mathbf{K}) \sim \int \boldsymbol{R}_{\sigma}(\Delta z) e^{-j\Delta K \Delta z} d\Delta z \tag{12}$$

Note that this simple equation does not include the ambiguity phase term e^{-jKz} where K is the wavenumber associated with the carrier frequency F and z is the distance to the scattering arena. There also is a $e^{-jK\Delta z}$ phase term. The influence of this is negligible if either the carrier frequency F or the scattering object is randomized slightly.

In summing up this section on the relationship between the scattering medium (the geophysical processes) and the measurable quantities of the EM field, figure 2 is presented.

We characterize the scattering arena in terms of a 4 dimensional wavenumber spectrum, 3 in space and one in time, $A(\mathbf{K},\omega)$. The power spectrum $\Phi(\mathbf{K},\omega)$ is obtained by computing $A(\mathbf{K},\omega)A^*(\mathbf{K},\omega)$.

When dealing with stochastic geophysical processes such as atmospheric turbulence or sea surface gravity wave phenomena, we shall use a simple expression for the wavenumber spectrum (the Fourier transform of the spatial autocorrelation function of the scattering cross-section)

$$\Phi(\mathbf{K},\omega) \sim K^{-n} \tag{13}$$

if we are dealing with inertial sub range turbulence $n = 11/3$.

We shall now seek methods that enable us to measure the spatial wavenumber spectrum. First, let us consider the functions in time domain bearing in mind that the spatial function $f(z) = f(c\tau)$ where c is the phase velocity of the EM wave and τ the time.

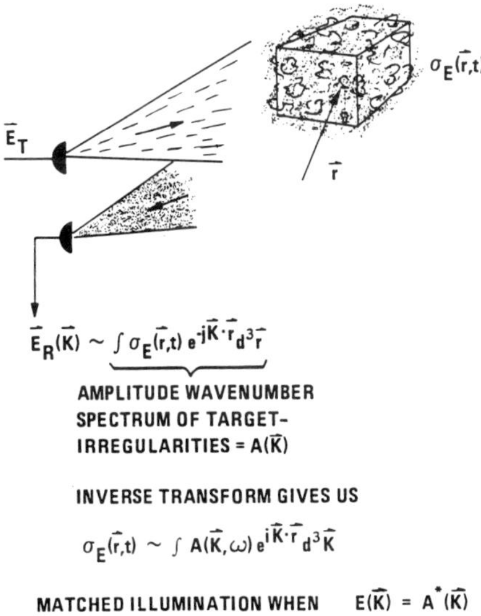

Fig. 2 *Any scattering object characterized by distribution in space of the scatters (can be decomposed into its Fourier components A(K). We will have constructive interference from all this scatterers if we choose a combination of illuminating sine waves $E_T(K)$ that couple to each of Fourier components A(K) of the scattering object.*

2.1.1 Characterization of the Scattering Arena by Manipulations in *Time Domain R(Δτ)*, Delay Function Considerations.

Using wide-beam antenna so that the multipath transmission is governed by the scattering mechanism rather than by the beam geometry, we first seek an expression relating path length

ℓ and the position in space of the scattering element, i.e., we require an expression relating ℓ and the scattering angle θ (figure 3). If d is the length of the chord between T and R. Simple geometric calculations give the required results:

$$\theta = 2\left|(\ell / d)^2 - 1\right|^{\frac{1}{2}} \tag{14}$$

If we transmit a short radio pulse, the power that reaches the receiver has traveled through a wide variety of different paths. By substituting for θ in the expression for the angular power spectrum $(P \sim \theta^{-n})$, we get the spectrum relating power and path length.

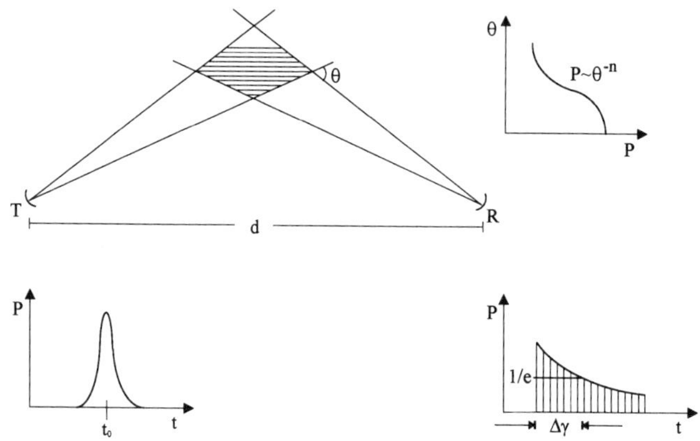

***Fig.* 3** *The delay spectrum is determined by the variation in path length.*

Expressing this spectrum in terms of the path length $\Delta\ell$, which is in excess of the minimum path length ℓ_o (i.e., writing $\ell = \ell_0 + \Delta\ell$), we find that the power spectrum referred to τ is given by:

$$\frac{P(\tau)}{P(0)} = \left(1 + \frac{8c\tau}{\theta^2}\right)^{-n/2} \tag{15}$$

Where we have transferred from space ℓ to time τ by $\ell = c\tau$.

The 10 dB width of this delay function P(τ) which is derived from the spatial wavenumber spectrum of the form $\Phi(K) \sim K^{-n}$ is then:

$$\Delta\tau_{1/10} = \frac{\theta^2 d}{8c}(10^{1/n} - 1) \tag{16}$$

Hence, the width of the temporal autocorrelation function $R(\Delta\tau)$ resulting from a K^{-n} wavenumber spectrum is given in equation 16. Now let us consider the correlation properties in the frequency domain $R(\Delta F)$ from which we can estimate the size distribution $\Phi(2\pi/L)$ of the scattering arena.

2.1.2 Characterization of the Scattering Arena by Manipulations in the Frequency Domain R(ΔF), Bandwidth Considerations

We can use the result of the simple network theory to obtain a simple approximate result. We know that the response to the delta function of a network is known as the impulse response V(τ) of the network. Furthermore, the Fourier transform of the impulse response is known as the transfer function F(ω) of the network. By multiplying this transfer function with its complex conjugate, we obtain the power spectrum that we are seeking. From the previous section we obtain the expression for the impulse response.

In order to obtain a simple first order estimate of the bandwidth from which we can characterize the scattering arena, we write:

$$\Delta F = \frac{1}{\Delta\tau} = \frac{8c}{\theta^2 d}(10^{1/n} - 1)^{-1} \tag{17}$$

The manipulation in the frequency domain $R(\Delta F)$ obtained by multiplying $\mathbf{V}(F)$ with $\mathbf{V}^*(F+\Delta F)$ enables us to couple to scales in the scattering arena of $\Delta L = 2\pi/\Delta K = c/\Delta F$ rather than scales $\ell = c/F$. This technique is often referred to as ΔK principles or frequency interferometry.

Finally, let us consider the spatial correlation properties of the scattered signal $R(\Delta L)$.

2.1.3 Characterization of the Scattering Arena by Manipulations in the Space Domain R(ΔL), Space Diversity Considerations

As shown above, this spatial field-strength correlation function is the Fourier transform of the angular power spectrum of the wave reaching the receiving antennas. Specifically, if the antennas are spaced horizontally, thus giving us information about the correlation properties of field strength along a horizontal direction, this correlation function is determined by the angle-of-arrival spectrum as measured in the horizontal plan. If the antennas are spaced vertically, it is the angle-of-arrival spectrum in the vertical plan that matters. Specifying the angle-of-arrival of a particular scattered wave by an elevation angle α (relative to the center line through T and R) and an azimuth angle β (relative to a great circle plane through T and R) we have in the same manner as above that a refractive-index spectrum $\Phi(K) \sim K^{-n}$ gives rise to an angular spectrum of form:

$$P(\alpha,\beta) \sim (\beta^2 + \alpha^2)^{-n/2} \tag{18}$$

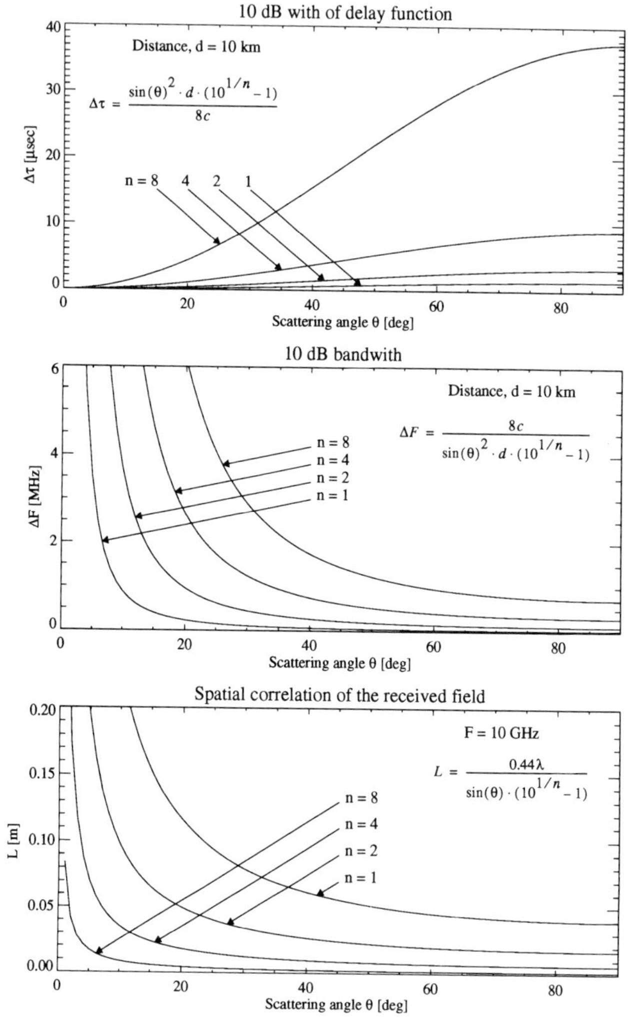

***Fig. 4** The correlation properties in time, frequency and space.*

The horizontal correlation of field strength is thus obtained by a Fourier transformation of P with respect to β, whereas vertical correlation is obtained from $P(\alpha)$ relationship.

A rigorous Fourier transform of equation 18 however lends itself to numerical computations only. In our case, we need a simple approximate expression. This can be obtained if we approximate $P(\alpha)$ and $P(\beta)$ by a $\sin(x)/x$ function, thus giving us a simple expression for the Fourier transform. This is a procedure well known in antenna theory. From antenna theory we know that if L is the width of the illuminating field strength distribution, the width of the resulting angular power spectrum (beam width) is given by:

$$\theta_{1/2} = 0.88\lambda / L \tag{19}$$

where λ = the radio wavelength.

From such considerations, we can obtain approximate expressions for the horizontal and vertical correlation distances:

$$L_H = \frac{0.44\lambda}{\theta\left(4^{1/n}-1\right)^{1/2}} \tag{20}$$

$$L_V = \frac{0.44\lambda}{\theta\left(2^{1/n}-1\right)} \tag{21}$$

2.2 Scattering From an Object that can be Characterized Deterministically as $\sigma(\mathbf{r})$

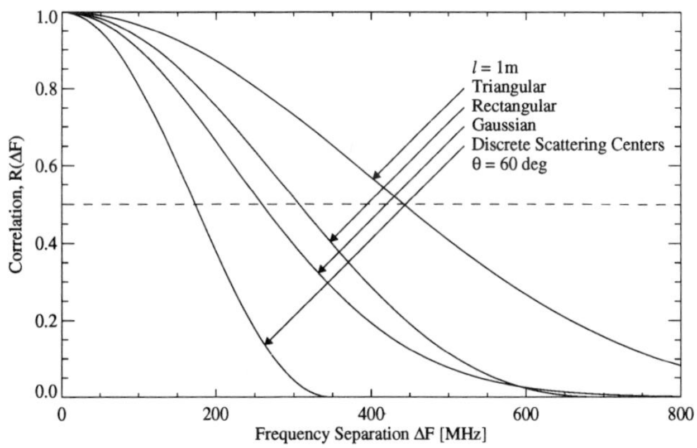

***Fig. 5** The autocorrelation function of different scattering distributions.*

For the sake of completeness of this simple approach aiming at characterizing a scattering arena by coding the signal in time, frequency and space in the way of a target adaptive matched illumination, we shall present some illustrative examples with reference to Fig. 2, where we have shown that there is a very simple relationship between the measured quantity $\mathrm{E}(K)$ and $\sigma(\mathbf{r})$. An illustration is given in Fig. 5 for a 1m long target of different shape.

2.3 Motion Pattern Considerations, Doppler

We have shown that if a rigid body is illuminated by a set of electromagnetic waves having different frequencies, information about the distribution in depth (along the direction of wave propagation) is obtained by studying the correlation properties of the scattered waves. Specifically, if two electromagnetic waves with frequency separation ΔF are illuminating the object, we obtain information about a particular irregularity scale L (spatial spectrum component $K = 2\pi/L$) where L is related to frequency separation ΔF, as follows:

$$L = \frac{c}{2\Delta F} \tag{22}$$

Let us now assume that a rigid body characterized by the delay function f(z) (distribution in depth of the scattering elements) is moving with velocity $\mathbf{V}$. We now want information about the Doppler shift to which this frequency ΔF is subjected. We base this calculation on the basic Doppler equation:

$$f = \frac{1}{2\pi}\mathbf{K}\cdot\mathbf{V} \tag{23}$$

Here $K = 2\pi/L$ where L is the scale-size to which the frequency pair ΔF is matched. The Doppler shift associated with the difference frequency ΔF is therefore:

$$f = \frac{1}{2\pi}\Delta\mathbf{K}\cdot\mathbf{V} = \frac{1}{2\pi}\frac{2\pi}{L}V\cos\phi = \frac{2\Delta F}{c}V\cos\phi \tag{24}$$

where ϕ is the angle between the wave vector **K** and the velocity vector **V**.

Hence, if we illuminate a moving object with two electromagnetic waves with frequency spacing ΔF (coupled to scale size $L = c/2\Delta F$), this frequency ΔF is subjected to a Doppler shift f which is proportional to ΔF and to the velocity V of the object. The power associated with the Doppler frequency f is the same as that associated with the frequency ΔF.

Thus, knowing the shape and velocity of the target (Fig. 5), the Doppler spectrum can be calculated.

3. Some Specific Illustrations Focused on Air, Sea and Land

In the preceding section the attention has been focused on a unified approach to the characterization of geophysical phenomena. We have shown that by the use of a set of coherent frequencies organized in a matched illumination fashion (synthetic pulse) and similarly spaced antennas (synthetic aperture) paying attention to amplitude and phase so as to get Doppler information, we can characterize the scattering arena (scattering object) in 3D shape and size of the scattering phenomenon and the velocity (Doppler) associated with each scale $\ell = 2\pi / K$ and $L = 2\pi / \Delta K$. We shall now use this simple and unfined approach to characterize geophysical phenomena (air, sea and ground).

As an introduction to this section it may from a tutorial point of view serve the purpose to "calibrate" our concepts by using objects of deterministic shape and size. For this purpose we use a "Gaussian shaped" air target F16 of length 16 meters and a rectangular ship of length 37.5 m. These are shown in Fig. 6 and 7.

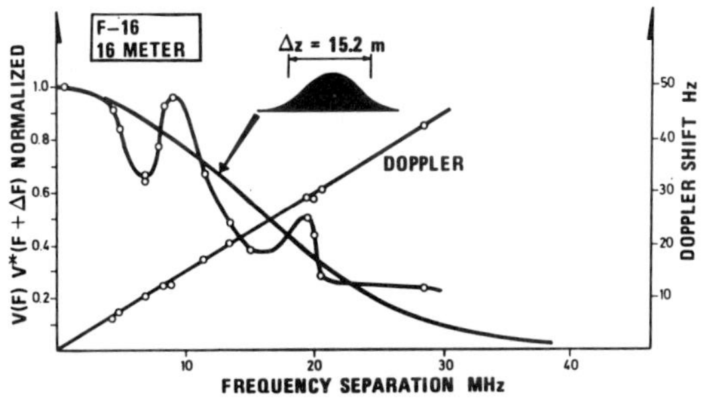

Fig. 6 Illuminating a fighter airplane with a suitable distribution of beat frequencies, the shape and the size can be established. Note that for = 7 MHz two different carriers (6 GHz and 4 GHz) were used giving the same R().

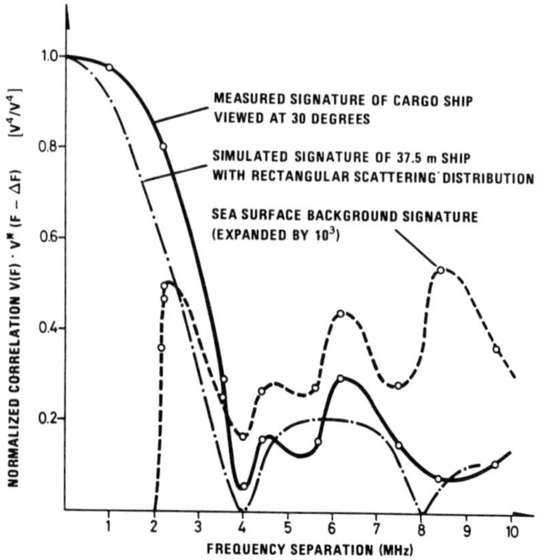

***Fig.** 7 The ship size (37.5 meters) is well determined experimentally. The experiments also suggest that the scatters are evenly distributed from bow to stern so as to given a signatures of form (sin x/x)^2.*

3.1 Characterization of Atmospheric Phenomena by a Bistatic System Coupling to Dominating Scales by Varying the Scattering Angle θ

The current overview aims at giving a broad outline of the physical principles involved and an estimate of the relative potential of the various methods aiming at measuring wind and temperature in the lower troposphere from ground up to a few kilometers altitude. To serve the purpose of this highlighting presentation, Fig. 8 is presented

- Vortexes from aircraft
- Downdraft behind mountains
- Wind shear
- Cross glidepath wind
- Wind profiles

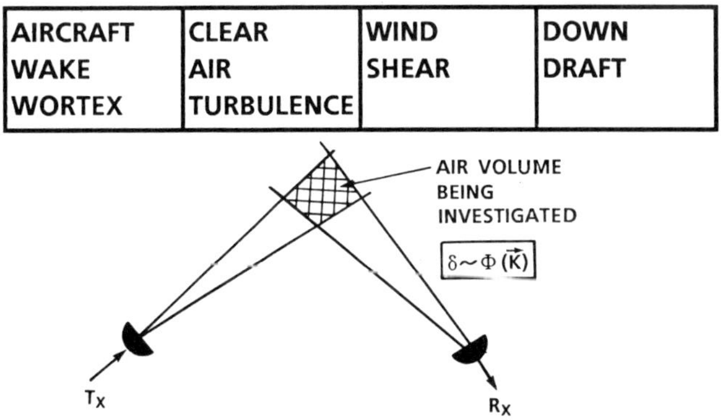

SIGNATURE DOMAINS

- **TRANSMISSION LOSS (SCATTERING CROSS SECTION)**
- **BANDWIDTH OF TRANSMISSION PATH**
- **DOPPLER SPECTRUM (MOTION PATTERN)**
- **SPATIAL CORRELATION OF SCATTERED FIELD**
- **FREQUENCY DEPENDANCE**
- **REFRACTION (BENDING OF RAY)**

Fig. 8 *Four different air dynamic classes can be characterized by 6 radio circuit parameters.*

The previous review, on which the current contribution is largely based, considered the influence on the quantities measured in trans-horizon propagation (correlation distance of field strength, bandwidth, pulse distortion, etc.) of the slope n of the refractive index spectrum, the refractive index gradient a, and the inhomogeneity factor v. The relationship between the measured quantities and n and v was calculated in a unified manner so as to enable us to compare the values of n deduced from the various experiments. Deducing then the spectrum slope n from various experiments, specifically beam swinging experiments, bandwidth measurements, measurement of vertical correlation distance of field strength, antenna coupling loss and two frequency experiments, it was found that the width of the essentially Gaussian n-distribution curves were very much the same for the different experiments. These distribution curves are shown in Fig. 9.

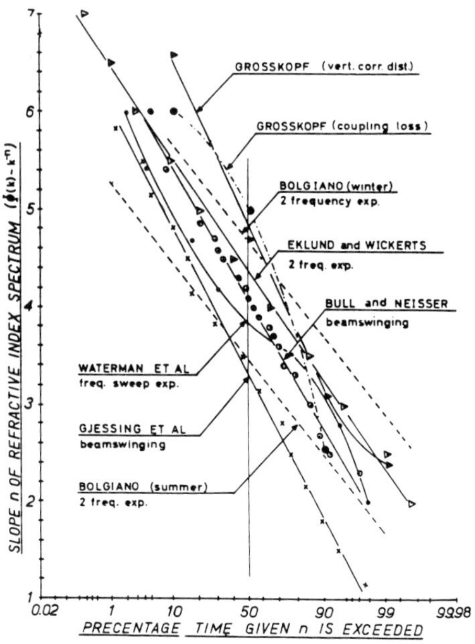

***Fig. 9** Probability distributions of the spectrum slope n as deduced from two-frequency and frequency sweep experiments.*

It should be emphasized that to review the rather large field under consideration within a reasonable number of pages and in order for the review to serve its purpose, we shall focus on typical phenomena.

From these sets of experimental verifications, we can draw the conclusion that the sets of basic equation based on more or less intuitively obvious physical concepts gives comparable results. From Fig. 9 we can draw the conclusion that the various degrees of freedom that we have at our disposal, all have the capability of characterizing the atmospheric structure in wavenumber domain $\Phi(K) \sim K^{-n}$.

Note also that the forward scattered signal gives us an extra degree of freedom, the scattering angle θ relative to the backscattered signal. For a scattering arena such as the sea surface or the turbulent atmosphere where the scattered signal:

$$P \sim (\sin\theta/2)^{-n} \tag{25}$$

the forward signal strength is enhanced by a factor $(\sin\theta/2)^{-n}$ relative to the backscattered signal. Thus for a scattering angle of some 20^0 we have a scattered signal that is 30 dB larger than that of the backscattered mode, for n = 4.

3.2 Characterization of Ocean Surface Structure by the use of Electromagnetic Waves

To illustrate the potential of our target adaptive matched illumination concept in relation to sea surface phenomena, let us focus the attention on 4 sea surface phenomena and confine ourselves to simple intuitively obvious phenomena before we present some experimental results:

a) Wind driven ocean gravity waves.

For deep water the dispersivity relationship is given by:

$$\omega^2 = gK \tag{26}$$

Phase velocity

$$v_p = \frac{\omega}{K} = \sqrt{\frac{g}{K}} \tag{27}$$

The phase velocity can not exceed the speed of the driver, the wind with speed V_{WIND} for fully developed sea and infinite fetch. Hence, we have a dominating phase velocity (and the corresponding wavenumber of the gravity waves K_0) given by:

$$v_p = V_{WIND} = \frac{\omega}{K} = \sqrt{\frac{g}{K}} \tag{28}$$

Hence, we shall have maximum wave energy $\Phi(K) \sim \textit{waveheight}\ H^2$ for a wavenumber $K_{SEA} = 2\pi/\textit{wavelength}$ given by:

$$V_{WIND} = \sqrt{\frac{g}{K_{SEA}}} \tag{29}$$

Hence

$$K_{SEA} = \frac{2\pi}{L_{SEA}} = \frac{g}{V_{WIND}^2} \tag{30}$$

From this $K_0 = g/V_{WIND}^2$ giving maximum wave height, the wave height H^2 is distributed roughly as K^{-4}.

As we have shown in section 2.1.2, this gives a 10 dB width of the autocorrelation function in the frequency domain $R(\Delta F)$ given by equation 17. This is illustrated in Fig. 10.

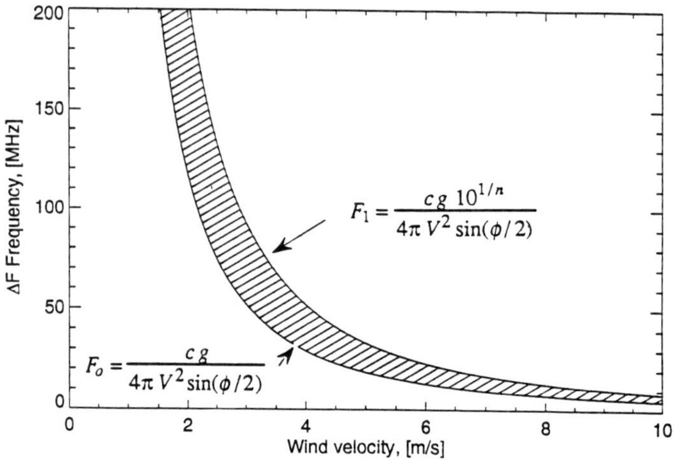

Fig. 10 *For a certain wind speed V there is a frequency (or beat frequency) band that couple to the sea surface structure for given scattering angle.*

The dispersivity (Doppler ω as a function of wavenumber K, see equation 26) is illustrated in Fig. 11. The wavelength H^2 as a function of wavenumber (wavelength) is illustrated in Fig. 12.

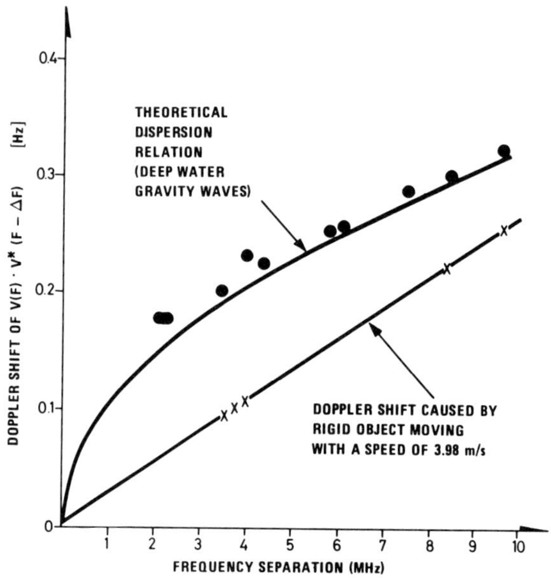

Fig. 11 *Doppler versus beat frequency for surface gravity waves and for a ship.*

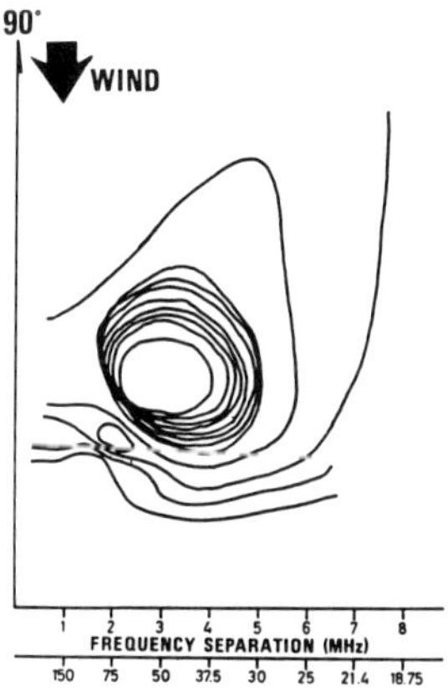

Fig. 12 *2D experimental results from radar experiments on surface gravity waves.*

b) Ship wakes

The ship wakes are viewed against the signature of the ship itself (Fig. 13) and against the background of wind driven gravity waves as characterized in the previous section. This is illustrated in Fig. 13 and 14.

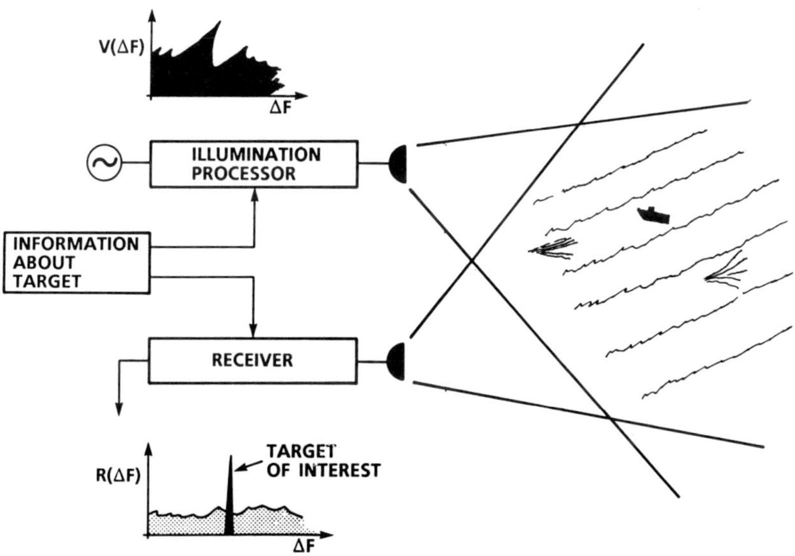

Fig. 13 *To detect and identify a small scattering cross-section phenomenon (gravity waves and internal waves, current, turbulence, ships and wakes) against a clutter background of no interest represents a considerable challenge.*

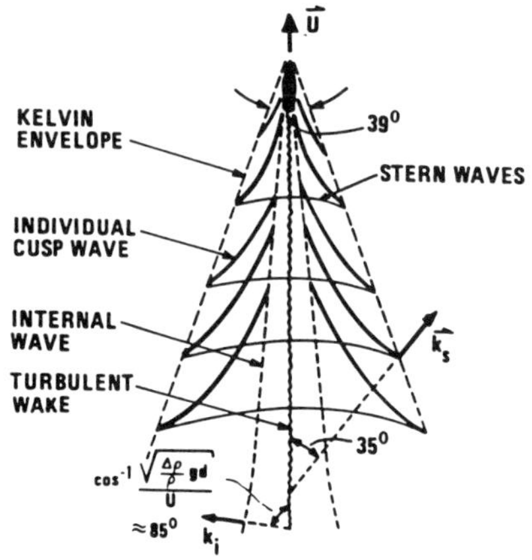

Fig. 14 *Many wave phenomena contribute to the ship wake signature: surface gravity waves, internal waves and turbulent waves.*

In summing up this section on sea surface phenomena, Fig. 15 is presented.

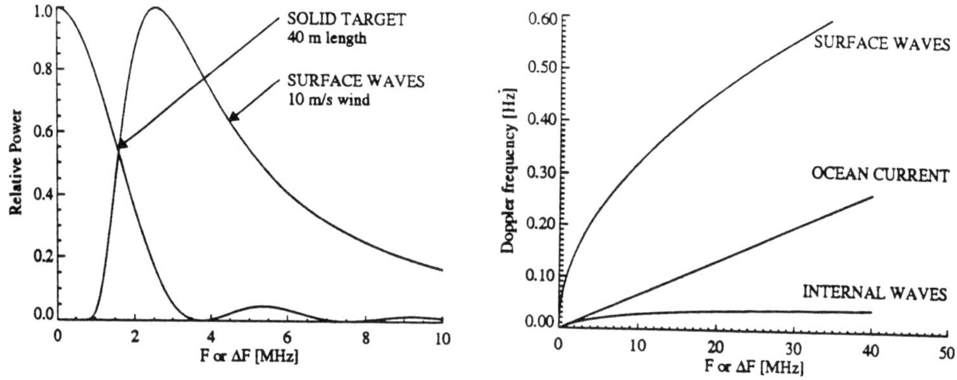

Fig. 15 *The various ocean phenomena manifests themselves clearly in space and time signatures.*

3.3 Characterization of Geological Structures by the use of Electromagnetic Waves

It is the endeavor of this section to characterize geological/glaciological structures using much the same technology as that for air and sea surface phenomena. The general concept is shown in figure 16. The transmitter illuminates a set of frequencies with total bandwidth ΔF_{MAX} giving a resolution in range for a monostatic configuration of

$$\Delta z = \frac{c}{2\Delta F} \tag{31}$$

The unambiguous depth is given by $\partial z = c/2\partial F$.

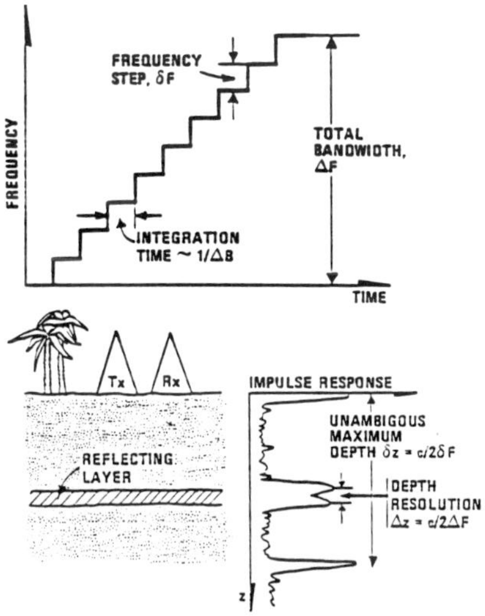

Fig. 16 The radar obtains the depth distribution through stepping the frequency and taking a Fourier transform.

We now characterize the geological structure in much the same way as that used for the atmosphere.

(a) We have a number of layers of a given thickness Δz.

(b) Within each layer we have an irregularity structure (turbulence, texture, conglomerate) that can be characterized as $\Phi(K) \sim K^{-n}$ and a variance in refractivity Δn.

This is illustrated in Fig. 17.

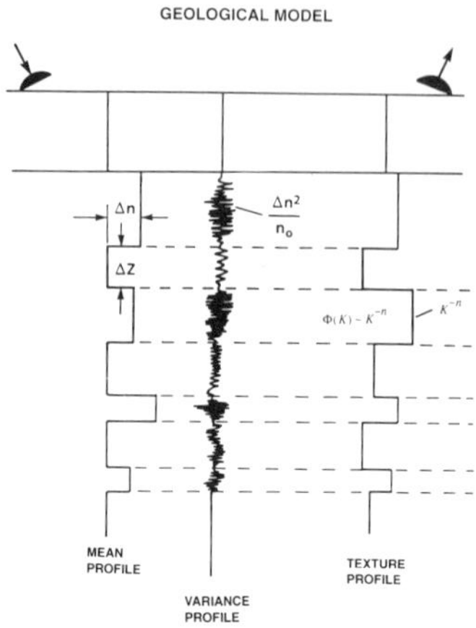

Fig. 17 The geological structure is modeled by a finite number of layers. Each layer is characterized by a given mean profile of wave velocity, by a variance profile and by a texture profile.

An example of the potential of this matched illumination technique is illustrated in Fig. 18. Hence, we have used 140 – 180 MHz illuminator stepping the frequency.

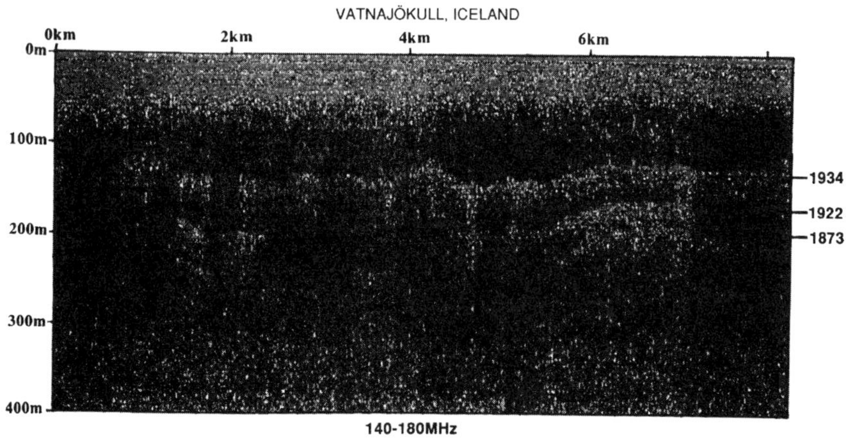

Fig. 18 The example is from the temperate glacier Vatnajökull, Iceland. Under the glacier there are volcano's which from time to time erupts and spread an ash layer over the surface. This will later be buried in the glacier. The profile show three such layers at different depths stemming from historically known eruptions. These layers can later be used to estimate the growth in the glacier between the different eruptions and thus maybe say something about the climate around the glacier [Hamram et al., 1991].

References

Apel, J., *Principles of Ocean Physics*, Academic Press, 1987.

Clifford, S. F. and D. E. Barrick [1978], "Remote sensing of sea state by analysis of backscattered microwave phase fluctuations," *IEEE Trans.*, AP-26.

Dysthe, K. T., [1981], "Havbølger og fysikk," *Fra: Fysikkens Verden*, No.4 and No.l.

Gjessing, D. T.,*Target AdaptiveMatched Illumination RADAR: Principles & Applications*, Peter Peregrinus Ltd., London, 1986.

Gjessing, D. T., *Adaptive Radar in Remote Sensing*, Ann Arbor Science Publisher Inc., USA, 1981.

Gjessing, D. T. [1967], "Radiophysical aspects of irregular structure in the atmosphere," in A. M. Yaglom (ed.) and V. I. Tatarsky (ed.), *Atmospheric Turbulence and Radio Wave Propagation*, Publishing House Nauka, Moscow.

Gjessing, D. T., *Surveillance by Electromagnetic Waves for Air-Water-Land*, Ann Arbor Science Publishers Inc., USA, 1978.

Gjessing, D. T. [1978], "A Generalized Method for Environmental Surveillance by Remote Probing," *J. Radio Sci.*, March/April.

Gjessing, D. T. and K. S. McCormick [1974], "On the Prediction of the Characteristic Parameters of Long-Distance Tropospheric Communication Links," *IEEE Trans. Cornm.*, 22, 9.

Gjessing, D. T. and J. Hjelmstad [1986], "Ocean waves and turbulence as observed with an adaptive coherent multifrequency radar," *J. Geophys. Res. Oceans*, January.

Gjessing, D. T., J. Hjelmstad, and T. Lund [1985], "Directional ocean wave spectra as observed with multi-frequency continuous wave radar", *Int. J. Remote Sensing,* 7, pp. 979-1008.

Hamran, S. E., D. T. Gjessing, J. Hjelmstad, and E. Aarholt [1995], "Ground penetration syntetich pulse radar: dynamic range and modes of aperture," *J. Appl. Geophysics*, 33, pp 7-14.

Hasselmann, K., et al. [1973], "Measurements of wind-wave growth and swell decay during the Joint North Sea Wave Project (JONSWAP)," *Ergänzungsheft zur Deutschen Hydroqraphischen Zeitshrift*, A/80, 12.

Ishimaru, A., *Wave Prapagation and Scattering in Random Media*, Academic Press, New York, 1978.

Panofsky, H. A. [1967], "Spectra of atmospheric variables in the boundary layer," *Radio Sci.*, 4, 12.

Phillips, O. M. [1974], "The equilibrium range in the spectrum of wind generated waves," *J. Fluid Mechanics*, 66.

Tatarsky, V. I., *Wave Propagation in a Turbulent Medium*, McGraw-Hil1, New York, 1961.

Waterman, A. T., D. T. Gjessing, and C. L. Liston [1961], "Statistical analysis of transmission data from a simultaneous frequency and angle scan experiment", *URSI Spring Meeting 1961*, Washington DC.

RADAR SYSTEMS IN IONOSPHERIC RESEARCH

Jürgen Röttger

1. Introduction

Since Earth's ionosphere is a highly dispersive medium for electromagnetic waves up to frequencies of some GHz, radio and radar methods have been extensively applied for scientific studies of this part of Earth's environment. The reasons for these studies are at least twofold, namely basic research to understand complex phenomena in our ambience and the communication oriented applications. A major contribution to our knowledge of the ionosphere stems from contributions of radar applications, operating in the mentioned frequency range. These will be briefly summarized in this tutorial by particularly taking into account most recent developments of systems, methods and their capabilities for ionospheric research. This paper is not a review of scientific achievements but a tutorial to explain briefly contemporary radar methods used for ionospheric research, flavored with a little scientific and applicational framework. We also have to limit ourselves to very few references, which are mainly of the most recent years. More details on scientific achievements can for instance be found in [*Dieminger et al., 1996*; *Kohl et al., 1996*] and will be published elsewhere in an extended paper version including figures, tables and more references.

When recapitulating the different types of ionospheric radars, it must be conceded that a specified nomenclature does not exist for the diverse radar methods. The given terms do not follow a unified standard definition, since they just had evolved during the development of these methods. We have used here the most common or perhaps acceptable terms to classify the relevant radars or radar methods. Further details of such systems can be found in [*Hunsucker*, *1991*] and in other contemporary publications. We will discuss the radar applications for studies of ionospheric and atmospheric structure and dynamics, plasma physics and ionospheric electrodynamics in more detail in subsequent sections.

The MF radars, also called Mesosphere-Lower-Thermosphere (MLT) radars, encompass partial reflection to determine electron density and collision frequency and wind-wave-turbulence measuring methods [*Reid et al.*, *1995*; *Roper and Brosnahan*, *1997*]. These radars cover the altitude region of about 60 km to 100 km. Meteor radars, covering 70 km to 120 km altitude, include techniques for studies of atmospheric / ionospheric [*Avery et al., 1990*] as well as astrophysical parameters [*Pellinen-Wannberg and Wannberg, 1996*; *Mathews et al., 1997*]. HF radars include the wide variety of the classical ionosondes, making use of total reflection, and their progressive digital evolutions to determine high-resolution profiles of ionospheric parameters [*Reinisch et al., 1997*]. They are operated from the ground for studies of the D-, E- and bottom-side F-region, and from satellites to observe the topside of the ionosphere [*Pulinets and Benson, 1999*] and are being designed for observing the magnetosphere [*Benson et al., 1998*]. HF radar encompass backscatter radars for distant observations of the ionospheric profile [*Russel et al., 1997*] and irregularities [*Villain et al., 1996*], imaging of ionospheric structure and dynamics, and its coupling to the magnetosphere [*Greenwald et al., 1995*; *Lester et al., 1997*] as well as the over-the-horizon (OTH) radars for detection and tracking of remote targets [*Headrick and Thomason, 1998*]. Ionospheric modifications, also called ionospheric heating [*Rietveld et al., 1993*], is done near ionospheric critical frequencies, i.e. in the HF band. It highly relies on radar methods for diagnostic purposes, operating in HF, VHF and UHF ranges [*Bond et al., 1997*]. They include the investigations of the modification of ionospheric background parameters and plasma turbulence [*Stubbe et al., 1992*] as well as the development of natural field-aligned [*Eglitis et al., 1998*] and artificial periodic irregularities [*Bakhmeteva et al., 1998*]. Heating facilities had been implemented as radars to try magnetospheric [*Hysell et al., 1997*] and mesospheric observations [*Karashtin et al., 1997*]. Mesosphere-Stratosphere-Troposphere (MST) VHF radars detect echoes from turbulence induced ionization irregularities [*Hocking, 1997a*] and dusty plasma layers in the D-region/mesosphere [*Cho and Röttger, 1997*]. The studies of the stratosphere and troposphere with MST radars are not subject of this tutorial [*Röttger and Larsen, 1990*]. It is also assumed that certain radar echoes from mesospheric ranges are related to electrical discharges [*Roussel-Dupre and Blanc, 1997*] and perhaps sprites.

Coherent scatter radars, ambiguously also called coherent radars, observe backscatter from ionospheric irregularities, which are aligned with Earth's magnetic field and for studies of the corresponding plasma instabilities [*Fejer and Kelley, 1980*; *Tsunoda, 1988*; *Haldoupis, 1989*; *Sahr and Fejer, 1996*]. The term field-aligned ionospheric irregularities (FAI) is frequently used, whereas most of these result from a spectrum of plasma waves, which are created by instabilities in the magnetized plasma. Incoherent scatter radars, detecting backscatter from the ionospheric plasma in thermal equilibrium, are the most powerful tools for studying the full profile of many ionospheric parameters, such as electron density, electron temperature, ion temperature and plasma velocity [*Foster, 1995*; *Nygren, 1996*; *Alcayde, 1997*; *Djuth et al., 1997*; *Heinselman et al., 1998*].

2. General Ionospheric Radar Principles

We will first discuss some features, which are common to all ionospheric radar techniques. Applications, which are special to certain systems, will be treated in the corresponding subsequent chapters.

2.1 Phase-Coherent Radar Systems to Observe Scattering and Reflection

Contemporary ionospheric radar systems are phase-coherent. The corresponding quadrature signal detection allows improvement of the signal-to-noise ratio, application of phase-coding, interferometry in the frequency and space domain as well as dedicated signal-processing routines. It is particularly required to obtain the full spectrum information of the scattered or reflected signals. It is not correct to equate coherent radars with coherent scatter radars, as is often done. We use the term coherent scatter when describing the process of scattering from coherent structures in the ionospheric plasma, such as irregularities with scales of the Bragg wavelength (i.e. half the radar wavelength for monostatic backscatter). Incoherent scatter results when the individual scatterers are randomly moving and distributed independently of each other, such as free electrons moving with thermal velocity in the ionospheric plasma. However, they are coupled in form of an ensemble to the ions in the plasma. The related ion-acoustic waves scatter radar signals; this means in strict terms that this scatter is coherent, too. These processes of coherent and incoherent scatter hold for radar frequencies, which are larger than the plasma frequency in the scatter volume. When the radar frequency is equal to the plasma frequency, the electron ensembles oscillate coherently with this plasma resonance frequency, which results in total reflection. Also partial reflection can result from gradients or irregularities in the plasma density profile.

2.2 Modulation Concepts of Ionospheric Radars

To obtain range information to the scattering or reflecting region in the ionosphere either the pulsed mode or the FM/CW (frequency modulated continuous wave) or chirp mode can be applied. The majority of the ionospheric radar systems apply pulse modulation, where frequently binary phase-codes and frequency hopping are employed to make use of the full average power of the radar transmitters. Particularities of coding schemes, such as double- or multi-pulse, Barker, complementary, alternating and random codes, adapted to ionospheric properties and the corresponding radar systems, will be discussed in the corresponding chapters. The FM/CW mode uses a time-varying frequency ramp (chirp), which covers the bandwidth corresponding to the range resolution. Continuous wave modulation allows this modulation to be used merely in bistatic or in monostatic systems, when ground and skywave can be sufficiently separated. Due to its relatively narrow instantaneous bandwidth it is less susceptible to interference than other pulse modulations. Continuous frequency sweeping can be replaced by random or deterministic frequency stepping, and spread-spectrum techniques are occasionally applied as well. The FM/CW technique is now mostly used in HF radar over-the-horizon applications.

New approaches have been performed recently which make use of existing radio transmitters rather than using an own radar transmitter [*Sahr and Lind, 1997*]. For this purpose it is required to convolve the directly received ground wave signal of a frequency-modulated broadcast transmitter with the signal of this transmitter scattered from the ionosphere. We may call such a system a passive bi-static coherent scatter radar. In certain aspects LF, MF or HF receiving systems, observing phase and amplitude variations as well as the diffraction pattern of signals from distant transmitters, which are not controlled by the receiver, can also be regarded as passive radar methods.

2.3 Ionospheric Radar Antennas and Transmitters

To obtain directional information, different kinds of antenna configurations are in use. The antennas of the ionospheric radar systems cover the wide range from systems of simple loop, dipole and rhombic antennas to giant phased-arrays or large-aperture parabolic dishes. MLT/MF radars use dipole antennas for transmission and frequently loop antennas for reception. Contemporary VHF meteor radars for measuring atmospheric parameters employ the spatial interferometer mode with a set of simple Yagi antennas. HF radars, namely digital ionosondes, employ wide-band vertically beaming delta, rhombic or log-periodic antennas for transmission and usually spaced loop antennas for reception. HF radars for over-the-horizon profiling of the ionosphere and ionospheric irregularity imaging use rhombics or arrays of log-periodic antennas, beaming into narrow azimuth sectors at low elevation angles. Monostatic as well as bistatic HF radar systems are common and interferometer configurations are often applied for reception. CW-Doppler HF sounding systems use simple dipole antennas with wide vertical beam angle, whereas HF heating facilities consist of large phased arrays with vertically directed beams at high gains in the order of 10-20 dBi. MST VHF radars use phased array antennas beaming in assorted close-to-zenith directions. The arrays, which can be subdivided for interferometer applications, consist of individual Yagi or co-axial collinear antenna elements, yielding a total gain usually in the order of 30 dBi. Coherent scatter VHF radars mostly use phased arrays and allow simultaneous multi-beam reception by interferometry applying Butler matrices. Incoherent scatter radars utilize phased arrays in the low VHF band. In the higher VHF and the UHF band large-aperture parabolic dish antennas or fully steerable dish antennas with antenna gains larger than 40 dBi are employed by incoherent scatter radars.

Transmitters are almost exclusively operated in pulsed-coded modulations. The MLT/MF radars, vertically beaming HF-radars and CW-Doppler systems operate at peak power levels up to a few hundred watts to a few kilowatts. Meteor radars and coherent scatter radars run transmitter power up to about ten kilowatts, whereas some trans-horizon and high-power MST radars apply up to some hundred kilowatts and incoherent scatter radars produce some megawatts of peak power.

The optimum frequencies of these radars are given by typical scales of the governing scattering and reflection processes, which range from a few MHz (MF) to some GHz (UHF). The detectability of the various systems is obviously determined by the frequency-dependent cross section of scattering irregularities or the critical frequency of the ionosphere, by the radar system parameters themselves and by external noise and interference. Whereas the detectability of MF radars and HF radars with typical transmitter power and antenna systems are frequently affected by external interference from conflicting transmissions in the MF and HF spectrum, the detectability of MST, coherent and incoherent scatter radars, utilizing frequencies between 40 MHz and 1.3 GHz, is given by the transmitter power, antenna aperture and the system temperature. The contributions to the latter are the receiver and the galactic noise power. Whereas the receiver noise increases with frequency, the galactic noise decreases with frequency. An optimum frequency range, where the total system temperature minimum occurs, is between 400 MHz and 500 MHz. Several of the high power incoherent scatter radars operate in this frequency band.

This is also the optimum frequency range for these radars, since one also has to consider here that the radar wavelength has to be larger than the Debye length of the ionospheric plasma.

When comparing the capabilities of different radar systems operating in the upper VHF and UHF band, a system figure of merit can be defined as: $F_m = P \cdot A \cdot T^{-1}$, where P is the average transmitter power, A the antenna aperture and T the system temperature. The system temperature should be taken without the external (sky) noise power, since this varies as function of frequency and sidereal time. For this reason, just the power-aperture-product P·A is often used. Typical values of F_m are larger than 10^6 Wm^2 K^{-1} for incoherent scatter radars and a maximum of 10^6 Wm^2 K^{-1} for MST radars, corresponding to a power-aperture-product of up to 10^{10} Wm^2.

2.4 Observation Principles of Ionospheric Radars

To cover larger areas and to measure velocity vectors, the Doppler method is utilized, where a narrow antenna beam is steered into different directions. It is also known as the Doppler beam swinging method (DBS). This method is always applied by the incoherent scatter radars, where the full spectrum information is measured at different, horizontally separated positions in the ionosphere. The spaced antenna method (SA), where the changing diffraction pattern on the ground caused by backscattering from aloft is measured, is applied to deduce the horizontal velocity, shape and coherence of the scattering layer by analyzing the cross correlation of signals received at spaced antennas. The SA method is used with the MLT/MF radars and becomes common also in MST radar systems. The interferometer method is a natural extension of the spaced antenna method. It utilizes the fact that the modern radar systems are phase-coherent and the amplitude and phase (or the quadrature signal components) are measured. Phase information at separated antennas allows the determination of the angle of arrival. It yields, with improved accuracy, similar parameters as the diffraction pattern method and the Doppler beam swinging method. It also allows digital beam forming and high-resolution radar imaging by employing a larger number of antenna modules and receiver channels. The advantages of this method are becoming more widely known and are accepted to constitute the features of contemporary radar systems for ionospheric and atmospheric research. The interferometer method is applied in many varieties in almost all radar systems, covered in this tutorial. Such mutations are radar imaging to deduce the visibility and brightness functions of scattering regions, digital beam forming (also called post-beam steering (PBS)) using spaced antennas and spatial domain interferometry (SDI). *Hocking* [*1997b*] has described advantages and limitations of such systems. Interferometry in the frequency domain (FDI) is used to detect thin layering, substituting implementation of shorter pulses. Interferometer applications have not been done with incoherent scatter radars where the scattering process itself would inhibit the required coherency of the diffraction pattern. Here the Doppler method with different antenna beam pointing angles is used. Another possibility measuring velocity and anisotropy is to apply multi-static observations using transmitting and receiving antennas at separated locations, such as the EISCAT tri-static UHF incoherent scatter radar system [e.g., *Westman et al.*, *1995*].

3. MLT/MF Radars

The mesosphere-lower-thermosphere (MLT) radars, also called MF radars, have their origin in the early applications of LF, MF and low HF radars to analyze partial reflections from the D-region by means of the differential phase and amplitude methods to deduce the electron density profile. The partial reflection of the radar waves was assumed to be from irregularities in the ionization, and it was soon recognized that this technique could be extended by measuring the mean drift and variability of these irregularities. Since the irregularities are strongly coupled to the neutral atmosphere due to the high collision frequency between neutrals and ions, this opened a possibility to study neutral dynamics of the mesosphere and lower thermosphere. For this purpose the spaced antenna method emerged as an optimum application, since sufficiently narrow beam antennas, which are needed for the Doppler method, cannot be achieved in this frequency range without very large-size antenna apertures. The spaced antenna method used with MLT/MF radars requires relatively small transmitting antennas with dimensions of a few wave-length and at least three simple spaced dipole or loop antennas for reception. The frequencies are usually in the range of 2-3 MHz, the transmitter peak powers are at some ten kW and the altitude resolutions are in the order of a few kilometers. The analysis of the variable diffraction pattern at the ground and the deduction of winds, waves and turbulence is being done by analyzing the received complex signals in the time domain by the full correlation analysis (FCA), in the frequency domain by the analogous full spectrum analysis (FSA) or by the imaging Doppler interferometry (IDI). There are numerous mutations of these analysis techniques. All of these established the appropriate applicability of the spaced antenna and interferometer technique with MLT/MF radars to study middle atmosphere dynamics and are accepted to yield essential ground-based contributions for understanding the global climatology of the mesosphere and thermosphere.

4. Meteor Radars

Meteoroids impinging on Earth's upper atmosphere ablate and eventually disintegrate. The major interaction with atmospheric particles takes place in the altitude region 70 to 120 km, where plasma trails are produced by the meteoroid. When interacting with Earth's atmosphere, meteoroids are called meteors. The head of the meteor as well as the ionized trail cause scattering and reflection of radar waves. Such observations have been used since about the early 1960s for studies of the scattering processes, the features of meteors and their trails and for the deduction of ionospheric and atmospheric parameters. The applied methods, aiming towards some different directions, namely astrophysics and atmospheric physics, have become known as the meteor radar technique. Meteor radars are particularly used to measure the wind velocity and the diffusion coefficient, which can yield the temperature. While the distance is obtained by range gating, the direction can be determined by using narrow beam antennas. This, however, has the disadvantage that the number of meteors is smaller than with wide angle antennas and unwanted meteor echoes can occur through antenna sidelobes. All contemporary meteor radars now use spatial interferometry to determine the angle of arrival with sufficient accuracy. Measuring range, azimuth and elevation allows unambiguous determination of the meteor echo location. Standard meteor radars usually operate in the frequency range 25 to 50 MHz. Typically general-purpose meteor radars for atmospheric studies run at less than 10 kW peak power and use a simple set of dipole or Yagi antennas. Also MST radars in the 40 - 50 MHz-range and MF and HF radars are used for such meteor radar studies. It has moreover been postulated that incoherent scattering from the ionization created by the meteor can explain VHF and UHF radar observations.

5. HF Radars

The HF radar technique has experienced a major rebirth after the original ionosonde technique, which was introduced some seventy years ago, seemed to have saturated. Also the trans-horizon oblique incidence sounding seemed to have lost belief in its importance. We will demonstrate here that these techniques have again developed exceptionally and gain highly valuable information on Earth's ionosphere and its coupling with the atmosphere and magnetosphere.

5.1 Digital Ionosondes

Vertical sounding of the ionosphere has been the predominant and quite indispensable technique yielding a wealth of information. The pulsed transmitters, called ionosondes, sweeping from low frequencies around 1 MHz up to some 20 MHz had essentially been used to measure profiles of virtual reflection heights from which the true height profiles of the ionospheric plasma frequency is deduced. Digital techniques were already employed to ionosondes about 25 years ago. The theory and the basic technique, yielding time of flight, polarization and absorption of signals as function of frequency, had been developed in extenso. Ionosondes operated from the ground can provide just profiles up to the ionospheric electron density maximum, since total reflection of the transmitted radio wave at the critical frequency prevents the transmission into the topside ionosphere. For studies of this upper ionosphere topside sounders in spacecraft, also operating according to the ionosonde principle, had been employed.

5.1.1 Ground-based systems

In the recent decades advanced techniques had been developed to measure more characteristics of the ionosphere as function of frequency such as wave polarization, Doppler frequency spectra, amplitude and phase as well as angles of arrival. Also quasi-automated on-line analysis techniques for ionospheric profiling have been designed and the ionospheric parameters are, thus, available in real time. The transmitter-receiver systems, providing multiple and more advanced information than before, have been miniaturized and operate at much less power than the initial ionosondes. This has become possible by application of coherent detection, frequency-shift and phase-coding for pulse compression, matched filtering, adapted coherent and incoherent integration, digital gain control, correlation techniques and intelligent frequency selection procedures to minimize mutual interference with other spectrum users. Several parallel system developments have taken place which led to some modern brands of digital ionosondes, such as the Dynasonde, the Digisonde Portable Sounder (DPS), the Canadian Advanced Digital Ionosonde (CADI), the Advanced Digital Ionosondes (IPS-71), the SCIPION or the Advanced Ionospheric Sounder (AIS). All these brands function according to basically equivalent concepts, such as demonstrated by the DPS. They are phase-coherent systems, computer-controlled by a digital sweep frequency synthesizer. The exciter generates phase and frequency modulated/coded signals, which are step-wise increased in frequency channels, which had been preselected by minimum interference criteria. Transmission in circular polarization requires two quadrature-phase power amplifiers at some 100 W peak power feeding two orthogonal linearly polarized antennas, which are vertically beaming rhombic, delta or log-periodic designs. The application of complementary code pulse compression can increase the effective power by some 20 dB. Usually a four element receiving array and four parallel receivers are used. This phased-array of receiving antennas allows to measure the polarization and angles of arrival of ionospherically reflected signals. Digital receiver gain control adjusts the gain to the varying signal levels. The signal is digitized at the intermediate frequency level and all processing of the complex data is done on-line in special DSP devices, usually performing the range gating, filtering, decoding, coherent integration, conducting fast Fourier transformation, and integration for each of the signals from the four antennas and the two linear polarizations. The group reflection height is determined by calculating the phase differences from two narrow-spaced frequencies. This method is more than an order of magnitude accurate than determining the group height from the leading amplitude edge of the echo pulse.

A major advantage of these modern systems is their capability to measure the angle of arrival by means of interferometry. This is usually done after the Fourier analysis, which allows determining the angle of arrival for each individual spectral component and is called Doppler sorting. This technique, providing angular and velocity information, is used to deduce the vertical and horizontal velocity of the ionospheric layers. The availability of complex signals from multi-receiver-channels also allows digital beam forming, which is antenna beam steering by transforming the digitized data. All these features, providing sky map images of the ionosphere are highly required, since the ionospherically reflected signals are mostly arriving from different parts of the ionosphere due its irregularity, patchiness and horizontal gradients. Sophisticated routines, which are basing on the standard methods of ionogram analysis, are automatically applied to calculate in real-time the true-height profiles of the ionospheric plasma frequency and control their quality.

5.1.2 Topside sounders

It is well known that ground-based ionosondes can only observe the ionosphere up to the height of the maximum electron density. The top-side ionosphere has to be studied by other means. For this purpose ground-based incoherent scatter radars are applied. The other possibility are digital ionosondes in satellites orbiting above the height of the electron density maximum of the ionosphere. This radar technique of top-side sounding has several advantages, such as the reduced man-made interference, since total reflection prevents reception of ground-based transmitters in the topside ionosphere. Topside sounding provides global coverage and allows studies of plasma physics in the local environment of the satellite. Basically top-side sounders apply the same technical principle as ground-based ionosondes, but there are some differences. One has to consider that the sounder is in a plasma environment, which affects the antenna matching and causes plasma resonances and wave cut-offs seen in the ionogram. One has to take into account the high speed of the satellite, which affects the accuracy of the sounding and makes skymap imaging fairly complicated. It is also required that the data have to be down-loaded to ground stations in regular intervals, which needs suitable telemetry. Receiving in a spacecraft the signal transmitted from the ground, or v.v., at frequencies beyond the critical frequency is a natural step to combine the ground-based and space-borne techniques. This method is called trans-ionospheric radio sounding and provides for instance further information on horizontal structures in the ionosphere.

The topside sounder technique had been introduced in the early sixties and more than ten topside sounders (e.g., Alouette, ISIS, Cosmos and EXOS etc.) have been operated since, yielding valuable information on the ionospheric parameters and space plasma physics. Whereas the ionospheric topside sounders observe the region around and below the satellite, emphasis has now shifted towards sounding the region above the satellite, namely the magnetosphere.

5.1.3 Magnetospheric sounding

The highly magnetized plasma and wave - particle interaction in the magnetosphere has almost exclusively been studied in-situ by spacecraft probes. The magnetosphere and ionosphere are highly coupled regions of Earth's environment. It is, thus, justified to discuss also the possibility of radar sounding techniques for studies of this region. A few attempts have been made to detect echoes from the magnetosphere with high power HF radars in Russia, which, if they were real, were assumed to be due ion-acoustic turbulence. Attempts for direct radar sounding of the plasma density, however, are scarce. Since the plasma density in the magnetosphere is as small as 10^6 m^{-3}, low frequencies have to be used for this sounding. Radar signals at such low frequencies

(VLF - MF range), transmitted from the ground, will usually be reflected by the ionosphere, unless they can enter a duct through an ionospheric trough. Consequently sounders above the ionosphere have to be applied for regular sounding. A new project, the Radio Plasma Imager (RPI), has been established, which is part of NASA's IMAGE mission to be launched in early 2000. The RPI consists of a sweep-frequency digital sounder covering the frequency range 3 kHz to 3 MHz. This system will image the echoing structures by determining the angle of incidence in addition to measuring Doppler spectrum and time of flight as function of operating frequency.

5.2 Over-The-Horizon Radars

Under this heading we summarize oblique incidence methods, namely the over(beyond)-the-horizon radar systems, which operate in point-to-point connection, in the ground (or sea) - backscatter or in the ionospheric irregularity backscatter mode. Such systems are naturally used for communication purposes or detection of remote targets. Conventionally the term Over-The-Horizon (OTH) stands for the latter method detecting remote targets [*Anderson and Abramovich, 1998*]. To gain information on the remote ionosphere certain data reduction and interpretation methods have to be applied and complicated ray-tracing techniques have been developed for this purpose, besides the aforementioned technical advances.

5.2.1 Oblique point-to-point sounding

The systems for ionospheric observations described in chapter 5.1 are monostatic, which utilize total reflection at vertical incidence into the ionized regions or scattering from ionization irregularities. Such systems can also be used in bistatic mode, where the receiver and transmitter are separated and the transmission is at oblique incidence. This is the standard method of point-to-point radio transmission, where receiver and transmitter have to be properly synchronized in frequency and time allowing observations of the ionosphere at remote locations. This technique applies the step frequency pulse-modulation or the frequency-modulated continuous wave (FM/CW) mode. Sophisticated pulse-compression and spread-spectrum modulations as well as multiple antennas and multi-channel receivers have been introduced to measure the time, frequency and angular dispersion on HF communication paths, and in the same instant providing information about the structure of the distant ionosphere [*Haines et al.*, 1997]. A new feature of these systems is that upon reception at the remote location the spread-spectrum signal is retrieved by correlation processing. Such techniques are particularly useful in areas where the ionosphere is highly irregular, such as in equatorial and auroral regions.

For completeness we mention here also the HF Doppler sounding systems, which operate multi-static in CW mode at fixed frequencies. They are sometimes imprecisely called radars, although they are not applying the true radar principle. They are successfully used for studies of ionospheric irregularities and wave signatures.

5.2.2 Backscatter from Earth's surface and remote targets

An obliquely transmitted wave reaches the Earth's surface after reflection in the ionosphere. Besides forward reflection a small amount of the radar wave energy is scattered due to roughness of the Earth's surface, waves on the ocean surface or other targets on the ground or in the air. The back-scattered wave, which had propagated back to the transmitter, is received at that location and yields information about the scattering properties of the remote target as well as on the passed ionosphere. The former is used in the OTH technique, which we will not discuss here, since it is primarily utilized to study ocean waves and identify remote target features. Propagation of the signals through the variable ionosphere cause uncertainties in determining parameters of remote targets and special methods, such as ray tracing and adaptive signal processing, are applied to mitigate these disturbing effects. Also digital ionosondes are employed for ionospheric sounding in support of these over-the-horizon radar measurements. Because of the small ground-backscatter cross-section the transmitter power and antenna gains have to be a few ten dB larger than those of oblique incidence point-to-point sounders. Attempts were very successful to deduce the ionospheric plasma frequency profile in the vicinity of the reflection region in the ionosphere, provided that the ionosphere is horizontally stratified. Ray tracing methods were introduced to notably overcome this limitation.

Atmospheric gravity waves imprint periodic variations in time and space of the electron density profile, which are called travelling ionospheric disturbances (TIDs). These in turn are detected in ground-backscatter ionograms due to focussing and defocussing of the radar waves. Model ray tracing has shown, that horizontal wavelength, phase velocity and period of TIDs can be deduced from the backscatter ionograms. The investigations of TIDs with ground-backscatter radars has gained increased interest in this decade after the introduction of the powerful HF coherent backscatter radars for polar cap studies.

5.2.3 Backscatter from the ionosphere

HF-Radars observing the polar cap ionosphere are known by their acronym SuperDARN (Super Dual Auroral Radar Network). They use backscattering from field-aligned F-region irregularities (FAI) to measure the plasma drift velocity. In addition to this direct scatter from the F-region these radars observe ground-backscatter and thus are also proper instruments to investigate F-region structures specific to the auroral region, such as plasma troughs and moving blobs and signatures of TIDs in the auroral ionosphere.

The direct backscatter process from the ionosphere, i.e. from FAI, falls into the category of coherent scatter. The ray bending of the HF radar wave due to ionospheric refraction is an essential requirement to meet the perpendicularity condition for this kind of scatter from F-region irregularities. Angle of incidence measurements, ray tracing methods, supportive digital ionosonde measurements and transponder ionosondes are helpful complements to determine the electron density profile on the propagation path and deduce the correct location of the scattering FAI. This is remote sensing of the ionosphere and magnetosphere, where plasma irregularities are used as tracers.

The SuperDARN systems have the capability to measure the polar ionosphere plasma convection pattern at high latitudes [*Bristow et al., 1998*]. The prototype radar, operated since 1983 in Goose-Bay, uses an electronically steered phased array and operates in the frequency range 8-20 MHz. Transmission is not in the sweep-frequency mode but on a selected fixed frequency. Beam angles are towards the pole over a 50 degrees azimuth sector. This radar has become the prototype of a whole set of HF radars looking into the northern and southern hemisphere polar caps. CUTLASS (Collaborative UK Twin Located Auroral Sounding System) is a new part of SuperDARN [*Lester et al., 1997*]. It includes two rows of log-periodic antenna arrays. One row, consisting of 16 antennas, is used for transmission and reception. Each antenna transmits 800 W peak power of coded multipulse sequences capable to provide full spectrum information. A second row of four antennas is used for reception only and applies interferometry to measure incidence angles in the vertical plane. Horizontal antenna beam steering into 16 azimuth directions with an angular width of a few degree is performed by an electronic phasing matrix of digital delay lines. Transmission is usually on fixed frequencies, which however can be changed quickly according to the ionospheric and interference conditions. Computer control and digital data acquisition delivers the standard parameters of backscattered power, Doppler velocity in radial direction and the spectrum width as function of range and angle. These systems are operated in the northern and southern hemisphere and can usually be remotely controlled and monitored. Velocity information from sets of radars with overlapping antenna beams yields the vector velocity. It was shown that these velocities compare well with velocities measured with incoherent scatter radars. The SuperDARN radars were applied also to analyze echoes from meteor trails [*Andre et al., 1998*], which opens a unique possibility to measure also lower thermospheric and mesospheric winds.

6. Ionospheric Heating and Diagnostics with HF Waves

The obvious fact, that ionospheric irregularities are required for studies with SuperDARN systems, can be further proved by creating such irregularities artificially by ionospheric modification. This modification of the ionosphere is done by high power radio waves in the HF range, which cause electron heating, parametric decay and many other kind of plasma instabilities resulting in plasma waves, striations, sheets, filaments and turbulence, i.e. irregularities. These in turn scatter radar waves, which are used to diagnose the plasma processes, which is done with HF, VHF and incoherent scatter radars. Another research tool using ionospheric heating in radar applications is the creation of artificial periodic inhomogeneities (API). This is resonance scattering from the standing wave pattern of the reflected heater pump wave. A probing HF-wave, which fulfills the Bragg condition with this standing wave, will be backscattered from this pattern. As long as the density irregularities in the standing wave pattern decay at long enough time scales, the probing radar wave can be transmitted after the pump wave has been switched off and will be scattered from this Bragg pattern. Analyzing the scattered radar wave can yield several ionosphere parameters such as electron density, electron and ion temperature, neutral atmosphere density and temperature, ion chemistry and vertical velocity, depending on the probed altitude region. Original work on these API and related irregularity patterns, studied or created by radar, were done at Jicamarca, Arecibo and SURA in Russia. These high-power heating facilities typically operate in the frequency range 4 - 12 MHz with CW-power up to 1000 kW and usually vertically pointing phased array antennas with gain in the order of 15 - 25 dB.

7. MST Radars

The technique of using sensitive clear air radars to investigate the upper atmosphere at altitudes from the troposphere to the mesosphere owes much of its success to the early days of ionospheric incoherent scatter observations. It was found that the Jicamarca radar, operating at 49.9 MHz, also detected echoes from the middle and lower atmosphere. In contrast to total reflection, which applies for the HF radars, scattering and partial reflection takes place in this VHF frequency range. These VHF radars operate in the monostatic mode. The capability of these radars to cover parts of the mesosphere and the stratosphere and the troposphere has lead to their name MST radars. We will discuss only the mesosphere observations with MST radars, since only these encompass the ionosphere. Whereas common MST radar echoes are resulting from neutral turbulence acting on an electron density gradient and usually are called turbulence echoes, a new brand of echoes had attracted wide attention in the last decade, namely the Polar Mesosphere Summer Echoes (PMSE).

MST radars operate in the 40 - 55 MHz frequency range, and, thus, are also called VHF radars. The three most powerful systems of this kind are the Jicamarca radar in Peru, the MU radar in Japan and the National MST Radar Facility in India. Also the SOUSY-VHF-Radar in Germany (and at the Arecibo Observatory in Puerto Rico), the Poker Flat radar in Alaska, and the VHF radar in Wales have been used for mesospheric studies. A large equatorial radar system is also proposed to be located in Indonesia. To detect mesospheric turbulence scatter high power-aperture products are required. The antennas with more than 30 dBi gain are large horizontally positioned fixed phased arrays with beams pointing close to vertical. Beam steering is done by phase delay lines, which are fixed or can be changed by switches. Antenna elements can be coaxial collinear or Yagi antennas. Partitioning of the antenna array into sectors allows spaced antenna applications. Such a partitioning is most versatile when also the transmitters and receivers are highly modular, such as at the MU radar operating on 46.5 MHz [*Fukao et al., 1990*]. A total of 475 crossed Yagi antennas, each fed by a single solid state transmitter module of 2.4 kW peak power, form this radar. The signals received at each Yagi can be merged in any combination into four receiver channels, allowing most flexible interferometer arrangements. The antenna beam of 3.6° width can be steered from pulse to pulse into 1657 individual directions. Transmitter peak powers of MST radars for mesospheric studies are in the order of 100 - 1000 kW and the average power is a few kW to some 10 kW. The largest radar of this kind with 290 m x 290 m antenna aperture is the Jicamarca VHF radar. The system figure of merit is about 10^6 Wm K^{-1}, where one has to consider that the sky noise is in the order of 2000 - 8000 K.

Spatial and frequency domain interferometry, digital beam forming and radar imaging to increase the three-dimensional resolution is applied. Standard range resolution for mesosphere observations is in the order of a few 100 meters. To achieve this resolution, pulse coding is inevitable. The complementary code has become the standard pulse compression modulation of the MST radars. This code consists of a pair of two code sequences, which are transmitted in successive interpulse periods. Complementary codes have the property that the sidelobes of their correlation functions are exactly equal, but opposite in sign. The coherent addition of the two correlation functions leads to the complete elimination of the sidelobes, provided the signal is coherent from pulse to pulse. The zero lag amplitudes add, since they have equal sign. Depending on the code length the equivalent transmitter peak power can, thus, be raised by several ten dB. MST radars are also operated in the frequency domain interferometer mode, where two frequencies within the system bandwidth are transmitted from pulse to pulse and their phases are compared. This allows determination of the position and thickness of thin layers within the pulse volume. Typical interpulse periods for mesospheric observations are in the order of 1 ms. The signal coherence time is usually more than two orders of magnitude longer, such that the condition for application of complementary codes are fulfilled. The long coherence time also allows coherent integration over many interpulse periods, and the decoding process can be done after the coherent integration. This substantially reduces the computational load. However, contemporary signal processing hardware makes coherent integration obsolete as done in the CLOVAR system [Hocking, 1997a]. It has the major advantage that the same system can also be used for measurements of processes, which have a short coherence time, such as meteor echoes.

The basic parameters measured with MST radar are deduced from the Doppler spectrum, such as the signal power, the Doppler shift and the spectral width, and spaced antenna methods are also used for this purpose. Furthermore the anisotropy of turbulence scattering layers can be determined. These parameters are used to study turbulence itself, where two methods are applied. One relies on the dependence of the scatter cross section on turbulence intensity and the background gradient of electron density, the other on the spectral width. From these measurements the eddy diffusion coefficients are deduced. Doppler shift or spaced antenna analysis yield estimates of the three-dimensional wind vector, which allows studies of mean winds, long-period waves, tides and gravity waves. Of particular importance is the deposition of momentum and energy by waves into the mesosphere, which strongly affect the mean mesospheric circulation. This can be measured with special opposite-beam pointing arrangements.

8. Coherent Scatter Radar Observations of Ionospheric Irregularities

It is notable that most of the observations to understand ionospheric irregularities result from backscatter radar experiments, which have become increasingly sophisticated. Furthermore, the inclusion of rocket measurements has turned out to be highly valuable, and satellite in-situ measurements in the F-region contributed additionally to the understanding of the irregularities. Theories and modeling has explained many features seen by the radars, although numerous open questions remain to be solved. The research on ionospheric irregularities is undoubtedly also stimulated by application requirements, since the irregularities can destructively hamper ground-to-ground and satellite communications. On the other hand, the scattering can also open additional communication channels in the HF and VHF ranges.

The term coherent scatter is commonly used, since the scatter is from structures in the ionospheric plasma density, which exhibit some temporal and spatial coherency on the Bragg scale of the probing radar. This is often seen in contrast to incoherent scatter, which is in itself coherent, too, since it is scatter from coherent structures due to ion- and electron-acoustic waves in the ionospheric plasma. In practice coherent scatter results from irregularities, created by structural modification of, or instabilities in the ionospheric plasma. Plasma irregularities are a synonym for a spectrum of plasma waves or turbulence, which are created by the plasma instabilities in the E- and F-region. Structural modifications are variations related to breaking waves and turbulence in the neutral atmosphere. This is certainly the case in the mesosphere (D-region) due to the high collision frequency between ions and neutrals. It is also believed that neutral turbulence affects the lower E-region. From the radar systems point of view, we can simply say that coherent scatter has normally a much larger cross section and narrower spectral width than incoherent scatter and is mostly highly anisotropic. In order to understand the radar system design, which has to be adapted to the scattering properties of the irregularities, we will briefly outline some of their essential features.

8.1 D-Region Irregularities

During the early operation of the Poker Flat MST radar in Alaska unusually strong mesospheric echoes were observed during the summer time. These echoes have since been detected by radars at frequency between 2.7 MHz and 1.3 GHz. Since the latter frequency corresponds to a Bragg scale of 0.12 m, it is impossible to explain such observations by turbulence scatter. These echoes preferably occur at high latitudes but are also reported at midlatitudes. After a long period of skepticism, whether these unusual echoes could be created by enhanced neutral turbulence, several more reasonable explanations were suggested, which base on the fact that the polar mesopause is very cold in summer and contains some water vapor. In the presence of meteoric dust and aerosols, this leads to the formation of heavy water cluster ions and ice particles, which affect the distribution of electrons in the ionospheric plasma. This results in particles dressed by electrons and is often called a dusty plasma. The effects are depletions in electron density at a scale of some kilometer and small-scale electron density variations. The latter in particular give rise to an strong increase in the radar scatter cross section.

Since these echoes are most pronounced in polar regions, they are called Polar Mesosphere Summer Echoes - PMSE. They are regarded to result from coherent scatter from highly structured irregularities, embedded in the ionospheric plasma and partially distorted by neutral turbulence. The origin and characteristics of these irregularities is still being discussed. The relation to heavy ions, aerosols or ice particles in the dusty plasma of the mesopause region seems to be accepted. The extension of scatter from irregularities at scales, which are much smaller than scales of neutral turbulence can be explained by the extension of the inertial subrange in the electron gas in the presence of large particles. These can reduce the ambipolar diffusion of electrons, which keep structures at shorter scales than neutral turbulence (the so-called Schmidt-number effect). It is also being discussed how charge accumulation on these large particles in the mesopause region could increase the scatter cross section. There is a clear frequency dependence of the scatter cross section, favoring the longer wavelengths. Whether this results from the same scattering mechanism over such an extended range of scales between some ten centimeters and several ten meters, is unclear. Due to the large scatter cross section at meter-scales VHF radars with relatively low power (some 10 kW peak) and small antenna aperture (1000 m^2) can be used for PMSE studies. Other radar system parameters are equivalent to those of MST radars.

8.2 E-Region Irregularities

Whereas neutral atmosphere and dusty plasma effects govern D-region irregularities, primarily electro-dynamic effects dictate irregularities in the E-region. In contrast to horizontal layering of D-region irregularities, which are essentially controlled by the neutral atmosphere, the E-region irregularities are predominantly elongated along the Earth's magnetic field. Radar observations of these irregularities are, thus, very sensitive to the pointing direction with respect to the magnetic field, i.e. the radar echo is strongest when the radar wave vector is perpendicular the Earth's magnetic field. Due to the connection of these irregularities to the direction of the Earth's magnetic field $\underline{B}$, the instability mechanisms and the radar observation methods applied in low, middle and high latitudes have to be treated slightly different. The term irregularities most frequently refers correctly to plasma waves, created by instabilities. The energy of these waves is distributed corresponding to spatial spectra. The radar echoes result from those spectral wave components, which fulfill the Bragg scattering condition. These E-region irregularities are basically created by two main plasma instabilities, the two-stream instability occurring when electrons and ions drift with different velocity, and the gradient drift instability, which occurs when there is an electron density gradient component perpendicular to the drift direction.

8.2.1 Equatorial Latitudes

In the equatorial region the Jicamarca VHF radar in Peru played a dominant role in the studies of E-region irregularities, since it is properly located close to the magnetic dip equator below the equatorial electrojet, allowing the antenna beam pointing almost vertical and perpendicular to the magnetic field. Its frequency 49.9 MHz is well adapted to prevalent irregularity scales and its large phased-array antenna, subdividable into modules, is most suitable for skillfully adapted, accurate beam pointing, incidence angle measurements and radar imaging.

Measurement of the full Doppler spectrum of the radar echoes from electrojet irregularities is mandatory. Since the spectra are replicating velocities in and of the irregularities (plasma waves), they have been categorized according to echo types. Type 1 echoes comprise a large mean velocity of about the ion acoustic velocity (usually about 350 - 400 ms^{-1}), a narrow spectral width and a large scatter cross section. They are due to the two-stream instability. Type 2 echoes show a mean velocity significantly smaller than the velocity of type 1, usually have a broad spectrum and a small cross section. They are assumed to be due to secondary waves creating gradient-drift instabilities. Plasma waves and turbulence in the equatorial electrojet show remarkable structures as measured with an altitude resolution better than 250 m. In addition to determining the vertical structure and velocity it is possible to evaluate also horizontal motions by applying simultaneous interferometer measurements. Emphasis on such observations had also been promoted in India, using a smaller VHF radar in Trivandrum and the National MST Radar Facility in Gadanki. Concentrated campaigns of radar and rocket observations of equatorial electrojet irregularities had been performed in order to measure electric fields, current density, plasma waves and irregularity spectra in-situ with rockets and compare these with concurrent radar observations. The most recent activity of this kind was the GUARA campaign in Brazil [*Pfaff et al., 1997*], where the CUPRI system (Cornell University Portable Radar Interferometer) was employed. This 46.9-MHz radar was particularly set up for interferometer measurements. Large-scale gradient drift waves in the equatorial electrojet can be detected with the spatial interferometer method using the JULIA radar at Jicamarca. Frequencies in the HF band had also been used for studies of the equatorial electrojet and coherent echoes from 150 km in the F1-region.

8.2.2 Mid Latitudes

Field-aligned E-region irregularities are also observed with VHF and HF radars in mid-latitudes. During the past decade, several radars have been used for mid-latitude E-region irregularity studies: The MU radar (46.5 MHz) in Japan, the Frequency Agile Radar (FAR) near Japan operating on 43.7 MHz and 24.5 MHz, the Chung-Li VHF-Radar (52 MHz) in Taiwan, the sporadic-E Scatter SESCAT bistatic CW system (50 MHz and 144 MHz) in Crete, and the Valensole HF radar in southern France. Similar to the FAR, which is a multifunctional approach for ionosphere remote sensing [*Tsunoda et al., 1995*], several portable VHF radars had been used at Arecibo and some other places. At least two major campaigns were performed for studies of midlatitude E-region irregularities, the Sporadic-E Experiment - SPOREX - in France and the Sporadic-E Experiment over Kyushu - SEEK - in southern Japan [*Fukao et al., 1998*]. Due to the Earth magnetic field geometry all these radar experiments are with obliquely pointing beam at some 30 - 50 degrees elevation angle. Since the scatter cross section of these echo types is moderately strong, a very high transmitter power and large antenna gain is usually not required. The signals have to be sampled at a fairly high rate, comparable to the radar interpulse period of a few milliseconds, or double or multipulse coding has to be applied to resolve Doppler velocities of several hundred meters per second. When a good range resolution is required, Barker coding would have to be implemented additionally to make use of the full average power of the radar transmitter. All the pulsed VHF radars apply spatial domain interferometry, and a few apply frequency domain interferometry. The Chung-Li VHF radar [*Huang et al., 1995*] has for example a separate ionosphere antenna system consisting of 3 rectangular arrays of 4 x 8 Yagi antennas pointing geomagnetically north at 41° elevation angle matching the perpendicularity to the Earth's magnetic field in the lower F-region. Three phase-synchronized transmitters are used together with three coherent receivers.

8.2.3 High Latitudes

What we have noted for low and middle latitudes holds in an even firmer manner for high (auroral) latitudes, namely that radar methods are the most essential tools for studies of E-region irregularities. This may result from early observations, that radio communication in auroral and polar regions can be highly disturbed and hampered, but on the other hand the scatter from ionospheric irregularities frequently reestablishes possibilities for communication. The high latitude E-region irregularity scatter is notably correlated with ionospheric disturbances and the visual aurora, such that these radar echoes were called radio aurora or auroral backscatter. Compared to the equator and midlatitudes there is a much larger complexity and variability in the scatter. This reflects the greater magnitude and variability of electric fields, currents and electron density due to the coupling with the magnetosphere.

The radars, which had been used in high latitudes, are covering the frequency range from the HF to the high UHF band. Deduction of the full spectrum is always done as well as interferometer and bistatic measurements and comparisons of observations on different frequencies. In addition to the two types of spectra dominating the equatorial and mid-latitude E-region backscatter, two further types were found. Type 3 characterizes a velocity lower than the ion-acoustic speed, whereas type 4 has a much larger velocity. We have to recognize that such a classification refers only to the observations of radar backscatter from the irregularities and that those spectral features are certainly affected by the radar system parameters. *Sahr and Fejer* [*1996*] argue that such nomenclature impedes the understanding of the irregularities and suggest another classification by statistically comparing the first and second moment of the spectra, namely the Doppler shift and the spectral width.

In the HF band E-region backscatter is observed with the SuperDARN systems. Most of the research, however, is done with VHF radars in the VHF band around 50 MHz and 140 MHz. At 50 MHz ionospheric refraction can still play a role and it is disputed whether this is a reason for the observation that the aspect angle dependence is not as pronounced in polar as in mid - latitude and equatorial regions. A 48.5 MHz Bistatic Auroral Radar System (BARS) is part of the CANOPUS program in Canada. The new Saskatchewan Auroral Polarimetric Phased Ionospheric Radar Experiment (SAPPHIRE-North) observes the region furthest north between the auroral oval and the polar cap. A scanning beam 50 MHz radar STARS (Syowa Station Auroral Radar System) is operated at the Japanese Antarctic station Syowa, where also two SuperDARN HF radars are located. This allows comparisons of VHF and HF scatter. The 50 MHz CUPRI system (Cornell University Portable Radar Interferometer) was operated in northern Scandinavia in combination with rocket measurements and the EISCAT UHF incoherent scatter radar in Tromsø, Norway. CUPRI measured backscatter from field-aligned E-region irregularities over EISCAT where the incoherent scatter radar measured the plasma parameters in the same region. Auroral backscatter is also observed with the 46.5 MHz and 93 MHz radars operated in northern Russia, which were combined with measurements with the EISCAT UHF incoherent scatter radar. Two further auroral backscatter radars are operated near 140 MHz, which are STARE (Scandinavian Twin Auroral Radar Experiment) and SABRE (Sweden and Britain Radar Auroral Experiment). These are dual-Doppler radars (not bistatic as sometimes quoted, since one radar does not receive the signal from its twin), where STARE has a transmitter and receiver in central Finland and central Norway, and SABRE in central Sweden and northern Scotland. Each radar transmits into 25° wide sectors and receives on a 16-element antenna array connected to a 16-port Butler matrix. Since the spectrum width and/or Doppler shift is frequently larger than the pulse repetition frequency, coded pulses (either double or multipulses) have to be used. The output signals of the Butler matrix, which are the analogue Fourier transforms of the input signals, are connected to eight receivers, allowing as many as eight adjacent 3.6° azimuth sectors to be sampled simultaneously. The 2 x 8 receiver beams of the STARE system overlap in areas in northern Scandinavia and those of the SABRE system south west of this area. This azimuth sectoring together with the range gating allows to detect backscatter from 20 km x 20 km bins. Spectral information of the echoes from these bins allows detailed studies of irregularities in an area 400 km x 400 km wide. The combination of radial velocity measured in these common viewing areas of the twin radars yields an estimate of the horizontal velocity component. When scattering irregularities exist, the plasma drift in the auroral E-region can be measured and the electric field $\underline{E}$ deduced, since the drift velocity is given by $\underline{E} \times \underline{B}$.

Observations of auroral E-region irregularities were also performed with the Millstone Hill 440 MHz radar, which is located at sub-auroral latitudes. An appropriate aspect angle to the magnetic field can be obtained at very low elevation angles of 4° to 10° at ranges around 850 km. This radar is extremely sensitive to coherent backscatter since it has an antenna gain of 44 dB and a few megawatts of peak power, which is required for the application of this radar for incoherent scatter observations. The narrow beamwidth of 1° allows proper mapping of the aspect sensitivity of scatter from E-region irregularities. Six kilometer range resolution, 10 s temporal resolution and 90 dB dynamic range was applied and the transmitter power had to be reduced to 10 kW peak to avoid receiver saturation. The peak cross section of the 34 cm irregularities was 10^{-9} m^{-1}, which is more than 90 dB above the incoherent scatter level. Using the EISCAT UHF (931 MHz) radar, observations of irregularities at 16 cm spatial wavelength were performed in the monostatic and bistatic mode, which allow measurements at different aspect angles. A special radar system had also been added to the EISCAT UHF radar, which consists of a low power CW-transmitter in Oulu, Finland. The reception is done by the EISCAT receiving sites in Sodankylä and Kiruna. This CW-system COSCAT (COherent SCATter experiment) had recently been expanded by introducing Barker coding to allow the determination of the spatial structure in the scattering region at some 10 ms time resolution and by improving the data acquisition to obtain full spectra at five second resolution.

A promising new approach for E-region irregularity studies was introduced by *Sahr and Lind* [*1997*], which uses TV broadcast transmitters in the frequency range 88 - 108 MHz. Signals of these transmitters scattered back from irregularities can be analyzed by cross correlating with the directly received groundwave signals. A perfect synchronization is required and the data rate is extremely high in the order of 250×10^3 complex samples per second. Real-time processing needs an extreme capacity of $10\text{-}20 \times 10^9$ floating-point operations per second. A range resolution of 1 km and velocity resolution of 1.5 ms^{-1} should be possible. First results obtained with this compact and low-cost passive bistatic radar system appear promising.

The radars covering wide frequency ranges and providing high time and space resolutions appear to remain essential tools, which need to be operated together with other diagnostic instruments, such as rockets and further ground-based optical remote sensing devices. The extensive radar observation has also created successful modeling and simulation and stimulated theoretical developments to understand the intricate plasma physics occurring in the ionospheric E-region.

8.3 F-Region Irregularities

The F-region is strongly coupled to the E-region, mainly due to the ionospheric electric field, which means there would be a certain relation of E-region to F-region irregularities. The background conditions, such as currents and electron density and its gradients are quite different, however. This causes differences in the instability mechanisms and the nature of field-aligned irregularities. The magnetic field geometry dictates the use of HF radars for high latitude studies, where only HF waves are sufficiently bent by ionospheric refraction to assure perpendicularity. At the equator vertically beaming HF and VHF radars can be used, since the magnetic field is horizontal. High and low latitude field-aligned irregularities had been studied intensively by radars, whereas only little is known of radar studies of mid-latitude F-region irregularities. Irregularities in the upper F-region which are of different kind and related to field-aligned currents are investigated by incoherent scatter radars.

8.3.1 High latitudes

Fejer and Kelley [*1980*] and *Tsunoda* [*1988*] have reviewed high-latitude irregularities. It is assumed that short-scale irregularities are produced by the gradient-drift interchange instability, where low density plasma is convectively mixed across mean plasma density gradients, by current-convective instabilities or by generation of electrostatic turbulence, where the limits between these processes appear not well separable. Which kind of instabilities play a role or dominate the backscatter from polar/auroral F-region irregularities remains still to be solved. This task appears far more complicated than studying mid- and low-latitude F-region irregularities, because HF radars have to be used in high-latitude studies which are affected by the ionospheric propagation and also have less spatial resolution capability.

Detailed characteristics of Doppler spectra of high latitude F-region are surprisingly rare as compared to E-region scatter. The SuperDARN radars usually measure the three first moments of the received backscattered signal, where particular emphasis is placed on the Doppler velocity to determine the plasma convection pattern. To improve the understanding of the scattering process and the nature of the scattering irregularities it would be required to have the full spectrum information available as well as observations with high spatial and temporal resolution. Applying a collective scattering approach, taking into account that the radar echoes are scattered from a distributed, non-uniformly moving volume, it could be shown that the collectively scattered signal correlation time is a statistical characteristics of the fluctuating fluid motion of the scatterers.

8.3.2 Mid-latitudes

Mid-latitude irregularities have not been a subject of intensive radar studies although irregular features are frequently observed as spread-F on vertical incidence ionograms. Such irregularities mostly occur in connection with traveling ionospheric disturbances (TIDs), which are resulting from neutral atmosphere gravity waves. Field-aligned irregularities of this origin have been studied at mid-latitudes with the 46.5 MHz MU radar [*Rishbeth and Fukao*, *1995*], also in combination with ionosondes. The unique multiple beam pointing capability of the MU radar allowed also to track patches of F-region irregularities.

8.3.3 Equatorial latitudes

Irregularities in the equatorial F-region are termed equatorial spread-F (ESF). This name stems from the appearance of spread traces on vertical incidence ionograms, which manifest strong horizontal and vertical gradients in electron density at some ten kilometers scales resulting from ionization depletions. Shorter-scale irregularities form on these gradients, which cause a blurring of the spread ionogram traces. Most of the research, leading to improved understanding of the instability mechanisms, the irregularities and the scattering of radio waves, resulted from observations with VHF radars operated close to the magnetic equator. Intensive work had also been done using transequatorial radiowave propagation in the HF and VHF band as well as analyzing scintillation of satellite signals and by satellite and rocket measurements. Besides some campaign operations with smaller VHF radars, the dominant radar instrument used for studies of ESF was and still is the 49.9 MHz VHF radar at the Jicamarca Observatory in Peru. More recently also the powerful 53 MHz VHF radar of the National MST Radar Facility in Gadanki, India, was applied [*Rao et al.*, *1997*].

The most spectacular phenomenon of ESF is the occurrence of strong coherently scattering irregularities in the topside F-region, which was first detected with the Jicamarca VHF radar. The observations of large upward moving plumes of irregularities clearly show the connection of the top-side spread-F with the bottom-side spread-F. Simultaneous measurements with satellites and incoherent scatter radar revealed that these plumes are related to depletions or bubbles of electron density. Such as for E-region irregularities, the fine structure of the F-region irregularities has been investigated by spatial interferometry [*Hysell and Woodman, 1997*]. The high anisotropy, which is consistent with a large horizontal coherence in magnetic west-east direction, allows fairly long baselines. The total baseline of such a radar interferometer at Jicamarca had a length of 560 m, and was separated into 5 submodules to yield all spatial lags at 50 m distance by choosing appropriate separations. It was shown that significantly improved angular resolution was achieved by increasing the baseline. Applying multiple spatial lags is a powerful means to obtain a dynamic high-resolution radar image of the structure of irregularities aloft. An advanced theory for such radar imaging introduced the brightness distribution and visibility function to radar observations [*Woodman, 1997*] and bases closely on methods applied in radio astronomy.

In contrast to the ESF, which is a nighttime phenomenon, there are other irregularities observed during daytime on HF and VHF, which cause the so-called 150 km radar echoes. These coherent echoes from field-aligned irregularities in the lower F-region can be utilized for studies of neutral wind variations by gravity waves and for measurements of the zonal electric field. A low-power system is used for these observations at the Jicamarca Radio Observatory [*Sarango et al., 1998*], which is called JULIA (Jicamarca Unattended Long-term studies of the Ionosphere and Atmosphere). It consists of an independent PC-based data acquisition system with a pair of 30 kW transmitters driving the Jicamarca phased array antenna, and is also applied in the interferometer mode. This coherent scatter system is uniquely suited for studying the day-to-day and long-term variability of equatorial irregularities. Using it as an MST radar, it is also suitable for climatological studies of the middle and lower atmosphere.

9. Incoherent Scatter Radars

The term incoherent scatter (IS) stands for the mechanism of scattering from electrons in the ionospheric plasma when the radar operating frequency is much larger than the plasma frequency. It stems from the initial assumption, that electrons scatter incoherently due to their random thermal motion. It is also called Thomson scatter, relating this term to the collective scatter of all the individual electrons in the volume probed by the radar. When the Debye length of the plasma is much smaller than the radar wavelength, ions are dominantly controlling the scattering process such that it becomes quasi-coherent. This generally holds for probing the ionosphere at radar wavelength of a few meters to some ten centimeters. The scattering is actually from the spatial component in the spectrum of the ion-acoustic and electron-acoustic waves in the ionospheric plasma, which match the Bragg scatter condition. The ion-acoustic waves causes the typical double-hump spectra of an IS signal from the F-region, which is called the ion line. In the lower E- and the D-region collisions between ions and molecules damp the ion-acoustic waves, such that the double-hump spectra merge into a single peak spectrum. The electron-acoustic waves cause the plasma lines, which are usually analyzed by off-setting the receiver oscillators by the plasma frequency. Commonly the Fourier transform of the ion-line spectra is used in most incoherent scatter radar applications. This transform, the autocovariance function, is usually, but not exactly correct, called autocorrelation function (ACF).

In a first approximation the scattered power is proportional to the electron density in the probed volume. The electron density, which is proportional to the square of the plasma frequency, can also be deduced by measuring the off-set frequencies of the plasma lines. The off-set and the widths of the up- and down-shifted ion lines are given by the ion and the electron temperature and the mass of the ions. The mean shift of the total incoherent scatter spectrum with respect to the radar carrier frequency is given by the bulk velocity of the plasma in the scatter volume. Applying proper analysis routines to fit measured spectra to theoretical spectra, calculated from the universal plasma dispersion relation, allows to determine the main ionospheric parameters: Electron density, electron and ion temperature and ion velocity. For this purpose it is, however, required to include assumptions or models of ion mass and composition as well as the ion-neutral collision frequency. By introducing reasonable a-priori postulations or iteration procedures, multi-parameter analyses of the latter parameters can be done. Another essential parameter is the electric field, which is deduced from the ion velocity measured in the F-region, where collisions can be neglected. The ion drift velocity is directly proportional to the vector product $\underline{E} \times \underline{B}$ of the electric field and the Earth's magnetic field. Due to the collisions, the ion velocity in the E-region is composed of the contribution by the drift due to the electric field and the neutral wind velocity. The combination of the electric field measurements at F-region heights and the ion drift in the E-region is used to determine the neutral wind velocity at E-region altitudes. In the D-region, where the ionospheric plasma is highly dominated by the collisions, the ion drift velocity is directly proportional to the neutral wind. From these basic parameters a multitude of deduced ionospheric parameters can be obtained, which usually requires further assumptions, models, independent measurements of other parameters with different instruments and dedicated computational routines. Such deduced ionospheric, thermospheric or mesospheric parameters are for instance: Neutral temperature, energy input and balance, E-region conductivities, precipitating particle energy spectra, recombination coefficients, field-aligned currents, ion flux etc. Detailed descriptions of some scientific revenue and the incoherent scatter radar technique can be found in the book edited by *Alcayde* [*1997*]. IS radars are also used as diagnostic tools for heating experiments [*Rietveld et al.*, *1993*].

The cross section of incoherent scatter is relatively small such that transmitters with up to a few megawatt peak power and antennas with gain above 40 dBi need to be applied. Incoherent scatter radars (ISR) are located in Jicamarca (operating frequency of 49.9 MHz) near Lima, Peru, at Arecibo (430 MHz) on Puerto Rico, at Millstone Hill (440 MHz) near Boston, USA, in Sondrestromfjord (1295 MHz) on Greenland, in Tromsø (224 MHz and 931 MHz), Norway, in Longyearbyen (500 MHz) on Svalbard, in Kharkov, Ukraine, and in Irkutsk, Russia. The Middle and Upper Atmosphere MU radar (46.5 MHz) near Kyoto, Japan, is also used as incoherent scatter radar. The radars in Tromsø and Longyearbyen are operated by EISCAT (EUropean Incoherent SCATter), an international association formed by six European countries and Japan. The EISCAT UHF radar on 931 MHz is the only tristatic system; it has receivers in Tromsø, in Kiruna, Sweden, and in Sodankylä, Finland. This allows to measure the three-dimensional vector velocity and the anisotropy of scattering processes. Plans exist to construct an incoherent scatter radar at the Polar Cap Observatory in Resolute Bay, Northern Canada. Its location close to the magnetic pole and the combination with the Sondrestromfjord and EISCAT IS radars will allow to observe a substantial portion of the polar cap ionosphere and its coupling with the magnetosphere.

The EISCAT Svalbard Radar (ESR) is the newest system, comprising an IS radar of the third generation [*Wannberg et al.*, *1997*]. The ESR operates at 1 MW peak power and a high duty cycle of 25%. The transmitter is modular using 16 television klystrons. The antenna is a turning-head design. Its main reflector of 32 m diameter, subreflector and feed horn are optimized for low sidelobes and a high gain of 42.5 dBi. The receiver has a noise figure less than 65 K. It is configured as a dual super-heterodyne device, where the second IF is sampled at 10 MHz. Further frequency translation, filtering and rate decimation is performed in the digital receiver backend, which allows very flexible and optimized adaption to the signal characteristics. The final on-line digital processing provides complex raw data samples or integrated complex autocovariance lag profiles. These are used for on-line or off-line analysis to obtain the basic ionospheric parameters. The ESR has the possibility to transmit and receive on eight frequency channels around the center frequency 500 MHz to observe the ion line. Plasma line capabilities will be added. Modulations can be coded in amplitude and phase.

The simplest form of ISR modulation is the uncoded long-pulse scheme, which is usually applied for F-region observations. The pulse length has to be longer than the correlation time of the echoes. Long un-coded pulses correspond to a coarse range resolution, which is acceptable as long as the scale height of observable parameters is larger than the range resolution. The scale height is several ten kilometers in the F-region such that the application of long-pulses, which match the correlation time, is justified. In the lower F- and the E-region the correlation time of the ISR signals is longer such that longer pulses would be needed. Since this is contradicting the shorter scale height at these altitudes, pulse-coding has to be applied to achieve the required range resolution. A possible method is the multi-pulse scheme, which consists of a set of amplitude modulated sub-pulses. The separation of the sub-pulses is chosen such that the required lagged products of the autocovariance function can be obtained. The minimum scheme is a double pulse, which just yields the mean Doppler shift. The range resolution is given by the duration of a single sub-pulse. In order to fill the gaps between successive sub-pulses, multi-pulse schemes are transmitted interlaced on separate frequency channels. When an even higher range resolution is needed, longer multi-pulse trains with shorter pulses would have to be implemented. This is unrealistic, since there are better methods to achieve this resolution. The sub-pulses can be Barker-coded and coded long-pulses or alternating codes can be employed. These modulation schemes apply amplitude and binary phase modulation. The ambiguity function of the Barker-code has sidelobes, which need to be considered when there are strong gradients of the scatter cross section. The coded long-pulse, characterized by a series of quasi-random binary phase changes within the pulse, has sidelobes, which can be reduced to acceptable levels by sufficiently long integration times. Most advanced systems apply the alternating codes, which consist of a cycle of binary phase-coded pulses. Their phase modulation changes deterministically from pulse to pulse. The decoding is done in the lagged profile domain according to a mathematically precise scheme. The lagged products of one cycle of pulses are finally added such that their sidelobes cancel completely. So far cycles of 16 pulses are applied, which are phase-modulated at 8 bits, which is the so-called weak alternating code. Strong codes consist of an equal number of pulses and modulation bits. An infinite number of alternating codes has recently been announced. It is required, though, that the scattering medium does not change during the cycle, which may not apply for too long cycles. We refer to *Nygren* [*1997*] for detailed discussions of such coding schemes. For F-region plasma-line studies also the frequency sweeping or chirp technique is applied [*Isham et al.*, *1996*], which is most useful for heating diagnostics. Here the frequency sweep is chosen to match the change of plasma frequency with height, which enhances the signal-to-noise ratio. Other modification of the ionosphere by spacecraft and rockets is also studied with ISR [*Bernardt et al.*, *1998*].

When the correlation time of the echo exceeds the time, which is required for the radar signal to travel to the scattering region and back to the receiver, the so-called pulse-to-pulse scheme has to be applied. This is the case for observing the D-region. Here complementary coding can be applied in principally the same way as in the MST radar technique. Depending on the observational goals, different modulation schemes have, thus, to be adopted. We also need to consider that there can be a superposition of incoherent and coherent scatter, such as during PMSE conditions. In the E- and F-region the plasma can be in a non-thermal state and enhanced ion-acoustic fluctuations can occur [*Forme et al.*, *1995*]. This dilutes the application of the standard incoherent scatter theory and opens wide fields of new aspects in plasma physics.

10. Summary and Outlook

We have to appreciate that radar systems are major tools for ground-based studies of the ionosphere and the upper atmosphere. We have summarized here the diversity of radar systems being applied for this research. The high state-of-the-art and continuous evolution of these systems in science and technology are an excellent proof for enriching the understanding of still many open questions and for envisioned new findings in geophysics and plasma physics.

This tutorial paper concentrates merely on short descriptions of the radar techniques whereas the scientific aspects could be discussed briefly only. These are compiled in an extended paper with tables, figures and complete references published elsewhere.

References

Due to space limitations it is not possible to incorporate all relevant literature in this list of references. Only some main references could be included here. An extended list of references covering the broad variety of this tutorial as well as some scientific revenue can be obtained from the author by contacting: <roettger@linmpi.mpg.de>.

Alcayde, D. (Editor) [1997], *Incoherent Scatter Theory, Practice, and Science*, *Techn. Rep. 97/53*, EISCAT Scientific Association, pp. 314.

Anderson, S. J. and Y. I. Abramovich [1998], "A unified approach to detection, classification, and correction of ionospheric distortion in HF sky wave radar systems", *Radio Sci.*, 33, 4, pp. 1055-1067.

Andre, D., G. Sofko, K. Baker and J. MacDougall [1998], "SuperDARN interferometry: Meteor echoes and electron densities from groundscatter", *J. Geophys. Res.*, 103, A4, pp. 7003-7015.

Avery, S. K., J. P. Avery, T. A. Valentic, S. E. Palo, M. J. Leary, and R. L. Robert [1990], "A new meteor echo detection and collection system: Christmas Island mesospheric wind measurements", *Radio Sci.*, 25, 4, pp. 657-669

Bakhmet'eva, N. V., V. V. Belikovich, E. A. Benediktov, V. D. Vyakhirev, N. P. Goncharov, A. V. Tolmacheva and G. S. Korotina [1998], "Studies of the ionosphere and neutral atmosphere using artificial periodic inhomogeneities in the ionospheric plasma", *Radio Sci.*, 33, 3, pp. 583-594.

Bernardt, P. A., J. D. Huba, W. E. Swartz and M. C. Kelley [1998], "Incoherent scatter from space shuttle and rocket engine plumes in the ionosphere", *J. Geophys. Res.*, 103, A2, 2239-2251.

Benson, R. F., B. W. Reinisch, J. L. Green, J.-L. Bougeret, W. Calvert, D. L. Carpenter, S. F. Fung, D. L. Gallagher, D. M. Haines, R. Manning, P. H. Reiff, and W. W. L. Taylor [1998], "Magnetospheric radio sounding on the IMAGE mission", *Radio Sci. Bull.*, 285, pp. 9-20.

Bond, G. E., T. R. Robinson, P. Eglitis, D. M. Wright, A. J. Stocker, M. T. Rietveld and T. B. Jones [1997], "Spatial observations by the CUTLASS coherent scatter radar of ionospheric modification by high power radio waves", *Ann. Geophys.*, 15, pp. 1412-1421.

Bristow, W. A., J. M. Ruohoniemi and R. A. Greenwald [1998], "Super Dual Auroral Radar Network observations of convection during a period of small-magnitude northward IMF", *J. Geophys. Res.*, 103, A3, pp. 4051-4061.

Cho, J. Y. N. and J. Röttger [1997], "An updated review of Polar Mesosphere Summer Echoes: Observation, theory, and their relationship to Noctilucent Clouds and subvisible aerosols", *J. Geophys. Res.*, 102, D2, pp. 2001-2020.

Dieminger, W. and G. K. Hartmann and R. Leitinger (Editors) [1996], *The Upper Atmosphere, Data Analysis and Interpretation*, ISBN 3-540-57562-6, Springer Verlag, Berlin, Heidelberg, New York., pp. 1014.

Djuth, T., M. P. Sulzer, J. H. Elder and V. B. Wickwar [1997], "High-resolution studies of atmosphere-ionosphere coupling at Arecibo Observatory, Puerto Rico", *Radio Sci.*, 32, 6, pp.

2321-2344.

Eglitis, P., T. R. Robinson, M. T. Rietveld, D. M. Wright and G. E. Bond [1998], "The phase speed of artificial field-aligned irregularities observed by CUTLASS during HF modification of the auroral ionosphere", *J. Geophys. Res.*, 103, A2, pp. 2253-2259.

Fejer, B. G. and M. C. Kelley [1980], "Ionospheric irregularities", *Rev. Geophys. Space Phys.*, 18, 2, pp. 401-454.

Forme, F. R. E., D. Fontaine and J. E. Wahlund [1995], "Two different types of enhanced ion acoustic fluctuations observed in the upper ionosphere", *J. Geophys. Res.*, 100, A8, pp. 14,625-14,636.

Foster, J. C. [1995], "Radar observations of magnetosphere-ionosphere coupling at mid and high latitudes", *J. Geomagn. Geoelectr.*, 47, pp. 801-812.

Fukao, S., T. Sato, T. Tsuda, M. Yamamoto, M. D. Yamanaka and S. Kato [1990], "MU radar: New capabilities and system calibration", *Radio Sci.*, 25, 4, pp. 477-485.

Fukao, S., M. Yamamoto, R. T. Tsunoda, H. Hayakawa and T. Mukai [1998], "The SEEK (Sporadic-E Experiment over Kyushu) campaign", *Geophys. Res. Lett.*, 25, 11, pp. 1761-1764.

Greenwald, R. A., K. B. Baker, J. R. Dudeney, M. Pinnock, T. B. Jones, E. C. Thomas, J.-P. Villain, J.- C. Cerisier, C. Senior, C. Hanuise, R. D. Hunsucker, G. Sofko, J. Koehler, E. Nielson, R. Pellinen, A. D. M. Walker, N. Sato and H. Yamagishi [1995], "DARN/SuperDARN: A global view of the dynamics of high-latitude convection", *Space Sci. Rev.*, 71, pp. 761-796.

Haines, D. M., B. W. Reinisch and G. P. Cheney [1997], "Simultaneous oblique sounding with Doppler analysis and low-probability-of-intercept communications using digital ionosondes", *Radio Sci.*, 32, 5, pp. 2065-2073.

Haldoupis, C. [1989], "A review on radio studies of auroral E-region ionospheric irregularities", *Ann. Geophys.*, 7, 3, pp. 239-258.

Headrick, J. and J. M. Thomason [1998], "Applications of high frequency radar", *Radio Sci.*, 33, 4, pp. 1045-1054.

Heinselman, C. J., J. P. Thayer and B. J. Watkins [1998], "A high-latitude observation of sporadic sodium and sporadic E-layer formation", *Geophys. Res. Lett.*, 25, 16, pp. 3059-3062.

Hocking, W. K. [1997a], "Recent advances in radar instrumentation and techniques for studies of the mesosphere, stratosphere, and troposphere", *Radio Sci.*, 32, 6, pp. 2241-2270.

Hocking, W. K. [1997b], "Strengths and limitations of MST radar measurements of middle-atmosphere winds", *Ann. Geophys.*, 15, pp. 1111-1122.

Huang, C. H., E. Kudeki, S. J. Franke, C. H. Liu and J. Röttger [1995], "Brightness distribution of midlatitude E region echoes detected at the Chung-Li VHF radar", *J. Geophys. Res.*, 100, A8, pp. 14703-14715.

Hunsucker, R. D. [1991], *Radio Techniques for Probing the Terrestrial Ionosphere*, Springer Verlag, Berlin, Heidelberg, pp. 293.

Hysell, D. L., M. C. Kelley, A. V. Gurevich, A. N. Karashtin, A. M. Babichenko, Y. M. Yampolski, V. S. Beley and J. F. Providakes [1997], "HF radar probing of the lower magnetosphere", *J. Geophys. Res.*, 102, A3, pp. 4865-4873.

Hysell, D. L. and R. F. Woodman [1997], "Imaging coherent backscatter radar observations of topside equatorial spread-F", *Radio Sci.*, 32, 6, pp. 2309-2320.

Isham, B., C. La Hoz, H. Kohl, T. Hagfors, T. B. Leyser and M. T. Rietveld [1996], "Recent EISCAT heating results using chirped ISR", *J. Atmos. Terr. Phys.*, 58, 1-4, pp. 369-383.

Karashtin, A. N., Yu. V. Shiyugaev, V. I. Abramov, I. F. Belov, I. V. Berezin, V. V. Bychkov, E. B. Eryshev and G. P. Komrakov [1997], "First HF radar measurements of summer mesopause echoes at SURA", *Ann. Geophys.*, 15, pp. 935-941.

Kohl, H. and R. Rüster and K. Schlegel (Editors) [1996], *Modern Ionosphere Science, A Collection of Articles Published on the Occasion of the Anniversary: 50 Years of Ionospheric Research in Lindau*, ISBN 3-9804862-1-4, EGS, Katlenburg-Lindau, pp. 551.

Lester, M., T. B. Jones, T. R. Robinson, E. C. Thomas, T. K. Yeoman, R. Pellinen, A. Huuskonen, H. Opgenoorth, M. Persson, A. Pellinen-Wannberg and I. Häggström [1997], "CUTLASS - A tool for co-ordinated satellite/ground based investigations of the solar terrestrial system", in *Satellite - Ground Based Coordination Sourcebook, ESA SP-1198*, ISBN 92-9092-391-1, pp. 191-202.

Mathews, J. D., D. D. Meisel, K. P. Hunter, V. S. Getman, and Q. Zhou [1997], "Very high resolution studies of micrometeors using the Arecibo 430 MHz radar", *Icarus*, 126, pp. 157-169.

Nygren, T. [1996], *Introduction to Incoherent Scatter Measurements*, Invers Oy, Finland, ISBN 951-97489-0-3, pp. 140.

Pellinen-Wannberg, A. and G. Wannberg [1996], "Enhanced ion-acoustic echoes from meteor trails", *J. Atmos. Terr. Phys.*, 58, 1-4, pp. 495-506.

Pfaff, R. F., J. H. A. Sobral, M. A. Abdu, W. E. Swartz, J. W. LaBelle, M. F. Larsen, R. A. Goldberg and F. J. Schmidlin [1997], "The GUARA campaign: A series of rocket-radar investigations of the Earth's upper atmosphere at the magnetic equator", *Geophys. Res. Lett.*, 13, pp. 1663-1666.

Pulinets, S. A. and R. F. Benson [1999], "Radio-frequency sounders in space", *Rev. Radio Sci. 1996-1999*, ed. W. R. Stone, in press, 1999.

Rao, P. B., A. K. Patra, T. V. Chandrasekhar Sarma, B. V. Krisjna Murty, K. S. V. Subba Rao and S. S. Hari [1997], "Radar observations of updrafting and downdrafting plasma depletions associated with the equatorial spread-F", *Radio Sci.*, 32, 3, pp. 1215-1277.

Reid, I. M., B. G. W. Vandepeer and S. C. Dillon [1995], "The new Adelaide medium frequency Doppler radar", *Radio Sci.*, 30, 4, pp. 1177-1189.

Reinisch, B. W., D. M. Haines, K. Bibl, I. Galkin, X. Huang, D. F. Kitrosser, G. S. Sales, and J. L. Scali [1997], "Ionospheric sounding in support of over-the-horizon radar", *Radio Sci.*, 32, 4, pp. 1681-1694.

Rietveld, M. T., H. Kohl, H. Kopka and P. Stubbe [1993], "Introduction to ionospheric heating at Tromsø - 1. Experiment overview", *J. Atmos. Terr. Phys.*, 55, 4/5, pp. 577-599.

Rishbeth, H. and S. Fukao [1995], "A review of MU radar observations of the thermosphere and ionosphere", *J. Geomag. Geoelectr.*, 47, pp. 621-637.

Roper, R. G. and J. W. Brosnahan [1997], "Imaging Doppler interferometry and the measurement of atmospheric turbulence", *Radio Sci.*, 32, 3, pp. 1137-1148.

Röttger, J. and M. F. Larsen [1990], "UHF/VHF radar techniques for atmospheric research and wind profiler applications", in *Radar in Meteorology*, ed. D. Atlas, Amer. Met. Soc., Boston, pp. 235-281.

Roussel-Dupre, R. A. and E. Blanc [1997], "HF echoes from ionization potentially produced by high-altitude discharges", *J. Geophys. Res.*, 102. A3, pp. 4613-4622.

Russel, C. J., P. L. Dyson, Z. Houminer, J. A. Bennett and L. Li [1997], "The effect of large-scale ionospheric gradients on backscatter ionograms", *Radio Sci.*, 32, 5, pp. 1881-1897.

Sahr, J. D. and B. G. Fejer [1996], "Auroral electroject plasma irregularity theory and experiment: A critical review of present understanding and future directions", *J. Geophys. Res.*, 101, A12, pp. 26,893-26,909.

Sahr, J. D. and F. D. Lind [1997], "The Manastash Ridge radar: A passive bistatic radar for upper atmospheric radio science", *Radio Sci.*, 32, 6, pp. 2345-2358.

Sarango, M. F., R. F. Woodman, B. B. Balsley and D. L. Hysell [1998], "First results from the Jicamarca unattended long-term studies of the ionosphere and atmosphere (JULIA) system", *Proc. 8th Workshop on Technical and Scientific Aspects of MST Radar*, SCOSTEP, Boulder, USA, pp. 310-313.

Stubbe, P., H. Kohl and M. T. Rietveld [1992], "Langmuir turbulence and ionospheric modification", *J. Geophys. Res.*, 97, A5, pp. 6285-6297.

Tsunoda, R. T., [1988], "High-latitude F region irregularities: A review and synthesis", *Rev. Geophys.*, 26, pp. 719-760.

Tsunoda, R. T., R. C. Livingston, J. J. Buonocore and A. V. McKinley [1995], "The frequency-agile radar: A multifunctional approach to remote sensing of the ionosphere", *Radio Sci.*, 30, 5, pp. 1623-1643.

Villain, J. P., R. Andre, C. Hanuise and D. Gresillon [1996], "Observation of the high latitude ionosphere by HF radars: Interpretation in terms of collective wave scattering and characterization of turbulence", *J. Atmos. Terr. Phys.*, 58, 8/9, pp. 943-958.

Wannberg, G., I. Wolf, L. - G. Vanhainen, K. Koskeniemi, J. Röttger, M. Postila, J. Markkanen, R. Jacobsen, A. Stenberg, R. Larsen, S. Eliassen, S. Heck and A. Huuskonen, "The EISCAT Svalbar Radar: A case study in modern incoherent scatter system design", *Radio Sci.*, 32, 6, pp.2283-2307.

Westman, A., T. B. Leyser, G. Wannberg and M. T. Rietveld [1995], "Tristatic EISCAT-UHF measurements of the HF modified ionosphere for low background electron temperatures", *J. Geophys. Res.*, 100, A6, pp. 9717-9728.

Woodman, R. F. [1997], "Coherent radar imaging: Signal processing and statistical properties", *Radio Sci.*, 32, 6, pp. 2373-2391.

THE MEASUREMENT OF WAVE DISTRIBUTION FUNCTIONS

Llewelyn R. O. Storey

1. Introduction

The wave distribution function (WDF) describes how the energy density of a random field of electromagnetic (EM) waves in a plasma is shared among the various waves present; a fuller definition will be given shortly. From an experimenter's standpoint, which will be adopted here, the measurement of WDFs can be viewed as a generalized form of radio direction finding. In non-scientific applications, direction finding generally concerns waves from point sources only and from only a few sources at a time. WDF analysis, which is the set of methods for measuring WDFs, has been concerned until now with waves from extended sources only. This paper argues that the scope of WDF analysis should be widened to include direction finding, in a unified set of methods that can analyze data on waves of either type.

The first step in describing the methods of analysis must be to state the problem that has to be solved, which means doing the following things: first, defining the raw data together with the processing that must be done to prepare them for analysis; second, defining the wave distribution function which is the desired output from the analysis; third, giving the relationship between the processed data and the WDF. The problem is how to invert this relationship so as to estimate the WDF from the processed data. In Section 2 this problem is stated in simple terms, that is to say, with simple assumptions made about the physics; the difficulties that arise when some of these assumptions do not hold are faced in Section 8.

Section 3 presents some techniques for direction-finding in space plasmas. First, the nature of the waves to which it can be applied is clarified in § 3.1. Then, assuming prior knowledge of how many waves are present, § 3.2 outlines the methods that can be used when there is only one wave, while § 3.3 deals with cases where there are more than one. Lastly, § 3.4 addresses the issue of how the number of waves can be determined.

In Section 4 the current techniques of WDF analysis are described in greater detail. The two most favored are the methods of maximum entropy and of maximum smoothness. Following a brief foreword in § 4.1, they are presented successively in § 4.2 and § 4.3.

After the methods of direction finding and those of WDF analysis have been described in Sections 3 and 4, Section 5 discusses how one might handle the difficult cases on the borderline between the domains in which these two sets of methods apply. Section 6 then explains how, before any analysis begins, the processed data might be classified so as to determine which is the right method; however, no software that does the classification has been published yet, and this is a gap in present-day methodology.

Indeed, there is a clear requirement for a large public-domain software package capable of analyzing any set of processed data, irrespective of the nature of the EM wave field that the data concern. It should classify them as mentioned above, apply the appropriate form of analysis, and display the results. Section 7 proposes a flow chart for this sequence of processes.

Data sets of less simple kinds may occur in practice. Examples include data from a cluster of spacecraft, and also data from a station on the ground. Section 8 examines these cases among others. Finally Section 9 presents the conclusions of this review paper, which is addressed to the community of actual and potential users of WDF analysis.

2. Problem Statement

2.1 Foreword

This paper is about the methods of data analysis that can be used for measuring WDFs, but any discussion of methods needs a physics base. The choice of method depends on the nature of the data, which depends in turn on the medium in which the waves are propagating, on the nature of the waves, and on the way in which their fields are being measured. Assumptions about these matters are made later, as and when required, but it may be helpful to list them all beforehand: the plasma is cold, collisionless, subject to a steady magnetic field, and uniform on a scale much larger than a wavelength; the wave fields are random, spread in frequency, and made of waves propagating in just one of the possible modes; several of their electric and magnetic components are being measured at a single fixed point in the plasma, over a period of time, in a narrow band of frequencies. These measurements are the raw data.

It must be stressed that these assumptions, though often fairly well satisfied in practice, are in no way conditions for the applicability of WDF analysis. They are being adopted for the time being as a simple framework in which to discuss the basic features of WDF analysis. In Section 8 some cases will be considered where these assumptions fail; then the methods of analysis have to be modified, but not basically.

2.2 Random Wave Fields

The raw data for WDF analysis are measurements of the electric and magnetic components of random EM wave fields. Any single elementary wave, a characteristic wave for the medium, is specified by its frequency, its wave-number vector (which is parallel to the wave normal), its polarization, amplitude and phase. A *characteristic wave* is a plane wave in which all the field components vary with position $\mathbf{r}$ and time t like $\exp[i(\mathbf{k}.\mathbf{r} - \omega t)]$, with the wave number $\mathbf{k}$ and the angular wave frequency ω satisfying the dispersion relation for the waves concerned. In the present context, a wave field is deemed to be random if it is composed of an infinite number of elementary waves with unrelated phases, and differing either in normal direction, or in frequency, or in both. These three possibilities distinguish three types of random wave field: in the first, all the elementary waves have the same frequency but are spread in normal direction, in the second they have the same normal direction but are spread in frequency, and in the third they are spread in both normal direction and frequency. Only the last two types will be considered here.

2.3 Wave Distribution Function

On the average, the energy density at any point in a random wave field is the sum of the energy densities of its component elementary waves, since these have no systematic phase relationships with one another. Therefore the average properties of the field can be described by a function that specifies, for each of the electromagnetic wave modes present, how the energy density is distributed among the component waves, and this is the wave distribution function.

The WDF can be defined in two ways: in one the component waves are specified by their wave number vectors $\mathbf{k}$, and in the other by their frequencies ω together with their wave-normal directions. This direction is that of the unit vector $\boldsymbol{\kappa} = \mathbf{k}/k$ where $k = |\mathbf{k}|$, and it may be specified conveniently with respect to the steady magnetic field vector $\mathbf{B}_0$ by two angles, namely a polar angle θ and an azimuth angle ϕ. The modulus k of $\mathbf{k}$ is related to the variables ω, θ, and ϕ by the dispersion relation.

Storey and Lefeuvre [*1974*, *1979*] defined the WDF by analogy with a distribution function for particles. In the first form of the definition, the function $\mathrm{f}_m(\mathbf{k})$ is the average energy density in the mode m, due to waves with the wave number $\mathbf{k}$, per unit volume of wave number space.

In the second form, $F_m(\omega, \boldsymbol{\kappa})$ is the same quantity, but due to waves with the frequency ω, per unit frequency range, and with the normal direction $\boldsymbol{\kappa}$, per unit solid angle of this direction. The function f_m or F_m also varies with the time t and with the position $\mathbf{r}$ of the point of observation in space, though here, for simplicity, these variables have not been listed among its arguments. Also, from here on, all the waves present will be assumed to be propagating in the same mode, so the subscript m will be omitted from f and F and from the symbols for related quantities.

2.4 Input Data

The raw data for use in determining the WDFs of random wave fields will be assumed, for the moment, to be measurements of the three components of the electric field **E** and of those of the magnetic field **H** at a fixed point in space, made over a time period T in a band of width $\Delta\omega$ centered on an angular frequency ω_0, with $\Delta\omega \ll \omega_0$. The postulate of a fixed observation point means that the wave frequencies are not Doppler shifted. In practice the number N of measured components of the EM field may be less than six. Accordingly, the raw data are N narrow-band gaussian random time series, and they will be assumed stationary over the period T.

Before the actual analysis can begin the raw data must be processed, a step that consists of using them to form the best possible estimates of the elements S_{ij} of the spectral matrix **S** of the measured field components, as functions of the frequency ω. The procedures for doing this are standard and will not be described here; see, for instance, *Priestley* [*1981*]. Needless to say, the resulting estimates differ from the true values of the elements. Some errors are due jointly to the stochastic nature of the field components, and to the limited intervals of frequency and time over which the measurements are made; errors of this kind, which are unavoidable, will be referred to as *fluctuations*. They need to be distinguished from errors due to interference or *noise*, such as the noise that arises within the receiving equipment, which can often be reduced by improving the techniques of measurement. Besides the data on the wave fields, ancillary data are needed on the plasma parameters, the vector of the steady magnetic field, and the attitude of the spacecraft.

2.5 Direct Problem

In developing WDF analysis, the first task was to solve the so-called *direct problem* of how to determine the spectral matrix for a given WDF. If only one wave mode is present, the expression for the matrix elements is

$$S_{ij}(\omega) = \frac{\pi}{2}\oint a_{ij}(\omega, \boldsymbol{\kappa}) F(\omega, \boldsymbol{\kappa})\, d\sigma \tag{1}$$

where $d\sigma = |d(\cos\theta)\, d\phi|$ is an element of solid angle centered on the wave-normal direction $\boldsymbol{\kappa}$, and the integral is taken over the full 4π range of solid angle (i.e., the "unit sphere"). If both wave modes are present, the term on the right-hand side of this expression is replaced by the sum of two similar terms, one for each mode. The kernel functions $a_{ij}(\omega, \boldsymbol{\kappa})$ in the integrals are different for the two wave modes; they embody all the relevant theoretical knowledge about the properties of the waves. General expressions for them, as functions of frequency, of the angles θ and ϕ, and of the properties of the plasma (which was assumed cold and collisionless) were given by *Storey and Lefeuvre* [*1980*]; an error in one of these expressions was corrected by *Lefeuvre et al.* [*1986*]. Recently *Storey* [*1998*] proposed that the spectral matrix and the integration kernels be redefined in a way that makes the kernels dimensionless without changing Equation (1).

Since, in Equation (1), the indices i and j both run from 1 to 6, it might seem that there are 36 independent kernel functions, corresponding to the 6 purely real auto-spectra and to the real and imaginary parts of the 15 complex cross-spectra that are the elements of the spectral matrix. For a cold plasma, however, four linear relationships exist between them. Three of these were given by *Storey and Lefeuvre* [*1980*], and the fourth follows from Equation (3.8) of a later paper by *Storey* [*1984*]. Thus, for a cold plasma, there are only 32 independent kernels.

2.6 Inverse Problem

The true problem of WDF analysis is the inverse of the one discussed above. The starting point is the set of Equations (1), but now the values of the quantities $S_{ij}(\omega)$ are given at one or more discrete frequencies, though as estimates subject to errors, and the problem is to determine the function $F(\omega, \boldsymbol{\kappa})$ at the same frequencies for the EM wave field in which the measurements were made; it will be considered here at a particular frequency ω_0.

An inverse problem may or may not have a unique solution. The problem of WDF analysis can belong to either category, depending on how many waves are present in the field. The issue of how small the number of waves must be for the solution to be unique will be addressed in the next section. The problems with unique solutions belong to the domain of direction finding, and all others, which have infinitely many possible solutions, to the domain of WDF analysis. These two forms of the inverse problem are discussed in Sections 3 and 4, respectively.

3. Direction Finding

3.1 Foreword

This section is concerned with random EM wave fields of the second of the three types defined in § 2.2 above. Such fields are composed of elementary waves that are spread in frequency over the full measurement band of width $\Delta\omega$ (see § 2.4), but all have the same normal direction. The resulting composite wave is a *quasi-monochromatic plane wave*. It is *plane* in the sense that its wave fronts are planes. The term *quasi-monochromatic* means that its electric and magnetic field components, measured as functions of time, are of the form $A(t)\exp\{i[\psi(t)-\omega_0 t)]\}$, where the amplitude A and the phase ψ vary randomly on a time scale of the order of $(\Delta\omega)^{-1}$, longer by far than a period at the center frequency ω_0 of the band. In the rest of this section and also in the next, the word *wave* will mean a quasi-monochromatic plane wave. The data considered in § 3.2 come from the field of a single wave, and in § 3.3 from a field made up of several waves.

3.2 Direction Finding for a Single Wave

Most of the methods for finding the normal direction of a single wave in a magnetoplasma are based on the fact that it is perpendicular at all times to the magnetic vector $\mathbf{H}$ of the wave, but in order for the direction to be fully determinable from measurements of the three components of $\mathbf{H}$, this vector must define a plane, i.e., the polarization cannot be linear. With circular or elliptic polarization, the normal direction can be estimated as the perpendicular to the plane of $\mathbf{H}$, albeit ambiguously. Two possible directions are found, exactly opposite to one another; measurements of one or more components of $\mathbf{E}$ can distinguish between them.

Theoretically, in the absence of noise, the measurement of direction would be error-free and instantaneous. In practice its accuracy is limited by errors in the measurements of the field components, and its rapidity by the transient response times of the receivers, but the key point is that in neither respect is the measurement affected by random fluctuations. The reason is that the fluctuations do not affect the orientation of the ellipse traced out by the tip of the vector $\mathbf{H}$: they do affect its size, so the mean square amplitude of the wave can be estimated only by averaging the fluctuations over time. However, if noise is present it affects the estimate of the wave-normal direction as well as the estimate of the wave amplitude, because in general the noise vector has a component perpendicular to the ellipse of $\mathbf{H}$ so it causes apparent variations in the orientation of the plane of this ellipse.

The most popular method of direction finding is probably that of *Means* [*1972*], in which the vector area of the magnetic polarization ellipse is computed from the imaginary parts of the cross-spectra of the three components of $\mathbf{H}$. Other methods in use for single waves include that of *McPherron et al.* [*1972*], which analyzes only the real part of the spectral matrix, and a method due to *Samson* [*1973*]. None of these methods needs the values of the plasma parameters. How well they perform in the presence of noise was examined by *Arthur et al.* [*1976*], who concluded that they were all about equally accurate.

If the statistics of the noise are known, the method of maximum likelihood (ML) can be used [*Buchalet and Lefeuvre*, *1981*]. This is the most accurate method since it analyzes all of the available data in the best possible way, unless the prior probabilities of the possible solutions are known too, in which case it is better to combine the likelihood function with the prior probability distribution in a process of bayesian inference [*Edwards*, *1972*]. However, unlike the previous three methods, it requires accurate knowledge of the plasma parameters; also it is relatively slow. In the other methods, the two angles that specify the direction of arrival are computed directly from the experimental data, and then, if necessary, the amplitude of the wave can be computed as well. The ML method, on the other hand, makes use of a theoretical model of a random plane wave with these two angles and the amplitude as its free parameters. Their values are adjusted by successive approximations so as to optimize the agreement between the theoretical spectral matrix and the data, according to the ML criterion. It is common in direction-finding work for the experimenter have a choice like this, between various methods that are rapid but sub-optimal and a unique one that is optimal but considerably slower.

3.3 Direction Finding for Several Waves

The ML method is equally applicable when the field is that of more than one wave, so long as there are not too many. Again, the wave-normal directions and the amplitudes of the several waves are estimated simultaneously, by maximizing their joint likelihood. Thus the number of variables in the likelihood function is three times the number of waves, and the more waves there are the slower is the computation. In their study of data from the GEOS 1 satellite, *Buchalet and Lefeuvre* [*1981*] used this method in several cases where two waves were present.

As in the case with a single wave, other methods exist that are more rapid but sub-optimal. The MUltiple SIgnal Characterization (MUSIC) method due to *Schmidt* [*1981*, *1986*] was among the earliest of this kind, and is still widely used. It belongs to the class known as *signal subspace* methods. Like the ML method, it estimates the normal directions by optimizing a function, but now the function concerned has just two independent variables, the angles θ and ϕ, so the whole process is much faster. It involves searching for several zeros of the function rather than for a single maximum. In the absence of noise there is a zero for each wave, at the corresponding pair of values of θ and ϕ. After the wave-normal directions have been found by this iterative method, the wave amplitudes can be computed forthwith. Once again, statistical fluctuations affect only the estimates of the amplitudes of the waves, not those of their directions, but noise affects the direction estimates as well.

A computer program that implements the MUSIC algorithm for direction finding on waves in the whistler mode has been written and was demonstrated at the Spring AGU meeting in 1993, but has not yet been published. Anyone can get this program by asking me for it.

3.4 Determination of the Number of Waves

Up to now, in discussing the methods for direction finding on one or several plane waves, the number of waves present has been taken as known, and it is time to see how this number can be determined. The standard procedures for doing so are based on the eigenvalues of the estimated spectral matrix. In the absence of noise, the vanishing of at least one eigenvalue is evidence that sufficiently few waves are present for direction finding to apply. The number of waves is given by the rank of the matrix, which is equal to the number of its non-zero eigenvalues; see § 3.8 of *Schmidt* [*1981*] and § 7.3 of *Johnson and Dudgeon* [*1993*]. If N components of the wave field are measured, then the estimated spectral matrix is of size $N \times N$ and it has N eigenvalues, so the largest number of waves for which the directions can be found simultaneously is $N-1$. In the presence of noise, the eigenvalues that would otherwise be zero have values dependent on the noise levels, and it may be hard to tell how many of them would be zero if there were no noise.

Various criteria have been proposed in attempts to answer this question objectively. In all of them, a measure of the quality of fit to the data is offset by a penalty term that increases with increasing complexity of the model, as represented by the number of its free parameters. *Akaike* [*1974*] was the first to propose one, but his criterion is inconsistent, which means that it does not always yield the right number of parameters in the limit of an infinitely large data set [*Kashyap, 1980*]. The *minimum description length (MDL)* criterion due to *Rissanen* [*1978*] is consistent and therefore preferable. Another consistent criterion was proposed by *Knockaert* [*1997*].

4. WDF Analysis

4.1 Foreword

The random EM wave fields considered in this section are of the third type defined in § 2.2: the elementary waves of which they are composed are spread in wave-normal direction as well as in frequency. In other words, a random field of this type is the sum of many quasi-monochromatic plane waves, *many* meaning any number equal to or greater than N, possibly even infinite.

If all the eigenvalues of the estimated spectral matrix are manifestly non-zero it follows that at least N waves must be present, but without some more information there is no way of telling how many. There may be N or any greater number, so the wave distribution function is not fully determined by the data. In such cases, the determination of the WDF by inversion of the set of Equations (1) has infinitely many different solutions, all equally compatible with the data.

None the less, unique estimates of the WDF can be made by the methods described below. In each of them, an assumption is made so as to pick out, from among all the possible solutions, the single one that is deemed the most reasonable. This process is known as *regularization*.

The common aim of these methods of regularization is to select a WDF free of all structure not forced upon it by the data. If the data were exact, the procedure for estimating the WDF at a particular frequency ω_0 would be to identify the non-negative function $\mathrm{F}(\omega_0, \boldsymbol{\kappa})$ that optimizes a certain functional of itself, subject to the constraints of compliance with the data, i.e., of being an exact solution of the set of Equations (1) with $\omega = \omega_0$. The functional concerned is an integral of some function over the unit sphere. The methods now in use fall into two classes: in one of them, the integrand is a function of F itself, while in the other it is a function of a derivative of F. These two classes, the maximum-entropy methods and the maximum-smoothness methods, are described below in § 4.2 and § 4.3.

When the data are subject to error, as in practice they always are, the general procedure for estimating the WDF is much the same, but with the difference that the WDF is usually not made to fit the data exactly. The fit to each item of data is made with a degree of flexibility determined by its variance. Otherwise, if the data were fitted exactly, highly inaccurate items would transmit their errors to the estimated WDF. It may even be expedient to do away with the least accurate items. In studying how errors in the data affect the results of WDF analysis, it is important to distinguish once again between the effects of statistical fluctuations and the effects of noise.

4.2 Maximum-Entropy Methods

The earliest application of the maximum-entropy (ME) method to WDF analysis was made by *Lefeuvre and Delannoy* [*1979*]. The functional they chose to use was the integral over the unit sphere of the function $\mathrm{F} \ln \mathrm{F}$, which has the same algebraic form as the thermodynamic entropy. They maximized it by the calculus of variations, thus obtaining the following parametric model for the WDF:

$$\mathrm{F}(\omega_0,\theta,\phi) = \exp[-1 + \sum_n \mu_n \mathrm{q}_n(\omega_0,\theta,\phi)\, d\theta\, d\phi] \tag{2}$$

The functions q_n are the real and imaginary parts of the integration kernels a_{ij} (see § 2.5). If N components of the EM wave field are measured, then there are N^2 such functions so the index n runs from 1 to N^2. The parameters μ_n are the Lagrange multipliers, and their values have to be adjusted so as to reconcile this model with the N^2 items of data, the real and imaginary parts of the estimated spectral matrix elements, as explained in § 4.1. A program for computing WDFs in this way was published by *Delannoy and Lefeuvre* [*1986*]; it was used for analyzing data from the satellites GEOS 1 [*Lefeuvre et al.*, *1981*; *Hayakawa et al.*, *1986*], ARCAD 3 [*Béghin et al.*, *1985*], and ISEE 1 [*Storey et al.*, *1991*].

This particular version of the ME method is open to the objection that the results are not independent of scale. If all the estimated spectral matrix elements are multiplied by some factor, the resulting WDF changes in shape instead of just being multiplied by this factor. The reason is that the entropy functional involves the logarithm of F, and F is not dimensionless. Accordingly *Oscarsson and Rönnmark* [*1989*] preferred to use an entropy functional in which the argument of the logarithm is F normalized relative to the integral of F over the unit sphere; this integral is the total wave energy per unit volume and unit bandwidth. More precisely, because these authors worked with the WDF in the form $\mathrm{f}(\mathbf{k})$, they made a similar normalization in that context. Their method of WDF analysis, as described by *Oscarsson* [*1994*], was used to analyze data from the satellites Viking [*Oscarsson and Rönnmark*, *1990*] and Freja [*Oscarsson et al.*, *1997*].

Two features of the maximum-entropy solution, apparent in Equation (2), deserve to be emphasized. First, the form of the solution is such that the WDF is automatically positive; this is one of its most significant properties. Second, entropy is not a criterion of smoothness. The basis functions used in constructing the maximum-entropy model are the real and imaginary parts of the integration kernels, so the solution for the WDF is a smooth function of the angles θ and ϕ only if the kernel functions are smooth.

4.3 Maximum-Smoothness Methods

In this alternative method, the functional that would be used if the data were exact is the integral of the function $|\nabla^2 F|^2$ over the unit sphere. It is a functional of the second derivatives of F with respect to the angles θ and ϕ. and is a measure of the roughness of F. The resulting solution is the wave distribution function $F(\omega_0, \theta, \phi)$ that minimizes this functional while being consistent with the data exactly. With inexact data, the functional to be minimized is a weighted sum of this one, and of another that expresses the degree of mismatch to the data. This approach to inverse problems is called *Phillips-Tikhonov regularization*. It was used by *Yamaguchi et al.* [*1994*] for the WDF analysis of data from a ground station (see § 8.4), then adapted for use with data from the Akebono (EXOS D) satellite by *Kasahara et al.* [*1994*, *1995*].

Functionals such as these cannot be minimized readily by the calculus of variations: the attempt to do so yields a differential equation for F that is no easier to solve. Hence they must be minimized directly, and the first step is to discretize the WDF, i.e., to replace the continuous function F by the set of its discrete values for very many closely spaced directions (θ,ϕ) over the unit sphere. This large set of parameters is then adjusted so as to minimize the functional.

This process yields the smoothest possible solution that is consistent with the data. It would make the WDF uniform if the data permitted. On the other hand, it does not compel the solution to be non-negative. Non-negativity could be imposed as a constraint on the process of minimization, but in most of the work published to date this does not seem to have been done.

4.4 Discussion

The problems of WDF analysis discussed in this section, and the problems of direction finding discussed in Section 3, are not wholly separate. It is possible to go from one class to the other by a continuous change in the definition of the problem. Consider, for instance, the case where the true WDF is comprised simply of one or of several $(< N)$ peaks. If the peaks are broad, then the WDF is a continuous function and the problem of inferring it from the estimated spectral matrix has no unique solution. Now imagine that the width of each of these peaks is made very small and its height very large while the volume under it is held constant, so that it approximates a delta function. Then the problem has effectively been turned into that of finding the directions and amplitudes of a few plane waves. Therefore an essential requirement of any general solution suggested for the WDF-analysis problem is that, in any such limit, it should reduce to the unique solution of the direction-finding problem.

In the case where the data are exact, this must be true for the maximum-entropy solution, it cannot be otherwise. When there is only one non-negative solution that agrees with the data, it is necessarily the one with maximum entropy. The same reasoning would apply to the solution of maximum smoothness, if the constraint of non-negativity were imposed upon it. In the absence of this constraint, however, it seems likely that the maximum-smoothness method would yield a relatively smooth solution that is negative for certain directions and therefore unacceptable.

If the data are in error because of fluctuations or noise, and the statistics of these errors are known, then in the limit just mentioned any valid solution of the WDF-analysis problem should go over to the known optimum solution of the direction-finding problem, namely the maximum-likelihood solution (or the solution of maximum posterior probability, if bayesian inference is being used). Neither the maximum-entropy (ME) nor the maximum-smoothness (MS) solutions have yet been shown to satisfy this requirement.

In evaluating and comparing the various methods of WDF analysis it is vital to distinguish between the algebraic and numerical solutions, i.e., between each formal solution of the inverse problem and the computer algorithm that implements it. A solution may seem effective and may satisfy the condition in the limit just mentioned, but the computer may fail to implement this solution properly, for instance because it uses an iterative algorithm that converges too slowly.

Some tests of the various methods have been made with artificial data. *Oscarsson* [*1994*] took a model WDF comprising a single narrow peak, from which he synthesized some data with a signal-to-noise ratio of 10 dB. Given these as input, his ME method reproduced the peak in the correct position, but slightly broadened. *Kasahara* [*1995*] took several model WDFs with one or two peaks, narrow or broad, produced synthetic data with a signal-to-noise ratio of 10 dB also, and analyzed these data both by an ME method and by an MS method. He found that the ME method was the better at reproducing narrow peaks, but that the MS method was less susceptible to noise. These conclusions should be regarded as provisional, since both classes of methods are open to improvement.

5. Borderline Cases

Sections 3 and 4 were concerned with the respective domains of direction finding and of wave distribution function analysis, but in § 4.4 it was pointed out that these domains are contiguous, and that consequently the results of WDF analysis should converge towards those of direction finding in any limit where a continuous WDF is changed so that it approximates, more and more closely, a delta function or a set of several $(< N)$ such functions. This is true in theory, but in practice some numerical methods, for instance those based on the maximum-entropy principle, have trouble in these borderline cases. In a parametric model such that of Equation (2), the WDF is modeled by forming the weighted sum of a set of smooth basis functions, and then distorting the sum by exponentiation. The parameters of the model, which have to be adjusted numerically, are the weights assigned to the basis functions. This procedure works well when the WDF is itself a smooth function, but in the approach to limits where the WDF becomes a set of delta functions, the values of two or more of the parameters tend to plus or minus infinity and this state of affairs leads to numerical difficulties.

In such a case, it is better to use a model with parameters that all stay finite in the limit. For example, *Santolík* [*1995*] has used one in which the WDF is modeled as a single peak or a set of several peaks of a given form, the parameters being, for each peak, the two angles defining its position, the volume under it, and its width. When the width of a peak becomes smaller than the scale of the angular variations of the integration kernels, the contributions from this peak to the spectral matrix elements become roughly proportional to the volume under the peak (that is to say, the total energy in the peak, per unit bandwidth), and are almost independent of the width. As the width tends to zero, the height of the peak tends to infinity but the volume under it stays finite, which is why the volume is a good parameter for use in a numerical model.

It remains to decide what analytic form should be given to each peak in the model, and as in § 4.4, the deciding factor is the need for continuity with the other two solutions. Continuity with direction finding is guaranteed: as its width decreases to zero at constant volume, each peak turns into a delta function irrespective of its form. Therefore the form should be determined by continuity with WDF analysis. In that case it will depend on what method of analysis is in use, maximum-entropy or maximum-smoothness. In various simple inverse problems defined by equations similar to (1), but with kernels that are simple analytic functions of the variables, it is easy to show that when the true function F is changed from a set of smooth peaks into a set of delta functions, the maximum-entropy solution (2) assumes an asymptotic form that behaves in the same way. The constants in the asymptotic ME solution are explicit functions of the data, at least when these are exact. I suspect that such solutions always exist, but have not yet discovered how to find them in the general case. If this conjecture proves to be correct, and if an ME method is in use, then the asymptotic ME solutions are the right analytic functions for modeling WDFs in these borderline cases: with a maximum-smoothness method, similar and possibly identical asymptotic solutions may exist, but they have not been investigated at all.

6. Data Classification

Three different methods of data analysis have been described in Sections 3 through 5, and the question arises as to how, given a set of data, one can tell which method is appropriate. A partial answer was given in § 3.4 in the context of direction finding, where the MDL criterion and some others were mentioned as ways of determining how many waves are present. Similar criteria can be used for helping to make decisions in each of the other methods of analysis, and for choosing between the three classes of method. These criteria have shortcomings, however. Firstly, they suppose that the wave data are corrupted only by fluctuations and noise. Other possibilities, for instance that they are subject to some error so gross that no solution for the WDF exists, are not covered, nor is the risk that supporting data such as the plasma parameters might be erroneous. Secondly, the way in which one decides between any two methods of analysis is to solve for the WDF by both methods, and then use the criterion to judge which solution is the better of the two. Obviously it would be preferable to have some means of classifying the data, to find out which of the different methods is the best for them, before the actual analysis begins.

One way of doing this would consist of plotting the wave data, in the form of the real and imaginary parts of the estimated spectral matrix elements S_{ij}, as the components of a vector in a multi-dimensional *data space*, having as many dimensions as there are items of independent data [*Storey and Chapron, 1988*]. Using the fact that the WDF cannot be negative, it can be shown that if the data are exact, then the data vectors corresponding to feasible solutions all lie inside a region bounded by a convex cone with its apex at the origin. The shape of this cone is governed by the kernel functions, which depend on the supporting data. Any set of nominally exact wave data for which the vector lies outside the cone cannot be analyzed to obtain a WDF: either they or their supporting data must be wrong. Data vectors that lie exactly on the cone correspond to fields consisting of one or several plane waves, few enough for the data to be analyzable by the methods of Section 3. If the data vector is well inside the cone then the methods of Section 4 are appropriate, or if it is only just inside, the methods of Section 5. When the data are affected by fluctuations and noise these conclusions are weakened, becoming matters of probability instead of being clear-cut. No software that performs this classification has been published yet.

7. Process Chart

If WDF analysis is ever to come into widespread use, a package of computer software must be developed that performs all of the foregoing processes swiftly, and as automatically as possible. The user should intervene only to make difficult choices, the decisive information for which is displayed by the computer. Figure 1 outlines a process chart for this essential software.

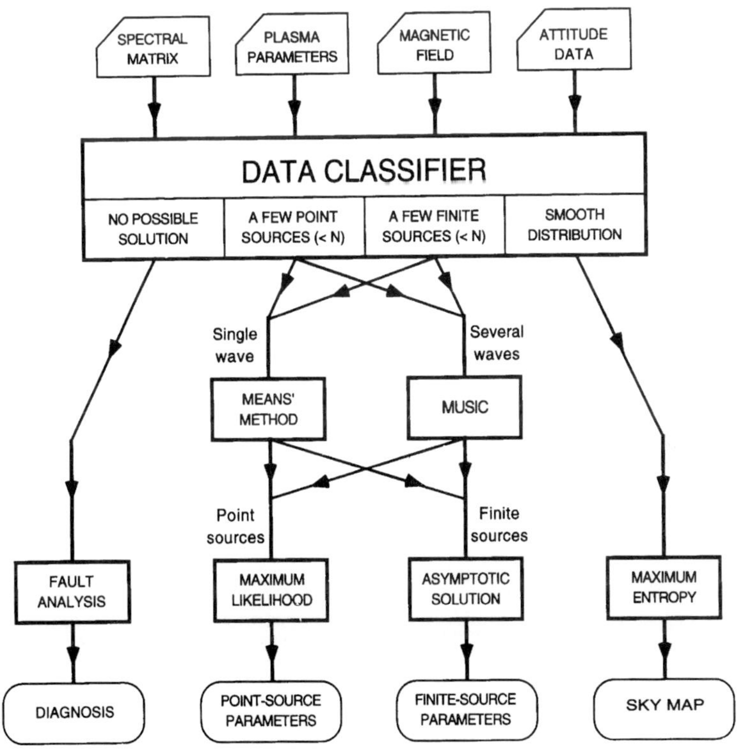

Fig. 1. *A process chart for wave distribution function analysis.*

The uppermost line represents the physical data that are the input for the analysis; they comprise the estimated spectral matrix, the plasma parameters, the vector of the steady magnetic field, and the attitude of the spacecraft. The first step in the analysis is one of classification, as described in Section 6. The spectral matrix elements are used to create the data vector, while the supporting data are used to compute the kernels which define the convex cone. The subsequent course of the analysis is decided by the position of this vector with respect to the cone.

If the vector is well outside the cone, then no feasible solution for the WDF exists, which means that some of the data must be badly wrong. The regions of data space within which the data vector can lie may be different for the different causes of error, and if so, its position there may be a clue to the nature of the fault. This diagnostic possibility needs investigation.

If the vector is only just outside, close enough to the cone that this fact can be attributed to random errors, then its most likely true position is exactly on the cone and the field is that of just one or of several plane waves, such as might be produced by a few point sources. In this case the methods of direction finding apply. In Section 3 it was noted that these fall into two classes: first, various methods that are fast but sub-optimal (e.g., Means' method if only one wave is present, or MUSIC if there are several); second, the maximum-likelihood method which is optimal but relatively slow. Hence the best approach might be to combine the two: first the directions and amplitudes are estimated roughly by one of the fast methods, the choice of which depends on the number of waves present, then these estimates are refined by the ML method as the figure shows.

In the borderline cases where the data vector lies just inside the cone, the field is such as might be produced by a few finite, extended sources and a similar process is applicable. As a first step, preliminary estimates of the parameters of a plane-wave model of the field are made in the way just described, and then a model with additional free parameters is fitted to the data by the ML method. This model is of a WDF comprising one or several peaks of finite width.

Finally, if the data vector lies well inside the cone, this means that the WDF is relatively smooth, not peaky, so it can be estimated by one of the standard methods, maximum entropy or maximum smoothness. The result can be displayed as a sky map on the unit sphere, as shown in the last line of Figure 1 which represents the outputs from the various processes. The full set of processes shown on this chart can be regarded as WDF analysis in a wide sense of this term.

8. Extensions

8.1 Foreword

Up to this point, the methods of WDF analysis have been discussed in their application to data of the simple, idealized type defined in § 2.1, all coming from a single spacecraft. However, the need to deal with real experimental data has motivated the development of various extensions of these methods, which will be described now. The next three sub-sections concern the analysis of data from a single spacecraft, from a cluster of several spacecraft, and from stations on the ground.

8.2 Data from a Single Spacecraft

The simplifications made in Section 2 include assuming the point of observation to be fixed in space, so the observed wave frequencies are not Doppler shifted. A second assumption, made in § 2.5, is that the dispersion relation for the waves is given sufficiently accurately by cold plasma theory. For data from a spacecraft in the magnetosphere, these two approximations are apt to fail in the same circumstances, namely when a large fraction of the energy density resides in waves propagating close to one of the plasma resonances; examples include whistler-mode waves close to the lower oblique or lower hybrid resonance, and electromagnetic ion cyclotron waves close to an ion gyro-resonance.

Oscarsson and Rönnmark [*1989*, *1990*] and *Santolik and Parrot* [*1996*] have shown how to avoid both these approximations, by taking account of Doppler shifts and by using warm-plasma wave theory. The Doppler shifts must be corrected for when computing the integration kernels. The WDF is then estimated in the normal way by one of the methods discussed in Sections 3-5. The result is an estimate is of the wave energy density per unit range of frequency, in the band of *observed* frequencies of width $\Delta\omega$ centered on a frequency ω_0 (see § 2.4), as a function of the wave-normal direction. For each direction, the *true* frequency of the wave in the frame of the plasma is known; it differs from ω_0 by a known Doppler shift. Next, this operation is repeated with data from a number of adjacent, and possibly overlapping, frequency bands. Then, for any given direction, the results from the different bands can be interpreted as points on a curve of the WDF versus true frequency. The value of the WDF at any true frequency in the whole range covered can be found by interpolation. By combining this information from the curves for many different directions, one can obtain the angular variation of the WDF at the chosen true frequency and can plot it as a sky map in the usual way. As regards the use of warm-plasma theory, the corrections that must be made to the dispersion relation to take account of the non-zero electron and ion temperatures are based on the work of *Rönnmark* [*1982*, *1983*]. If WDF analysis is ever used for analyzing data on the field of Alfvén waves in the solar wind, these two factors, especially the Doppler shift, will be of major importance instead of being just minor corrections.

8.3 Data from Several Spacecraft

When data from a single spacecraft are analyzed for the WDF, the resolution attainable in the direction of propagation is severely limited. There is a well known relation between the size of an antenna and its beam width, or angular resolution: an antenna of characteristic dimension d, used at a wavelength λ, has a beam width of the order of λ/d in radians. The antennas carried by a single spacecraft, be they electric dipoles or magnetic loops, are usually much smaller than a wavelength so they have cosine reception patterns. With such poor directivity, the best that can be done is to take all the available data on the wave field by measuring all six of its components, and then to analyze them in the best possible way. This involves using all other relevant data, in particular measurements of the plasma parameters needed for calculating the dispersion relation and other properties of the waves; the fact that the WDF cannot be negative is most helpful also. By these means, WDF analysis achieves some degree of "super-resolution", but none the less it is clear that the directional resolution could be greatly improved if the wave data were taken over a longer baseline than can be had with a single spacecraft, and clusters of several spacecraft offer this possibility.

In preparation for the ISTP Cluster mission, which involves four spacecraft, much thought has been given as to how the resulting wave data could be analyzed; see *Cornilleau-Wehrlin et al.* [*1997*] and the references therein. In principle, the methods of WDF analysis as described above could be used with these data. The basic equation (1) still applies, but now the spectral matrix **S** is larger, being that of all the measured field components. It includes the cross-spectra of all of the pairs of components of which one is measured on one of the spacecraft and the other on another. In such cases, the expressions for the kernels a_{ij} take into account the vector distance separating the two spacecraft. These particular elements of the estimated spectral matrix are the data with potential for improving the directional resolution.

A different method of analysis has been developed by *Pinçon and Lefeuvre* [*1988*, *1991*], who foresee that Cluster may encounter waves or wavelike disturbances for which the dispersion relation is wholly or partly unknown. With electromagnetic waves, this situation arises whenever some of the relevant properties of the plasma are unknown; for instance, if the ionic composition is not measured, the dispersion relation is uncertain for EM waves with frequencies of the order of the ion gyro-frequencies. This means that, for such waves, the modulus k of the wave-number vector $\mathbf{k}$ is not known as a function of the frequency ω and of the wave-normal direction. Nor is their total energy density known as a function of these two variables and of the amplitudes of the field components. To be exact, the unknown quantity is the contribution from the kinetic energy of the plasma particles: the electromagnetic energy density, i.e., the sum of the energy densities of the electric and magnetic fields, is calculable if all six field components are measured. These authors therefore introduce a field-energy distribution (FED) function specifying how the EM energy density alone is distributed with respect both to the frequency ω and to the wave number vector $\mathbf{k}$ $(= k\boldsymbol{\kappa})$. The modulus k, no longer given as a function of ω and of $\boldsymbol{\kappa}$ by a dispersion relation, is now another independent variable on which the FED depends. Their procedure for estimating the FED, from data taken by a cluster of spacecraft, is based on one devised by *Capon* [*1969*] for analyzing data from an array of seismometers on the ground; satisfactory results have been obtained in tests with artificial data.

8.4 Data from a Ground Station

Electromagnetic waves that emerge from the ionosphere, and are then observed at stations on the ground, also can be studied by WDF analysis. However, the data taken at a single ground station differ in the following respects from those taken by a single spacecraft: only three components of the EM wave field can be measured; the waves coming down from the ionosphere are observed in the presence of their reflections from the ground; lastly, their polarization is uncertain. These points are examined below.

At low frequencies, to a fair approximation, the ground can be regarded as being a perfect conductor. This implies that the two horizontal components of the electric field of a wave, and the vertical component of its magnetic field, all vanish at its surface. Hence only three non-zero components remain to be measured, namely the vertical component of the electric field and the two horizontal components of the magnetic field. The restriction to three components limits the possibilities of the analysis: for instance, the normal directions of no more than two plane waves can be determined at the same time, for the reasons given in § 3.4. With fewer data, on the other hand, analyzing them is a lighter task.

Because of these boundary conditions at the Earth's surface, every down-coming wave is accompanied by an up-going reflected wave of equal amplitude, a fact that has to be taken into account when deriving the integration kernels. With only three field components measured, the spectral matrix has dimensions 3×3; therefore there are only nine kernels, the expressions for which were given by *Shimakura et al.* [*1992*]. For consistency with previous work on direction finding, they were expressed as functions of the direction of arrival (DOA) of the down-coming wave, which is the opposite of its wave-normal direction. The DOA is specified by its inclination angle θ to the zenith, and by its azimuth angle ϕ which has the east as origin. The integration kernels depend also on the polarization of the down-coming wave.

One drawback of data from a ground station, as compared with data from a spacecraft, is that the polarization is not known theoretically as a function of the direction of arrival. To some degree this can be overcome by measuring the polarization at times when the field appears to be just that of a single down-coming wave plus its reflection. Given many such measurements, it might be possible to find an empirical relationship between the polarization and the angle θ. In practice, however, there is much variability, the polarization depending not solely on this angle but also, no doubt, on the unknown conditions in the lower ionosphere from which the wave has emerged. Seemingly, in WDF analysis, the best approach is to assume the average polarization for each value of θ, while recognizing that the variability is a source of error. *Shimakura et al.* [*1992*] assumed the polarization to be right-handed circular for all θ, and the errors that arose due to the departures from this value were evaluated by *Shimakura and Hayakawa.* [*1992*]. A maximum-entropy method of analysis was used in this initial work, but was replaced later by maximum-smoothness methods using Phillips-Tikhonov regularization [*Yamaguchi et al., 1994*; *Iwama et al., 1995*; *Hirari and Hayakawa, 1996a*], one of which was implemented by means of a neural network [*Hirari and Hayakawa, 1996b*].

9. Conclusion

In this review paper, the measurement of wave distribution functions in space plasmas has been discussed mainly in what was meant to be the simplest possible context, as defined in Section 2. The reason for doing so, even though work in the field has moved on to more advanced problems such as those described in Section 8, is that this simple one has not yet been fully solved, in the sense that no computer program exists that can be used for estimating the WDF in all cases. A program of this kind would have to include some of the methods used for direction finding. In Section 7, a possible outline of such a program was presented in the form of a process chart, but several of the blocks in this chart represent operations for which not even the basic algorithms have been developed, not to mention any software. The chief gaps concern the procedures for classification, both the initial data classification shown at the top of the chart, and also the tests, mentioned in Section 3 but not shown on the chart, that may be needed to help the user to decide which of several possible models fits the data best. The entire program could be developed most readily in the simple context of Section 2, and later extended so as to broaden its scope.

Acknowledgments

I thank the referee, T. Oscarsson, for his valuable comments, and Y. Kasahara, M. Parrot, J.-L. Pinçon and O. Santolik for their advice and help. This paper is based on a lecture given at the University of Iowa in June 1997, at the invitation of D. A. Gurnett.

References

Akaike, H. [1974], "A new look at the statistical model identification," *IEEE Trans. Automatic Control*, AC-19, 6, pp. 716-723.

Arthur, C. W., R. L. McPherron, and J. D. Means [1976], "A comparative study of three techniques for using the spectral matrix in wave analysis," *Radio Science*, 11, 10, pp. 833-845.

Béghin, C., et al., [1985], "Experimental evidence of field-aligned ELF plasma ducts in the ionospheric trough and in the auroral zone," in *Résultats du Projet ARCAD 3 et des Programmes Récents en Physique de la Magnétosphère et de l'Ionosphère*, Toulouse, Cepadues Editions, pp. 517-527.

Buchalet, L. J. and F. Lefeuvre [1981], "One and two direction models for VLF electromagnetic waves observed on-board GEOS 1," *J. Geophysical Research*, 86, A4, pp. 2377-2383.

Capon, J. [1969], "High-resolution frequency-wavenumber spectral analysis," *Proc. IEEE*, 57, pp. 1408-1418.

Cornilleau-Wehrlin, N., et al. [1997], "The Cluster spatio-temporal analysis of field fluctuations (STAFF) experiment," *Space Science Reviews*, 79, pp. 107-136.

Delannoy, C. and F. Lefeuvre [1986], "MAXENTWDF — A computer program for the maximum entropy estimation of a wave distribution function," *Computer Physics Communications*, 40, pp. 389-419.

Edwards, A. W. F. [1972], *Likelihood*, Cambridge University Press.

Hayakawa, M., M. Parrot, and F. Lefeuvre [1986], "The wave normals of ELF hiss emissions observed onboard GEOS 1 at the equatorial and off-equatorial regions of the plasmasphere," *J. Geophysical Research*, 91, A7, pp. 7899-7999.

Hirari, M. and M. Hayakawa [1996a], "A Bayesian regularization approach to ill-posed problems with application to the direction finding of VLF/ELF radio waves," *IEICE Trans. Communications*, E79-B, 1, pp. 63-69.

Hirari, M. and M. Hayakawa [1996b], "A neural network for the DOA of VLF/ELF radio waves," *IEICE Trans. Communications*, E79-B, 10, pp. 1598-1605.

Iwama, N., et al., [1995], "GCV-aided linear reconstruction of the wave distribution function for the ground-based direction finding of magnetospheric VLF/ELF waves," *J. Electromagnetic Waves App.*, 9, 5/6, pp. 757-782.

Johnson, D. H. and D. E. Dudgeon [1993], *Array Signal Processing: Concepts and Techniques*, Englewood Cliffs, NJ, Prentice Hall.

Kasahara, Y. [1995], *Study on ELF/VLF Emissions in a Multicomponent Plasma Observed by the Akebono Satellite*, Kyoto University, Ph. D. thesis.

Kasahara, Y., H. Kenmochi, and I. Kimura [1994], "Propagation characteristics of the ELF emissions observed by the satellite Akebono in the magnetic equatorial region," *Radio Science*, 29, 4, pp. 751-767.

Kasahara, Y. et al., [1995], "Propagation characteristics of auroral hiss observed by Akebono satellite," *J. Geomagnetism and Geoelectricity*, 47, pp. 509-525.

Kashyap, R. L. [1980], "Inconsistency of the AIC rule for estimating the order of autoregressive models," *IEEE Trans. Automatic Control*, AC-25, pp. 996-998.

Knockaert, L. [1997], "Maximum *a posteriori* maximum entropy order determination," *IEEE Trans. Signal Processing*, SP-45, 6, pp. 1553-1559.

Lefeuvre, F. and C. Delannoy [1979], "Analysis of a random electromagnetic wave field by a maximum entropy method," *Annales des Télécommunications*, 34, 3-4, pp. 204-213.

Lefeuvre, F., et al., [1986], "Rapid determination of the sense of polarization and propagation for random electromagnetic wave fields. Application to GEOS-1 and AUREOL-3 data," *Annales Geophysicae*, 4A, pp. 457-468.

McPherron, R. L., C. T. Russell, and P. J. Coleman [1972], "Fluctuating magnetic fields in the magnetosphere, 2, ULF waves," *Space Science Reviews*, 13, pp. 411-454.

Means, J. D. [1972], "Use of the three-dimensional covariance matrix in analyzing the polarization properties of plane waves," *J. Geophysical Research*, 77, 28, pp. 5551-5559.

Oscarsson, T. E. [1994], "Dual principles in maximum entropy reconstruction of the wave distribution function," *J. Computational Physics*, 110, pp. 221-233.

Oscarsson, T. E. and K. G. Rönnmark [1989], "Reconstruction of wave distribution functions in warm plasmas," *J. Geophysical Research*, 94, pp. 2417-2428.

Oscarsson, T. E. and K. G. Rönnmark [1990], "A combined wave distribution function and stability analysis of Viking particle and low-frequency wave data," *J. Geophysical Research*, 95, A12, pp. 21187-21202.

Oscarsson, T. E., et al., [1997], "Toward a consistent picture of the generation of electromagnetic ion cyclotron ELF waves on auroral field lines," *J. Geophysical Research*, 102, A11, pp. 24369-24386.

Pinçon, J. L. and F. Lefeuvre [1988], "Characterization of a homogeneous field turbulence from multipoint measurements," *Advances in Space Research*, 8, 9-10, pp. (9)459-(9)462.

Pinçon, J. L. and F. Lefeuvre [1991], "Local characterization of homogeneous turbulence in a space plasma from simultaneous measurements of field components at several points in space," *J. Geophysical Research*, 96, A2, pp. 1789-1802.

Priestley, M. B. [1981], *Spectral Analysis and Time Series*, London, Academic Press.

Rissanen, J. [1978], "Modeling by shortest data description," *Automatica*, 14, pp. 465-471.

Rönnmark, K. [1982], "WHAMP – Waves in homogeneous anisotropic multicomponent plasmas," Kiruna Geophysical Institute, Kiruna, Sweden, Report No. 179.

Rönnmark, K. [1983], "Computation of the dielectric tensor of a maxwellian plasma", *Plasma Physics*, 25, (6), pp. 699-701.

Samson, J. C. [1973], "Descriptions of the polarization states of vector processes: applications to ULF magnetic fields," *Geophysical J. Royal Astronomical Soc.*, 34, pp. 403-419.

Santolik, O. [1995], *Etude de la Fonction de Distribution des Ondes dans un Plasma Chaud*, University Doctorate thesis, University of Orléans, France.

Santolik, O. and M. Parrot [1996], "The wave distribution function in a hot magnetospheric plasma: the direct problem," *J. Geophysical Research*, 101, A5, pp. 10639-10651.

Schmidt, R. O. [1981], *A Signal Subspace Approach to Emitter Location and Spectral Estimation*, Stanford University, California, Ph. D. thesis.

Schmidt, R. O. [1986], "Multiple emitter location and signal parameter estimation," *IEEE Trans. Antennas Propag.*, AP-34, pp. 276-280.

Shimakura, S. and M. Hayakawa [1992], "Wave distribution functions of magnetospheric VLF waves with multiple field components: the effect of the polarization model in the integration kernels on the reconstruction of wave distribution function," *IEICE Trans. Fundamentals*, E75-A, pp. 1014-1019.

Shimakura, S., et al., [1992], "On the estimation of wave energy distribution of magnetospheric VLF waves at the ionospheric base with ground-based multiple electromagnetic field components," *J. Geomagnetism Geoelectricity*, 44, pp. 573-590.

Storey, L. R. O. [1984], "The energy density of electromagnetic waves in a cold magnetoplasma," *J. Plasma Physics*, 32, 2, pp. 309-318.

Storey, L. R. O. [1998], "Revision of the basic equations of wave distribution function analysis," *Annales Geophysicae*, 16, pp. 651-653.

Storey, L. R. O. and S. Chapron [1988], "Conditions for the existence and uniqueness of solutions to a constrained linear inverse problem," *Inverse Problems*, 4, pp. 249-271.

Storey, L. R. O. and F. Lefeuvre [1974], "Theory for the interpretation of measurements of the six components of a random electromagnetic wave field in space," in M. J. Rycroft and R. D. Reasenberg (eds.), *Space Research XIV*, Berlin, Akademie-Verlag, pp. 381-386.

Storey, L. R. O. and F. Lefeuvre [1979], "The analysis of 6-component measurements of a random electromagnetic wave field in a magnetoplasma — I. The direct problem," *Geophysical J. Royal Astronomical Soc.*, 56, pp. 255-270.

Storey, L. R. O. and F. Lefeuvre [1980], "The analysis of 6-component measurements of a random electromagnetic wave field in a magnetoplasma — II. The integration kernels," *Geophysical J. Royal Astronomical Soc.*, 62, pp. 173-194.

Storey, L. R. O., et al., [1991], "Initial survey of the wave distribution functions for plasmaspheric hiss observed by ISEE 1," *J. Geophysical Research*, 96, A11, pp. 19469-19489.

Yamaguchi, M., et al., [1994], "A new direction finding method of magnetospheric VLF/ELF radio waves using the linear regularization and generalized cross-validation," in H. Kikuchi (ed.), *Dusty and Dirty Plasmas, Noise and Chaos in Space and in the Laboratory*, New York, Plenum, pp. 405-414.

RADIO STARS: THE HIGH SENSITIVITY FRONTIER

A. Russell Taylor

1. Overview of Radio Continuum Emission from Stars

Concerted effort for more than two decades has resulted in the detection of several hundred stellar radio sources. The pace of radio detections increase dramatically with the commissioning of the Very Large Array in the early 1980's when the combination of sensitivity and angular resolution made possible routine imaging at the arc-second level and sub-mJy sensitivity. The loci of 441 radio-detected stellar systems are shown on a Hertzsprung-Russell Diagram (HRD, brightness M_V vs. colour B-V) in Fig.1. Radio emission is associated with virtually all stages of stellar evolution. We clearly see the main sequence from luminous OB stars down to very low luminosity M stars with a noticable clump in the G-K subgiant/giant area, mostly due to RS CVn binary systems and pre-main-sequence (PMS) stars. The giant and supergiant branches are visible, with concentrations in the red giant/supergiant area, mostly associated with the ionized winds of cool stars with hot, ionizing companions (symbiotic stars). Despite the apparent plenitude of radio detections across the HR diagram, this diagram reflects strong selection bias and detection limitations. The radio luminosity increases drastically with increasing absolute brightness (decreasing M_V). However, since the space density of optically luminous stars is very low, these stars are on average distant objects, with only very radio-luminous examples being detected by radio telescopes.

The radio luminosity distribution with position in the HR diagram is graphically illustrated in Fig 2. The 2-D surfaces represent an approximation to the envelope of the peak radio luminosity detectable as a function of M_V and B-V (binning, peak search, and median smoothing was applied for this representation). The right panel represents only single stars (or components in binaries that are not tidally interacting). There is an apparent correlation between optical and radio luminosity, with stars on the lower main sequence also representing the lowest radio luminosities. However, the mechanism giving rise to the radio emission varies over the HR diagram, with largely non-thermal processing dominating an the lower main sequence and thermal process (free-free radiation) in high luminosity OB stars and cool, red giants.

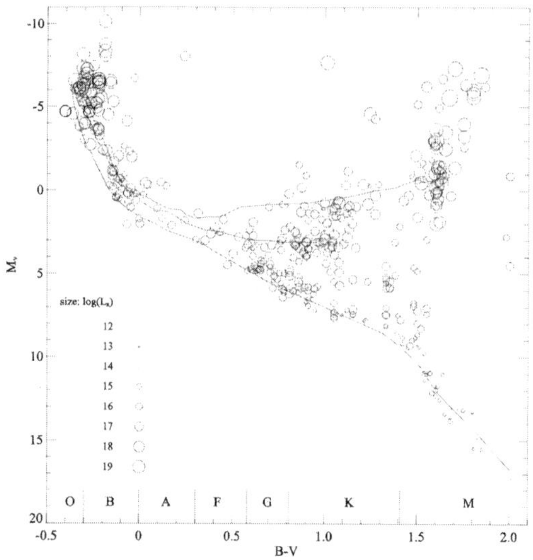

__Fig. 1.__ The Radio HR Diagram showing the loci of 441 radio detected stellar systems [Guedel, 1999]. The stellar classes are colour-coded, and the symbol sizes indicate the radio luminosity. Only detections within 1– 10 GHz have been considered. The radio star catalogue by Wendker [1995] has been used, complemented by (mostly Hipparcos) distances and B-V colours, with some colour corrections due to emission lines (WR stars) or to extinction.

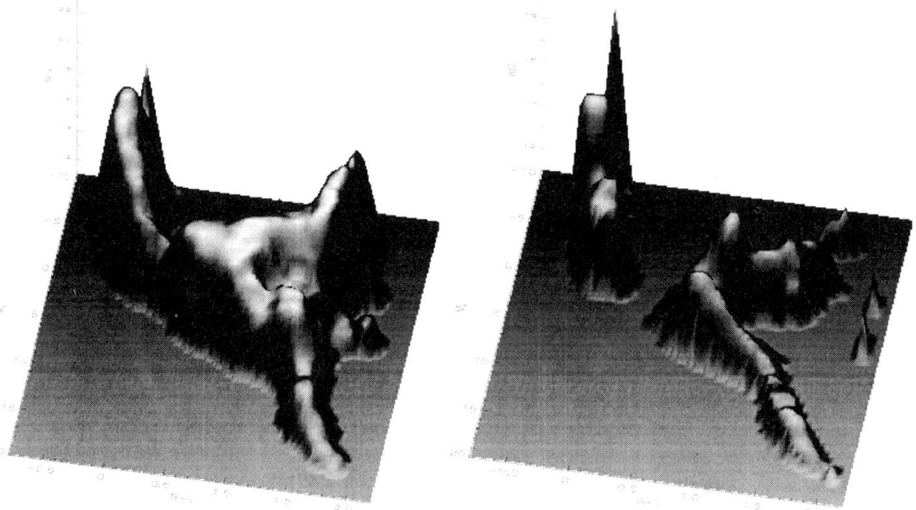

__Fig. 2.__ The distribution of radio luminosity with position in the Hertzsprung-Russell Diagram. The envelope of peak radio luminosities is shown as a 2-D surface. The left panel includes all stars (see figure 1). The left includes only single stars (or components in binary systems that are not tidally interacting). Figure is from Guedel [1999].

With very rare exception, the stars in these plots represent rare phenomena. Except for the most nearby supergiants, "normal" stellar photospheric emission cannot be detected with current sensitivities. Detectable radio emission is virtually always associated with energetic stellar phenomena, such as outbursts, enhanced magnetic activity or copious, ionised mass loss. In hot, young stars, radio emission stems from radiation driven mass loss during the short stellar lifetimes, and from electrons accelerated in shocks that form in strong stellar winds. In the cool-giant area, many sources are binaries, where the binary interaction leads to radio emission in the shocks of colliding winds or photoionized regions from the ultraviolet radiation of a hot, compact companion. Chromospheric emission is also detected from some red giant stars. Along the low-mass end of the main sequence, non-thermal radiation from energy release associated with magnetically active chromospheres is the dominant emission mechanism. The range of radio emission mechanisms and stellar class can be broadly categorised based on the characteristic power law, α, of the radio continuum spectrum, ($S_\nu \propto \nu^\alpha$). This is summarized in Table 1.

Table 1. *Class of Radio Continuum Emitters Characterized by Spectral Index*

	Spectral power law (α)	**Stellar Class**
1	+2	*Thermal emission* from stellar photospheres
2	+0.6 to +1	*Thermal emission* (Bremsstrahlung) from circumstellar environments and stellar ejecta in the form of winds (WR and OB stars), shells (novae, planetary nebulae), and jets (symbiotic stars, PMS stars)
3	0 to –1	*Non-thermal* synchrotron and gyrosynchrotron emission from flare stars, active binary systems, pre-main sequence (PMS) stars, and x-ray binaries
4	–1 to –2	*Non-thermal* emission from pulsars

While the past twenty years has provides us with this picture of stellar radio characteristics. For all these stellar types we have only a very poor census of properties. Our current understanding of these phenomena is based largely on investigations of a few bright members of

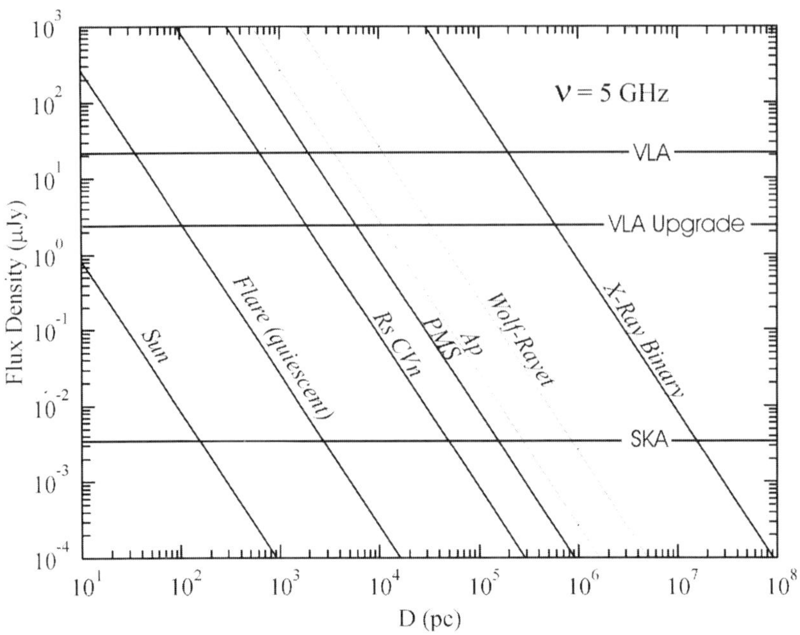

Fig. 3. *Flux density at ν=5 GHz versus distance for a number of non-thermal radio stars. The continuum sensitivity (3σ) after an integration time of 8 hours is shown for the VLA, the planned upgraded VLA and the SKA (after Seaquist [1996]).*

the class. Figure 3 shows the expected flux density versus distance for non-thermal emitters in category 3 of Table 1, along with the 3σ sensitivity of the current Very Large Array, the upgraded VLA and the proposed Square Kilometer Array (SKA) after 8 hours of integration. Only the very brightest radio stars, the luminous outbursts associated with x-ray binary systems, are now detectable throughout the Galaxy. Similarly, Figure 4 shows the flux density-distance plot as a function of observing frequency for stars at the other end of the spectral index class, the photospheres of Mira-like red giant stars. The vast majority of known radio stars in all classes are contained within a sphere of a few hundred parsecs radius from the Sun, and the numbers in each category typically total a few tens. Clearly fundamental advances will require a more complete census of the radio emission from stars, and samples largely free of selection effects. The increase sensitivity of the SKA will allow the detection of many classes of stars throughout the entire galaxy and in the Magellanic Clouds. Instead of a few hundred radio stars currently catalogued, hundreds of thousands of stars will be detectable. The radio properties for whole stellar populations will be measured, allowing meaningful studies of the relationship between radio phenomena and other stellar properties and evolutionary states.

Quiescent radio emission from flare stars will be detectable out to several kpc, and typical RS CVn systems, PMS stars, peculiar magnetic stars, and non-thermal components of WR winds would be detected over the entire Galaxy and the Magellanic Clouds. X-ray binary systems and brighter WR stars would be detectable in M31. For most of these classes, only a handful of objects are now known. With the SKA, in addition to studying large and complete samples of these populations, we will be able to trace their spatial distributions in star formation regions and clusters and within the Galaxy as a whole, thereby relating radio properties to characteristics such as age and environment.

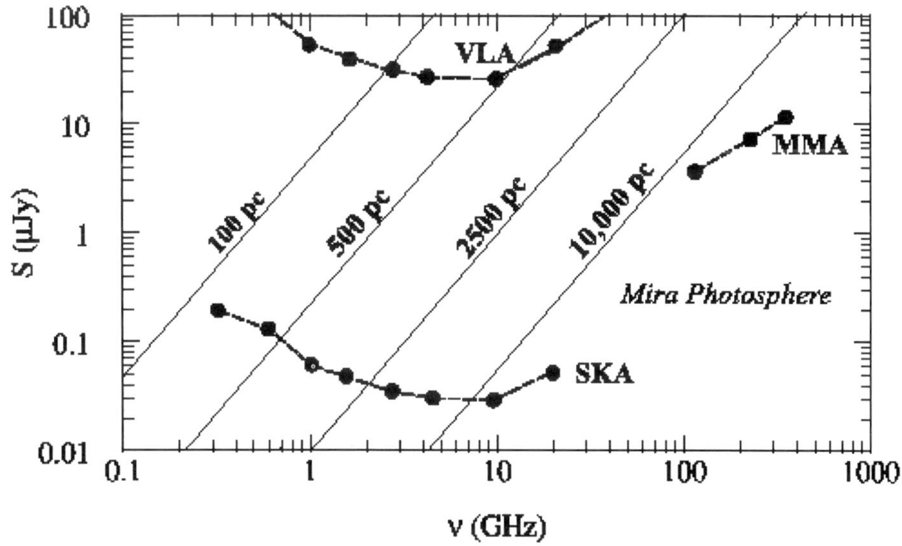

Fig. 4. The flux density expected for a Mira-like (late M) red giant star as a function of frequency and distance (after Seaquist, [1996]). The 3σ sensitivity after 8 hours of integration is shown for the VLA, the SKA and the MMA.

2. Imaging the Surfaces of Stars

Imaging the surfaces/atmospheres of other stars is one of the major frontiers in stellar astronomy.

The largest stars on the sky are also the coolest, thus red giant and supergiant stars have been the targets of attempts to resolve stellar disks at radio wavelengths. Both red giant and supergiant stars lose mass at a prodigious rate of $\sim 10^{-7}$–10^{-6} $M_\odot$ yr^{-1}, driven by radiation pressure on dust grains [*Kwok, 1975*]. The large radio photospheres give rise to detectable thermal radio emission. From a sample of 6 stars, *Reid and Menten* [*1997*] showed that the radio flux density spectrum of late-M giant stars such as Mira follows the expected $S \propto \nu^2$ dependence at centimetre wavelengths, suggesting that their radio photosphere has a sharp opacity edge. *Reid and Menten* [*1997*] partially resolved the radio photosphere of the red giant W Hydra, and find an average diameter of 80 ± 15 mas, corresponding to ~8 AU, and a brightness temperature of 1500 ± 570 K. The radio photosphere is approximately twice the size of that measured in the optical, and the temperature 30% lower than the effective temperature. The *Reid and Menten* [*1997*] data indicated that this relationship is generally true for red giant stars.

Early-M supergiant stars have a flatter radio flux density spectrum of $S \propto \nu^{1.0\text{-}1.3}$ as observed for Betelgeuse [*Newell and Hjellming, 1982*] and Antares. *Lim et al.*, [*1998*] have imaged the radio photosphere of Betelgeuse at 7mm using the VLA. The results are shown in figure 6. As shown in the left panel, the radio photosphere is about twice the diameter of the optical disk, and appears to be asymmetric. Because of the density gradient in the extended atmosphere of the star, longer wavelengths sample large radii. Measurements taken simultaneously at cm-wavelengths partially resolved the stellar atmosphere, and directly measure its radial temperature profile. At 7mm the temperature approach the optical values. Above the optical photosphere, temperature decreases with height. The measurement of the temperature profile allows the radius of dust grain formation to be determined, and models of mass outflow to be tested.

We are just beginning the exciting enterprise of imaging the surfaces of stars. What will the future bring? The higher angular resolution need to resolve the surfaces of a large sample of stars must be coupled with higher surface brightness sensitivity. At radio wavelengths, where the Rayleigh-Jeans approximation to the Planck radiation law is valid, the brightness temperature of an emitter with total flux density S and angular diameter θ is given by the simple expression

$$T_b = 1.97 \cdot S \frac{\lambda^2}{\theta^2}\ \mathrm{K},$$ where S has units of mJy, λ cm and θ arc seconds.

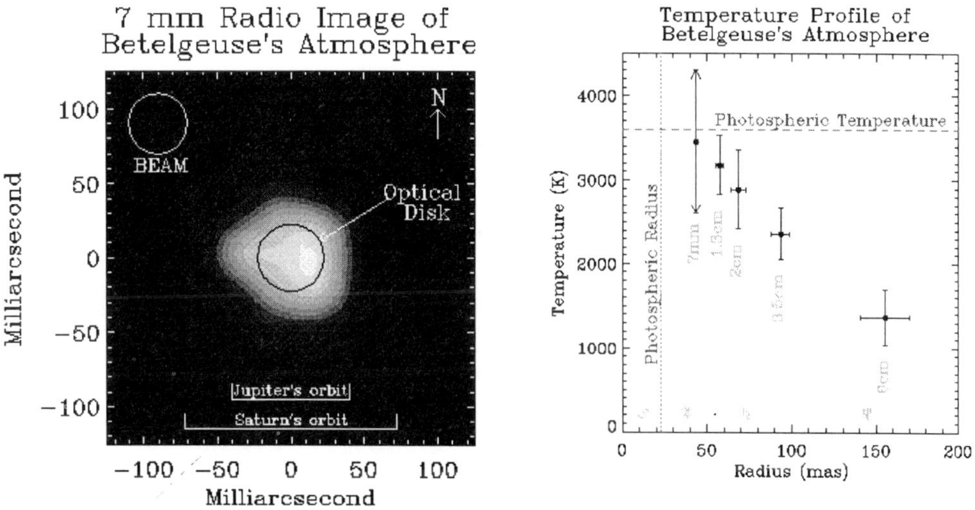

***Fig. 5**. A false color image of Betelgeuse made at a wavelength of 7 mm with the VLA (left panel). Simultaneous observations at longer wavelengths partially resolve the radio photosphere, which increases in size with increasing wavelength, and are used to deduce the radial temperature profile of the atmosphere (right panel). From Lim et al., [1998].*

Figure 5 shows plots of T_b versus θ at wavelengths of 6cm and 2cm. On these plots, lines of constant flux density appear as diagonals. The two dashed lines show the angular radius of a source required to produce flux densities of 1 mJy and 1 μJy as a function of the brightness temperature of the source. The dark solid lines characterise the imaging capabilities of the most powerful existing radio telescope arrays and the SKA. The horizontal portion shows their maximum resolution, and the diagonal portion shows the minimum detectable flux density. For a radio source with a given angular radius and brightness temperature to be resolved by a particular radio telescope array, it must lie above the solid line. The current suite of most sensitive radio telescope facilities occupies the upper right portion of this diagram, and is able to image only sources with either very high brightness temperature or large angular size. For thermal radio sources (below 10^4 K) the maximum attainable resolution is ~0.1 arcsec (with th VLA). The effective temperatures and angular diameters at 100 pc of stars on the main-sequenc (V), giant (III) and supergiant (I) branch are indicated by the gray lines. Nearby stars at the ti of the supergiant branch are just resolvable by MERLIN (at λ = 6 and 2 cm) and the VLA (at λ = 2cm). To make inroads toward the smaller angular diameters, the giant step in sensitivit provided by the SKA is required. The area of parameter space opened up by the SKA i indicated by the shaded area. The surface of supergiants and many giant stars will be resolved.

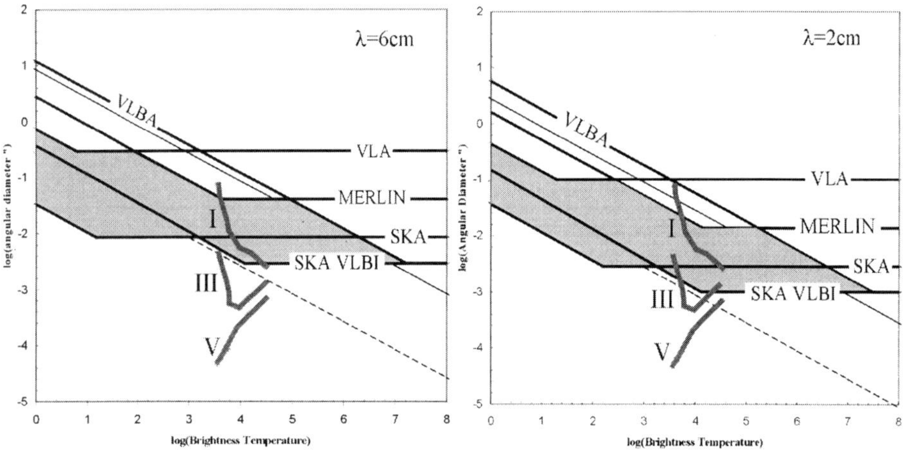

Fig. 6. *Brightness temperature versus radio source angular size at λ 6cm (left) and λ 2cn (right). The dashed line is the locus of radii and brightness temperatures for a source with tota flux density of 1μJy. The sensitivity and resolution limits are plotted for the world mos powerful existing interferometers, and for the SKA. The effective temperatures and angula radii at 100 pc are shown for stars on the main-sequence (V), giant (III) and supergiant (I branches of the HR diagram. The area of (T_b, θ) parameter space that will be opened up by th SKA will allow for the first time routine imaging of the surfaces of nearby stars.*

The SKA will have an especially significant impact on studies of red supergiant sta atmospheres. For those nearer than 2 kpc, the SKA will be able to directly measure th temperature and structure of the stellar atmosphere over a large range of heights. Comparison of the atmospheric structure at different heights will be vital for understanding both the physica properties and dynamics of the stellar atmosphere, and (combined with optical photospheric images) will provide powerful constraints on the mechanisms responsible for driving the atmosphere outward.

By the time the Square Kilometer Array (SKA) is constructed, a number of powerful optical and IR interferometers (OIRI) will also be in operation. Consideration of angular resolution suggests that the SKA and OIRI will be quite compatible, providing similar resolution at widely dispersed wavelengths from the IR wavelength range 5 - 50 microns and the radio range of 1 to 20 cm. Because of the common angular resolution and the common sensitivity to thermal sources the most obvious science overlap is in the study of stellar surfaces. This type of imaging should allow extensive investigation of the vertical structure of the photospheres and, via long term monitoring of surface and stellar wind features, make it possible to correlate surface activity with changes in the stellar wind. Such studies of the atmosphere-wind interface will not only lead to a better understanding of the wind phenomenon but may also shed light on the dynamo that drives surface activity [e.g. *Guedel et al., 1997*]. Additionally, such data place important constraints on convection models leading to a better understanding of energy transport in these stars.

3. Detecting Stellar Analogues of the Sun

The non-flaring Sun has been studied at metric/decimetric/centimetric wavelengths in appreciable detail. The nature of the emission varies strongly during the magnetic activity cycle; at maximum, most of the measured flux stems from active regions. At wavelengths ≤6 cm, active region emission consists of bright compact sources and a diffuse halo. The former, strongly sunspot-related component with a brightness temperature of a few times 10^6 K, is attributed to optically thick gyroresonance emission of coronal plasma in rather strong (~1 kG) magnetic fields [e.g., *Lang et al., 1987*; *Vourlidas, Bastian and Aschwanden, 1997*]. The halo component is often due to optically-thin thermal free-free emission from the corona, but also from the cooler transition region and the chromosphere.

During flares, the Sun can be a source of copious radio (*gyrosynchrotron*) emission for up to tens of minutes. The emission is produced by accelerated, mildly relativistic electrons spiralling around the magnetic field lines. This radiation often reaches its peak flux (transition

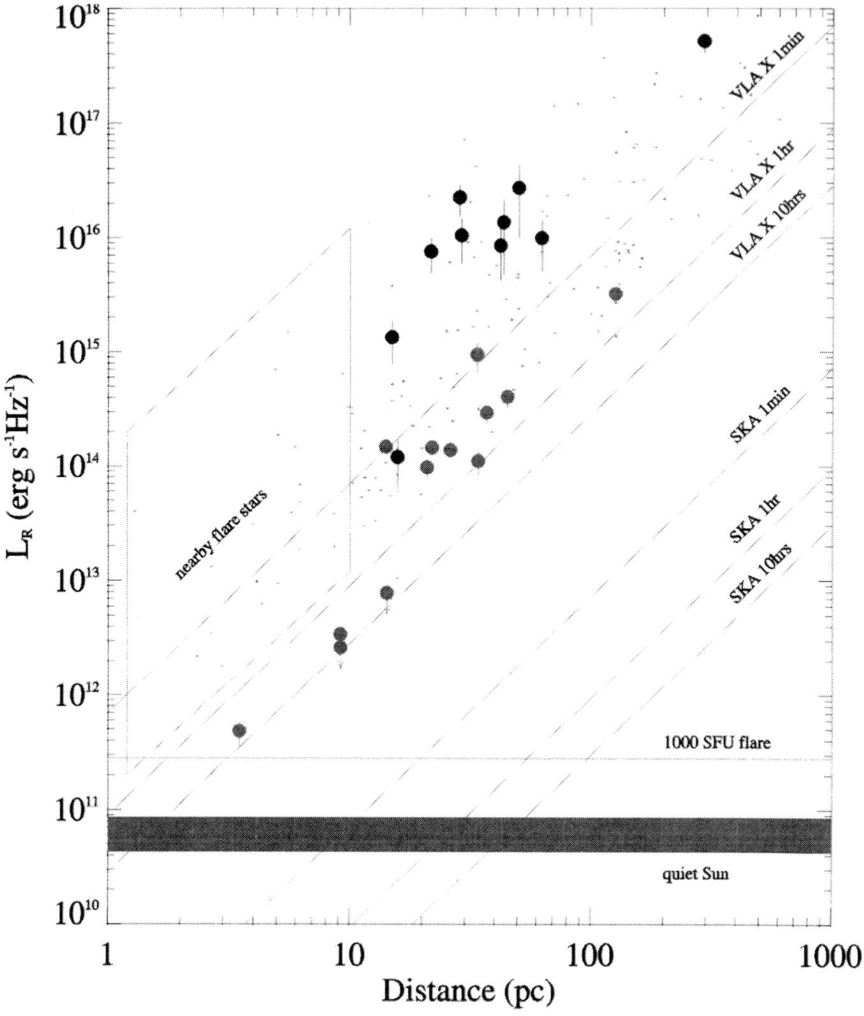

Fig. 7. *F/G-type MS stars so far detected. Many M dwarf flare stars fall within the schematic quadrangle at left, within 10pc, [White et a!., 1989]. Various other radio detections [Wendker, 1995] are plotted as small crosses. The average radio luminosity at 8 GHz of the non-flaring Sun is indicted by the bar at bottom, and the luminosity of a typical solar radio flare is marked by the dotted line. The diagonal lines represent the 3σ detection limits for three different continuum integration times (1 min, 1 hr, 10 hrs), both for the VLA (8.4 GHz, 100 MHz bandwidth; upper set) and the projected SKA (maximum bandwidth; lower set).*

from optically thick to optically thin) in the 1–10 GHz region. Its spectrum contains rich diagnostic information: The electron energies can be derived from the optically-thick portion, magnetic field strengths from the turnover frequency, and the electron energy distribution from the optically-thin part. Many of these diagnostic tools have been applied to easily observable, strong stellar flares, albeit with considerably smaller temporal and spectral (and no spatial) resolution.

The non-flaring radio luminosity of the Sun at several GHz is $\log L_R \approx 10.7$ [erg s^{-1} Hz^{-1}] (principally due to gyroresonance and bremsstrahlung emissions) and therefore well below the base level in the logarithmic 3-D diagrams in Fig. 1. No single radio star with L_R equalling that of the Sun has ever been detected. The lowest-luminosity radio star found so far, the F5 subgiant Procyon at a distance of 3.5 pc, is at $\log L_R \approx 11.7$, an order of magnitude higher than the Sun [*Drake, Simon and Brown, 1993*]. Presently, the VLA could detect a source like the Sun marginally out to 1.5 pc after 10 hours of integration time. The SKA, in contrast, will detect solar twins at 50 pc distance! This is illustrated in Figure 7 which shows radio detections of active stars as a function of distance form the Sun. Extrapolating from the *Gliese and Jahreiss*, [*1991*] Catalogue of Nearby Stars or *Allen* [1973], volume densities of stars in the solar neighbourhood, we infer that 30,000 stars (99% of them being cool MS stars) are potentially within reach of a telescope like the SKA if the solar activity level is taken as a lower limit for all MS stars. Clearly, the SKA will open a completely new window to "normal' stars. Numerous physical problems so far studied only on the Sun can be investigated on a wide range of stars and stellar types.

4. Stellar Astrometry and Planet Detection

Recent reports of extrasolar planets have revolutionised the field of planetary formation because of the Jupiter-mass planets that appear to exist in close proximity to their stars [e.g. *Mayor and Queloz, 1995*]. Such planet-star configurations may be common. Unfortunately, no extrasolar candidate has yet to be confirmed with independent techniques. Most such candidates have been uncovered with the radial velocity technique. However, given the high angular resolution of the SKA such systems are ideal for astrometric detection.

As noted in section 3, the SKA will detect the non-flaring Sun out to ~50 pc. Figure 8 shows that the binary motion of the Sun due to the orbit of Jupiter would be measurable out to several 10's of pc. The ability of the SKA to detect normal radio emission from solar-type stars would therefore allow the detection of Jovian planets by astrometric observations of the parent stars. Such studies would yield a census of Jovian planets in the Solar neighbourhood and statistical constraints on the range of orbital separations and masses.

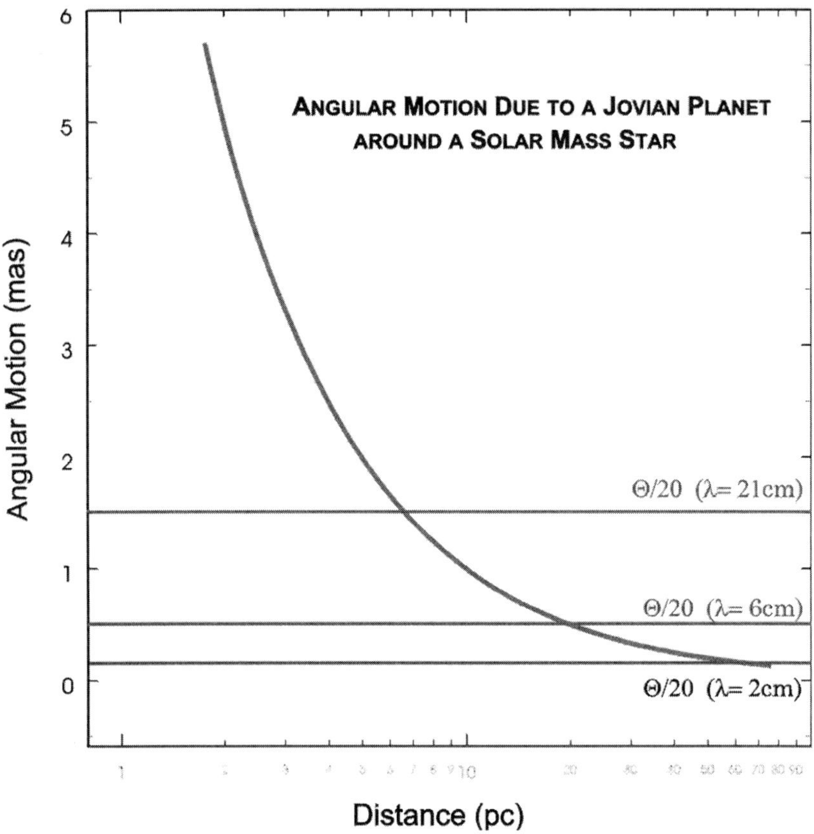

Fig. 8. The angular motion of a one solar-mass star due to the presence of Jupiter as a function of distance. The astrometric accuracies of the SKA at λ21 cm, λ6 cm and λ1.5 cm are shown by the horizontal lines. At λ1.5 cm the presence of Jovian planets would be detectable up to several tens of pc distance.

5. Conclusion

Stellar radio astronomy is in urgent need of a new instrument as powerful as the proposed SKA. About 20 years ago, the VLA became operational; it set new standards in stellar radio astronomy by discovering many stellar classes routinely for the first time and thus yielding access to new physics. Stellar radio astronomy has now reached the point where another major increase in sensitivity is required for fundamental advances.

Acknowledgements

The author wishes to thank Jeremey Lim and Manuel Guedel for material provided in preparation of this manuscript.

References

Allen, C.W., [1973], *Astrophysical Quantities*, 3rd Edition (London: Athlone).

Aarons, J., R.S. Allen, and T.J. Elkins, [1967], "Frequency Dependence of Radio Star Scintillations," *J. Geophysical Research*, 72, 11, pp. 2891-2902.

Gliese, W. and Jahreiss, H., [1991], "Third Catalogue of Nearby Stars", Astronomishes Rechen-Institut, Heidelberg, Germany, Software version.

Guedel, M. et al., [1997], *Astrophysical J.*, 483, 947.

Lang, K.R., et al., [1987], *Astrophysical J.*, 322, 1044.

Lim, J., Carilli, C.L., White, S.M., et al., [1998], Nature, 392, 575.

Mayor, M. and Queloz, D., [1995], Nature, 378, 355.

Newell, R.T. and Hjellming, R.M., [1982], *Astrophysical Journal Letters*, 263, 85.

Reid, M.J. and Menten, K.M. [1996], in *Radio Emission from the Stars and the Sun*, ed. A.R. Taylor and J.M. Parades, p. 87.

Seaquist, E.R., [1996], in *High Sensitivity Radio Astronomy*, eds. N. Jackson and R. Davis, p. 23.

Vourlidas, A., Bastian, T.S. and Aschwanden, M.J., [1997], Astrophysical J., 489, 403.

Wendker, W. [1995], *Astronomy & Astrophysics Supplements*, 109, 177.

White, S.M., Kundu, M.R. and Jackson, P.D. [1989], *Astronomy & Astrophysics*, 225, 112.

AN ASSESSMENT OF THE BIOEFFECTS INDUCED BY POWER-LINE FREQUENCY ELECTROMAGNETIC FIELDS

Russel J. Reiter

1. Introduction

Over the last 10 years, research into potential biological consequences of electromagnetic (EMF) field exposure has yielded a veritable quagmire of apparently contradictory findings. Scientists as well as regulators have been frustrated by the failure of seemingly similar studies done under what appear to be the same experimental conditions to produce dissimilar outcomes. Indeed, it generally appears that the discrete control of exposure parameters, i.e., reducing transients, producing "clean" fields, etc., has tended to eliminate effects that were seen when "dirty" fields were used. Even this, however, has not been proven.

A vast array of effects due to EMF exposure have been reported in both in vitro and in vivo studies. The inconsistency of the findings, the so-called Cheshire cat phenomenon, while not totally discounting their existence has certainly brought their significance into question. The following survey examines some of the most intensively investigated areas in EMF research, but the reader is reminded that for every effect reported, it is probably counterbalanced by a negative finding even though space limitations may have precluded an extensive discussion of both sides of the issue.

2. Electric and Magnetic Fields and Their Sources

The exposure of increasingly larger populations of humans to higher mean levels of extremely low frequency (ELF, 0-3,000Hz) EMF is a consequence of the electrification of residential and commercial buildings, particularly in the well developed countries of the world. The strength or intensity of these man-made fields is substantially greater than the naturally occurring ambient electric and magnetic fields which are in the range of about $10^{-4}Vm^{-1}$ and $10^{-13}T$, respectively. Additionally whereas the naturally occurring magnetic fields are commonly dc (static), those imposed by man are frequently complex sinusoidal (ac) fields. While concerns about these fields have a long history, they were particularly highlighted by a report published in 1979 that childhood cancer may be increased as a consequence of such exposures [*Wertheimer and Leeper, 1979*]. That EMFs can exert biological effects albeit not necessarily health effects is consistent with their widespread use in clinical medicine when they have been shown to promote neural regeneration and bone repair [*Bassett, 1993*]. The field strengths used for these purposes, however, are typically greater than those encountered in residential and commercial environments. Thus, for the purpose of the current report, the publications related to these findings will not be discussed.

The EMFs to be considered here are those derived from power distribution systems which in the United States are 60Hz and much of the remainder of the world are 50Hz. Typically, the electric and magnetic field strengths measured under energized power transmission lines are in the range of 1-10kVm^{-1} and 1-10μT. By contrast, in most homes background electric fields are 1-10Vm^{-1} although they can increase to 200Vm^{-1} near small, electrical appliances [*Merchant et al., 1994*]. For heated water beds and electric blankets, both of which have been proposed to induce biological effects [*Wertheimer and Leeper, 1986*], electric fields can range up to several kVm^{-1}. Background magnetic fields are usually between 0.01 and 1μT while appliances can generate field strengths of 0.1-100μT. Among appliances, highest magnetic field strengths derive from hair dryers, microwave ovens and vacuum cleaners. Near a personal computer the average field strength is about 10μT. Electric fields are produced wherever there is a voltage and are easily shielded by most objects; conversely, magnetic fields are directly proportional to current flow and have a high level of penetrance, i.e., they are not easily filtered. Magnetic field strengths do, however, decrease rapidly with distance from the source of the field. Thus, at a distance of 1m, the magnetic field strength from household appliances is barely detectable.

3. Epidemiological Studies Related to Cancer

Observations from epidemiological reports have been a major factor in driving laboratory research on the bioeffects of ELF-EMFs. The seminal report was that of *Wertheimer and Leeper* [*1979*] who noted that the incidence of cancer was higher in children living in homes where EMF exposure was presumed to be higher than usual. The actual estimates of the field strengths to which the children were exposed were indirect, being based on wire code configurations of the homes occupied by the children included in this case-control study. Wire code configurations were used as a surrogate method for estimating long term exposures to EMF; they are classified as either high-current or low-current configurations and they are based on the location of transformers, placement of service drops, and age of the wires, i.e., pre- or post-1956. In this case, children who died of cancer were more likely to have lived in the presence of a high wire code configuration and therefore they were presumably exposed to higher EMF strengths throughout much of their life. In the report by *Wertheimer and Leeper* [*1979*], the cancer types that were noted included leukemias and lymphomas, with the incidence of these cancers being 2-3 fold greater than in children living in homes with low-current wire code configurations. The implication of the findings was that residential exposures to power line frequency EMF increase cancer incidence in children.

Prompted by the observations of *Wertheimer and Leeper* [*1979*], many other epidemiologists undertook investigations to uncover potential health effects as a consequence of exposure to sinusoidal EMFs. The endpoint in most of these studies was the frequency of cancer, of various types, in both children and adults. While many of the residential studies suggested a slight increase in cancer risk as a result of power line frequency exposures, there was certainly not universal agreement on this point [*Savitz et al., 1988; Severson et al., 1988; Jackson, 1992; Feychting and Ahlbom, 1993; Linet et al., 1997*]. Many of the investigations were well performed and controlled although the studies were constrained by the usual limitations confronting epidemiologists, e.g., unknown confounders, selection bias, appropriate control subjects, use of surrogates as measures of field exposures, etc..

The conclusions of the recently published NIEHS Working Group Report [1998] included the statement that there is limited evidence to implicate residential exposure to power line frequency magnetic fields in the cause of cancer in children; this included cancer of the lymphoid organs and central nervous system, the two organ systems that have received the largest focus in epidemiological reports.

Studies of residential exposures in relation to cancer in adults have been as numerous as those in which children were the center of inquiry. The most frequently mentioned cancers that were noted in these investigations were leukemias [*Preston-Martin et al., 1993; Lovely et al. 1994*], breast cancer [*Vena et al. 1994; Li et al., 1997*] and tumors of the central nervous system [*Feychting et al., 1993; Verkasalo et al., 1996*]. These reports included investigations of 3 types of studies in which the use of electrical appliances as the source of the magnetic field exposure were considered; studies in which the field strengths were calculated either from the distance from transmission lines, from historical loads on lines or from power line configurations; and investigations in which direct measures of fields in the homes of the study subjects were made. None of the reported associations between cancer (either leukemia, breast cancer or brain cancer) and residential exposures to magnetic fields provided unequivocal evidence for a cause-effect relationship. One of the major problems with the residential studies in adults is often the lack of consideration of occupational exposures of the study groups. This omission could have led to substantial misclassification of the total exposure of some of the individuals.

By far the largest group of epidemiological studies related to EMF exposure and cancer involves exposures in occupational settings [*ORAU Committee Report, 1992; Kheifets et al., 1997*]. The quality of these studies in controlling for potential confounders and estimating exposures have improved steadily and in recent years and some very large, well done studies have been performed. Thus, *Tynes et al.* [*1992*], *Sahl* and colleagues [*1993*] and *Savitz and Loomis* [*1995*] carried out large cohort studies to determine the effect of EMF exposure on cancer of all types in adults working in environments where the exposures were expected to be elevated. Also, in a nested-case control study, *Theriault et al.* [*1994*] examined the occurrence of cancer in relation to magnetic field exposure at three large utility companies. None of these thorough studies provided conclusive evidence that the EMF exposures experienced by workers in these studies caused an unusually high incidence of any cancer type.

Besides the reports summarized in the previous paragraph, a large number of other well done studies have been performed on workers in jobs where EMF exposure was considered to be elevated and where the incidence of specific types of cancers were of interest. Most often the studies were directed at either leukemia [*Floderus et al., 1993; London et al., 1994*], cancers of the brain [*Theriault et al., 1994; Harrington et al., 1997*], breast [*Guenel et al., 1993; Floderus et al., 1994*], or lung [*Theriault et al., 1994; Savitz et al., 1997*]. As with the other investigations, it is apparent that individuals working in occupations where their exposure to electricity is high may have a slightly increased risk for cancer. The data on the whole, however, are inconclusive. Meta-analyses have been generally not very informative in helping to clarify whether residential or occupational EMF exposure contributes significantly to cancer burden [NIEHS Working Group Report, 1998]. Such marginal findings create problems for health care professionals and regulators who are attempting to judge whether EMF exposure is a genuine hazard for the individuals in those occupations and present difficulties in setting recommendations on maximal daily exposures.

4. Electromagnetic Field Exposure and Cancer in Experimental Animals

There have been several long-term studies to evaluate whether animals that live under high EMFs for prolonged periods experience a higher incidence of tumor growth than do animals not exposed to such fields. The first of these was reported by *Margonato* and co-workers [*1995*] and involved large groups of male Sprague-Dawley rats exposed to 50Hz fields for 32 weeks; the field strength in this study was 5μT and exposure was administered for 22 hours each day. The conclusion of this study was that there were no harmful effects, either cancer or otherwise, as a result of this exposure which was comparable to the magnetic field strength under a high power line.

There have been at least three similar studies where rodents were exposed to magnetic fields of longer duration, i.e., greater than 32 weeks. Both *Mandeville et al.* [*1997*] and *Yasui* and colleagues [*1997*] used rats and exposed them to magnetic flux densities between 2 and 5,000μT for up to two years. These were complete studies which included a comprehensive anatomical evaluation of tissues from animals that died during the course of the study or were killed at its conclusion and in neither experiment were consistent exposure-related pathologies found. A final even more complete investigation which is yet to be published in the peer-reviewed literature was conducted under the auspices of the US National Toxicology Program [*NTP, 1998*]. In this case both rats and mice receive a 60Hz linearly polarized magnetic field exposure of either 2, 200 or 1,000μT continuously or 1,000μT intermittently (1 hour on/1 hour off) for 2 years. Like the previous long-term exposure studies, these workers concluded that increased carcinogenicity was not a consequence of magnetic field exposures even when they were maintained essentially throughout the life time of the animals.

That these long term magnetic field exposures did not initiate additional tumor growth in rodents is consistent with a large body of evidence showing that such fields are not directly genotoxic [*McCann et al., 1993*]. Thus, it is generally accepted that the energy transfer from ELF EMF to cells is insufficient to directly damage the genome, i.e., DNA. It is possible, however, that alterations in cellular processes such as the prolongation of free radical half lives by magnetic fields [*Roy et al., 1995*] could result in indirect DNA damage and, therefore, an increased incidence of cancer. This indirect mechanism was used to explain the increased DNA damage in neurons following a 100μT field exposure for 2 hours; using the microgel electrophoresis method commonly referred to as the comet assay, Lai and Singh [1997] found both single and double strand breaks under these exposure conditions. The presumption for free radical involvement was emphasized by the fact that the administration of an antioxidant, melatonin [*Reiter, 1997a*], overcame the effects of the magnetic fields on chromosomal breakage [*Lai and Singh, 1997*]. This report is the first of this kind and has yet to be confirmed in another laboratory. Using methods different from the comet assay, other workers have not been able to demonstrate that sinusoidal magnetic fields alter the structure of nuclear DNA or chromosomal aberrations at field strengths comparable to those associated with high power lines [*Scarfi et al., 1994; Tofani et al., 1995; Lagroye and Poncy, 1997*].

With the exception of the paper of *Lai and Singh* [*1997*] there is virtually no evidence to suggest that EMFs by themselves damage DNA. This would imply that they are incapable of initiating the primary events required for tumor induction. This conclusion has often been reached by those working in this field and, at this point, it is widely accepted [*McCann et al., 1993*].

5. Electromagnetic Field Exposure and Tumor Promotion in Experimental Animals

5.1 Mammary Cancer Promotion Studies

While studies related to EMFs and cancer initiation have been predominantly negative, the experiments dealing with the role of power line frequency exposures on cancer promotion have yielded somewhat less consistent results. Many of these studies dealt with various models of mammary gland cancer and were stimulated by *Stevens* [*1987*] who proposed the first testable model by which EMF exposure may promote the growth of initiated tumor cells in breast tissue (Fig. 1).

The first report in this series was published by *Beniashvilli and co-workers* [*1991*] who exposed female rats in which mammary cancer had been initiated with the chemical carcinogen *N*-methyl-*N*-nitrosourea [MNU] to a 50Hz, 20μT field for the lifetime of the animals. This exposure caused a higher incidence of mammary tumors, a larger total number of tumors and a shorter latent period to tumor onset than in the non-exposed control rats. In the only replication of the *Beniashvilli et al.* [*1991*] study, *Anisimov* and colleagues [*1996*] exposed rats to a 50Hz, 20μT field throughout their lifetime; these animals were given MNU at 3 week intervals. This combination of treatments caused a non-significant rise in mammary tumor incidence but a significant reduction in the latency period for tumor development; comparisons were made with MNU only treated rats. While these two studies yielded similar results and provided evidence that magnetic field exposure increases the likelihood of tumor growth in rats initiated with a chemical carcinogen, a thorough evaluation of the results is compromised by a paucity of methods and/or experimental details provided in these reports.

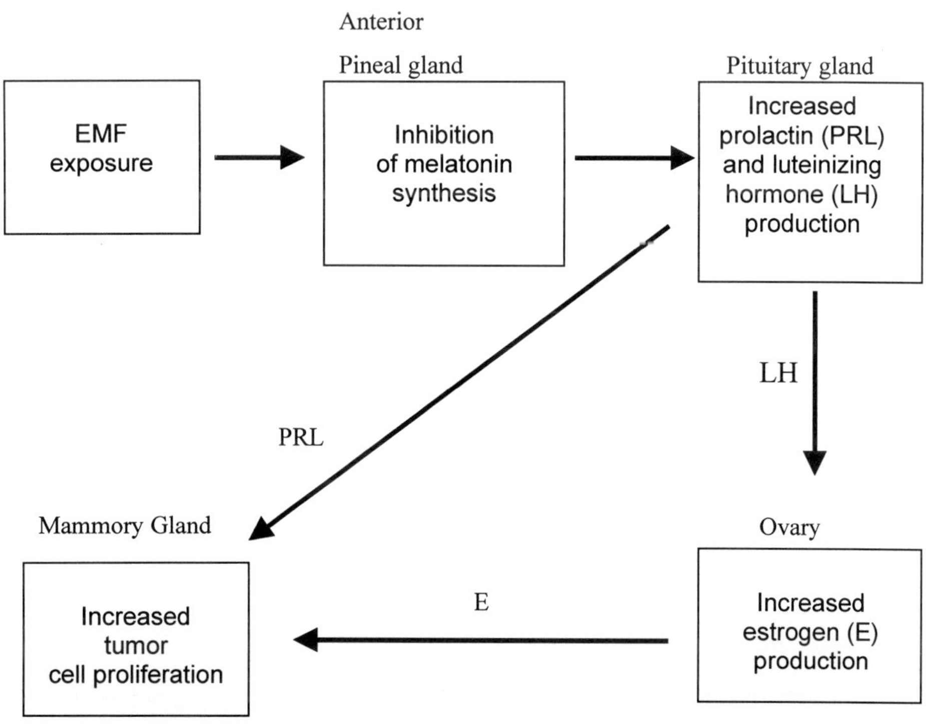

Fig. 1.*The Stevens' [1987] theory states that magnetic field exposure depresses pineal melatonin production which is followed by a reduction in serum melatonin levels. The drop in melatonin lifts its inhibitory influence on prolactin (PRL) and luteinizing hormone (LH) secretion from the pituitary leading to increased serum levels of these hormones. LH increases estrogen secretion from the ovaries and this hormone, in combination with PRL, stimulates proliferation of initiated mammary epithelial cells leading to an increased incidence of observable cancer. This theory has been tested in a variety of experimental settings with inconsistent outcomes.*

Perhaps the most comprehensive studies of the influence of magnetic field exposure on the growth of initiated mammary tumors are those of *Löscher and Mevissen* [*Löscher et al., 1993, 1994; Mevissen et al., 1993, 1998*]. Rather than NMU this group used dimethylbenzanthracene (DMBA), given intragastrically, to initiate mammary cancer. In their initial report, *Mevissen et al.* [*1993*] followed DMBA administration with exposure of the rats to a 50Hz, 30mT horizontal magnetic field for 13 weeks. In this study the group size was not large (33-36 rats each) (e.g. when compared to other studies of 99 animals per group) and the results indicated an increased incidence of tumor burden in the MF exposed animals. Replicate experiments with smaller groups of rats did not confirm this and later studies in which the rats were exposed to a flux density gradient (50Hz, 0.3-1.0μT] also failed to show that fields had a positive impact on tumor incidence [*Löscher et al., 1994*]. Based on these studies it was the opinion of the authors that high flux densities may act as a promoter or co-promoter of mammary cancer.

Using a similar experimental design in which large groups of rats were treated with DMBA and subjected to a 100μT field, the palpable tumors in breast tissue were larger and greater in number (relative to that of non-exposed rats) although their number in rats with tumors was not increased. The implication of these findings was that a 100μT field increases the growth of DMBA-induced mammary gland tumors [*Löscher et al., 1993*]. On the basis of a follow-up study where all tumors were subjected to a complete histopathological examination the same authors again concluded that magnetic field exposures may alter tumor numbers and may accelerate tumor growth but the authors were not dogmatic in this conclusion; thereby the fields presumably functioned as a tumor promoter (*Baum et al., 1995*). In large studies where the authors tried to establish a dose-response relation with magnetic field exposure, the outcomes of the experiments did show some statistically significant evidence of an association (*Mevissen et al., 1996a, 1996b, 1998*).

Collectively, this series of studies by the *Löscher/Mevissen* group are the most complete thus far performed in terms of the role of magnetic field exposure in functioning as a promoter or co-promoter of induced mammary tumors. While some measures suggested a positive effect of the fields on tumor growth, this was not always confirmed across experiments and there were a variety of tumor parameters that were not changed by the field exposures. Overall, the evidence does not provide compelling evidence that power frequency magnetic fields influence tumor growth and development. This is consistent with a recent report by *Ekström* and co-workers [*1998*] who used similar endpoints in a study in which rats were treated with DMBA and subsequently exposed to a 50Hz, 250 or 500μT field intermittently applied; their conclusion was that the fields were inconsequential in terms of tumor growth. Reference to NTP study report (now paper in press) in Carcinogenesis omitted here.

5.2 Skin Cancer Promotion Studies

Skin tumors in mice provide a well defined and accepted model in which to study multistage carcinogenesis and the potential effects of promoters on tumor growth. The first study that utilized this model in the context of a magnetic field investigation was that of *McLean* and colleagues [*1991*]. In this case, the DMBA-induced skin tumors were treated with the phorbol ester 12-0-tetradecanoyl phorbol 13-acetate (TPA), a tumor promoter. When the mice were subsequently exposed to a 60Hz, 2000μT magnetic field tumor growth was not influenced. A second study by the same group [*Stuchly et al., 1992*] found slight differences in various parameters of skin tumor growth but only some of the changes reached statistical significance. This group persisted and conducted a third 52 week study using the same skin carcinogenesis model and concluded that magnetic fields exposure does not function as a co-promoter in the well-established skin cancer model [*McLean et al., 1997*].

Rannug and co-workers [*1993, 1994*] also used the skin tumor model to investigate the effects of magnetic field exposures on tumor development; in this case the mice were studied throughout their lifetimes and, at death, all mice were subjected to a completed necropsy and histopathological examination of the skin tumors. The outcomes of neither of these studies support the hypothesis that magnetic field exposures promote skin cancer in SENCAR mice; in these studies the flux densities used were 50 and 500μT. *DiGovanni* study on skin cancer in SENCAR mice omitted.

5.3 Other Tumor Models

As noted in the work summarized above, the majority of studies to date in relation to magnetic fields as a potential co-promoter in experimental cancer models have investigated their effects in initiated epithelial breast cells or in skin epithelial cells. There are, however, other models that have been used in an attempt to define the role of magnetic field exposure on cancer cell growth.

In a series of investigations in which hepatic cancer was initiated in rats by giving them *N*-nitrosodiethylamine (NDEA) their subsequent exposure to 50Hz magnetic fields with flux densities ranging from 0.5 to 500μT caused low amplitude and inconsistent changes in hepatic tumor growth. When the authors evaluated the collective data from this series of studies it was their conclusion that they did not provide evidence of either a promotional or co-promotional effect of the field exposures [*Rannug et al., 1993*].

The published literature also contains two reports in which leukemia/lymphoma models were used to examine the co-promotional effects of magnetic fields. In one of these studies, leukemic cells were injected into Fischer rats [*Sasser et al., 1996*] while in the recent report the initiating agent was DMBA [*Shen et al., 1997*]. In the experiment conducted by *Sasser* and co-workers [*1996*] the transplantation of leukemia cells from diseased, aged Fischer rats to young males was followed by the growth of tumors in the recipient animals. The exposure of the experimental animals to a 1mT, continuous 60Hz field did not significantly alter the course of the disease process relative to that in ambient field-exposed control rats.

These observations are consistent with those of *Shen et al* [*1997*] wherein one-day-old mice pups were initiated with an injection of 35μg DMBA. When half of these animals were thereafter exposed to a 50Hz, 1mT magnetic field the percentage of rats with thymic lymphomas and lymphomatous leukemia was the same as animals only exposed to background magnetic fields. Thus, in neither the study of *Sasser* and colleagues [*1996*] nor in that of *Shen et al* [*1997*] was their any evidence of a promotional effect of magnetic field exposures on leukemia/lymphoma tumor models in two rodent species.

6. Epidemiological Studies Related to Neurodegenerative Diseases

Besides cancer, neurological diseases that occur as a result of the significant loss of neurons in the central nervous system have been examined recently as to their potential relationship to EMF exposure. Of particular interest has been Alzheimer's disease (AD) and other forms of dementia [*Davanipour et al., 1997*]. In reference to the former condition, *Sobel and Davanipour* [*1996*] proposed that it may be aggravated by EMF exposure while *Feychting* and co-workers [*1998*] implied that such fields may relate to the development of all forms of dementia.

Several epidemiological studies have considered EMFs as a predisposing factor for AD. *Sobel et al* [1995] used three case-control data sets to investigate "sporadic" AD. In these studies, field exposures were classified as high, medium or low by an industrial hygienist. The most common occupations that involved medium to high magnetic field exposures (average spot measurements of 1.93μT) were seamstresses and tailors. In these groups, a significant association between AD and magnetic field exposure was found for women (OR=3.0; 95% CI=1.7-8.9) and for women and men combined (OR=2.9; 95% eI=1.6-5.4). A follow-up study by the same group [*Sobel et al., 1996*] came to a similar conclusion, i.e., occupations with high magnetic field exposure (in this case tailors) have higher than the usual incidence of AD.

In two additional studies by *Savitz* and co-workers [*1998a, 1998b*] a non-significant association was found between AD and career EMF exposure in one study while in the second one a weak, but significant relative risk for AD for men working in electrical occupations was found. In this report no association between electrical occupations and Parkinson's disease was observed.

In a comprehensive study of Swedish twins, *Feychting et al* [*1998*] noted a significant association between all dementias and exposure to magnetic fields in excess of 0.2μT. In this study, an association between magnetic field exposure and AD was also found. The risks for these conditions were reportedly increased further when fewer than 10 years had elapsed between the last job and disease onset. The study is limited by a small number of cases, particularly for AD, and the lack of autopsy confirmation of the diagnosis of AD.

While the findings related to high occupational EMF exposure and neurogenerative diseases, particularly AD, have piqued the interest of experimentalists and clinicians, at this point the data are not compelling in support of an association. One of the major problems is in the diagnosis of AD itself which cannot be made reliably in the absence of autopsy material, and in the majority of the studies summarized in this section that material was either not available or was not considered. Additionally, the complex nature of genetic and possibly epigenetic factors which cause dementia mitigates against epidemiological studies consistently uncovering a minor relationship that could possibly exist between magnetic field exposures and neurodegenerative diseases.

7. Bioeffects of EMF Exposures in Animals which may Relate to Cancer and Neurodegenerative Diseases

Clearly, the epidemiological evidence supporting a link between power line EMF exposure and negative health outcomes is scanty and weak; nevertheless, there are some biological consequences of such exposures that have been validated. The following section summaries some of these effects and considers the data that either support the findings or reject them.

7.1 EMFs as Genotoxic Agents

It has been theorized that EMFs may act as direct mitogens much like do conventional carcinogens. Many experimental endpoints have been used in a variety of studies to determine whether EMF exposure causes DNA damage; these methods included assays of point mutations and analyses of alterations such as micronuclei formation, single and double strand DNA breaks, translocations and estimates of cells with an inappropriate chromosome complement. The results of a few of these studies are recounted below.

When human lymphocytes were exposed to either continuous sinusoidal 50Hz or 60Hz fields or to pulsed fields no detectable chromosomal aberrations or inhibition of DNA repair mechanisms were noted [*Rosenthal and Obe, 1989; Cadossi et al., 1992; Nafziger et al., 1993*]. One the other hand, primary human lymphocytes were reported to exhibit increased micronuclei formation when exposed to 32Hz, 75 or 150 μT fields but only when a parallel 42μT static field was concurrently applied [*Tofani et al., 1995*]. While there was also one report claiming that chromosomal defects increased in human amniotic cells following their exposure to a 50Hz continuous or intermittent 30μT field [*Nordenson et al., 1994*], attempts to confirm this observation have failed [*Galt et al., 1995*].

There are other examples where various types of magnetic field exposures seem to have altered DNA structure but in virtually all cases the findings have not been replicated. Also, in some cases the controls were inadequate raising a question concerning the validity of the original findings. While chromosomal aberrations due to magnetic field exposure could be used to explain the alleged increased cancer risk in some epidemiological reports, there is virtually universal agreement that magnetic fields are not capable of directly damaging DNA [*McCann et al.,1993*]. Even if such damage did occur, it would be unlikely to explain the reported increased incidence of neurodegenerative diseases that have been described.

7.2 EMFs as an Inhibitor of Melatonin

One subject that has received considerable interest in terms of an EMF effect that could explain the slightly increased cancer risk in the epidemiological reports is that such field exposures reduce pineal melatonin synthesis and secretion [*Reiter and Richardson, 1992*]. This field of research was stimulated by the theory put forth by *Stevens* [*1987*] as summarized above (Fig. 1). To reiterate, the theory states that EMF exposures reduce the production of endogenous melatonin leading to a rise in estrogen and prolactin secretion which in turn stimulate the growth of breast cancer cells thereby resulting in a higher incidence of at least mammary gland cancer as a result of such exposures.

Two lines of investigation have flourished as a result of this theory. In one set of studies, animals, including humans, were exposed to biologically relevant EMFs to test their efficacy in reducing melatonin. While some of these studies yielded information suggesting a reduction in some aspect of melatonin production or secretion [*Reiter, 1997*], the findings are by no means consistent. Thus, on many occasions no change in melatonin was seen; this is particularly true when humans were investigated in terms of their exposure to magnetic fields [*Selmaoui et al., 1996, Graham et al., 1997*].

The second line of investigation prompted by the theory of *Stevens* [*1987*] was that EMF exposure would increase mammary tumor development in rats given a chemical carcinogen. As already mentioned above, the studies of *Löscher and colleagues* [*1993, 1994*] are especially noteworthy for their thoroughness and attention to detail. Despite this, the results of these studies varied between and sometimes within the same report. Whereas the findings generally suggest that power line frequency magnetic field exposures increased induced cancer incidence in rats and that this may be associated with a lowering of endogenous melatonin production, the findings do not permit this unequivocal conclusion because of outcome variability.

From all the published reports, the one consistent finding seems to be that when EMFs influence melatonin production, the change is always in the direction of inhibition [*Reiter, 1993*]. Explaining the outcomes of the substantial number of studies in which no changes in melatonin were observed, however, has become difficult.

7.3 EMFs as Free Radical Modifiers

Free radicals are molecules that have an unpaired electron in their outer orbital, a feature whi makes them more highly reactive in biological systems. While there are some beneficial actio of free radicals, on the whole their actions are generally considered to be devastating to cel because they destroy molecules such as DNA, proteins and lipids.

There is a modicum of data which suggests that the reactions of free radicals are modifi in the presence of magnetic fields. Typically, free radicals have a very short half-life (on t order of fractions of seconds) which, if prolonged, can increase the likelihood of them damagi a macromolecule. According to theoretical calculations and biological results reported l several groups [*Harkness and Grissom, 1994; Grissom, 1996; Brocklehurst and McLauchla 1996*], magnetic field exposures may indeed increase the availability of free radicals to plund essential molecules because the fields prolong their half-lives. This was most compelling demonstrated by *Roy and colleagues* [*1995*] who specifically showed that a 60Hz, 0.1m magnetic field increased free radicals measured in activated rat neutrophil preparations. If th were to occur in vivo, increased molecular damage would be expected. If the specific molecul damaged were DNA then an increased incidence of cancer could be expected [*Cerutti, 1994*].

Perhaps of additional interest in this regard is that melatonin is a free radical scaveng [*Reiter, 1995; Reiter et al., 1997a*], i.e., it neutralizes them and thereby reduces the subseque damage. This being the case, a worst case scenario for EMF exposure may be increased radic half-lives combined with reduced melatonin levels. While this theory is attractive and has bee used to explain the findings in one report [*Lai and Singh, 1997*], it for the most part remains a untested theory [*Reiter, 1997*]. Its attractiveness as a possibility also stems from the fact that fre radicals not only damage DNA which can lead to cancer, but they are also part of th degenerative processes of a number of neurodegenerative diseases [*Reiter, 1998*]. Thus, if the are influenced by EMFs they could theoretically explain both the increased cancer risk and th higher incidence of age-related neurodegenerative conditions reported in the epidemiologic investigations.

7.4 EMFs as Modifiers of Intracellular Responses

There are a large number of biologically relevant intracellular changes that have been reported a a consequence of the exposure of cultured cells to EMF. Whether these relate to the healt consequences reported in the epidemiological studies is not clear, but the basic nature of th changes would be expected to have significant effects on the physiology of normal and transformed cells. The changes of major interest relate to field effects on calcium flux and intracellular free calcium measurements [*Bawin and Adey, 1976; Coulton and Barker, 1993*], the activity of the enzyme ornithine decarboxylase (ODC) [*Byus et al., 1987; Litovitz et al., 1991*], uridine uptake and incorporation [*Greene et al., 1991; Phillips et al., 1992*], and the activities of the early response genes, *fos* and *jun* [*Goodman and Shirley-Henderson, 1991; Lin et al., 1994*]. As with this field of research in general, replication studies related to magnetic field effects at the cellular level have not always yielded consistent findings. This disturbing feature has made some question the significance of the numerous cellular changes that have been observed.

8. Concluding Remarks

A survey of the numerous reports and diverse changes that have been reported in terms of the health effects and biological consequences of EMF exposure cannot be adequately dealt with in the context of a short review such as this. The interested readers is urged to consult far more comprehensive treatises on this subject [*National Research Council Report, 1997; NIEHS Working Group Report, 1998; Brainard et al., 1999*].

In evaluating the data published to date it is difficult to draw any firm conclusions as to whether there are health effects of exposure to power line frequency EMFs. In the case of the epidemiological reports, the evidence in general is "limited" or "inadequate" for making judgments. "Limited" evidence means that the effect is driven by a clear finding in a single study for a single endpoint or there are unexplained problems with the data set. "Inadequate" evidence can mean several things: 1), there is insufficient data for making a judgment; 2), data suggest a positive link but they cannot be interpreted as suggesting a causal link; 3) the data suggest a negative effect but cannot be interpreted as suggesting no effect; and 4), the data are contradictory and no clear pattern is discernible. For the epidemiological data summarized in this survey, in individual cases each of these issues may apply.

For mechanistic studies, data may be classified as "weak" "moderate", or "strong". Judging from the published data, it would seem most appropriate to conclude that mechanistic studies probably provide moderately strong data to support a causal relationship between EMF exposure and biological effects. Certainly no one set of mechanistic studies stand out in terms of being the likely mechanism to explain the epidemiological data. A conservative interpretation of the results reported to date would be that EMFs in the power frequency range may in some cases induce measurable biological changes; however, whether they are indirectly carcinogenic or accelerate neurodegenerative diseases remains questionable. Unfortunately, as a result any public-health decision by necessity would be based on limited and inconclusive evidence of an increased health risk associated with EMF exposure. Additionally, decisions concerning the value of continued expenditures for research in this area likewise rest on uncertain data.

References

Anisimov, V.N., O.V. Zhukova, D.S. Beniashvilli, et al. (1996), "Effect of the light regime and electromagnetic fields on carcinogenesis of the mammary gland in female rats," *Biophysics* 41, pp. 817-823.

Bassett, C.A. (1993), "Beneficial effects of electromagnetic fields," *J. Cell. Biochem.*57, pp. 387-393.

Baum, A., M. Mevissen, K. Kamino, et al (1995), "A histopathological study on alterations in DMBA-induced mammary carcinogenesis in rats with 50Hz, 100μT magnetic field exposure," *Carcinogenesis* 16, pp. 119-125.

Bawin, S.M., and W.R. Adey (1976), "Sensitivity of calcium binding in cerebral tissue to weak environmental fields oscillating at low frequency," *Proc. National Academy Sci. USA* 73, pp. 1999-2003.

Beniashvilli, D.S., V.G. Benianishvilli, and M.Z. Menabde (1991), "Low frequency electromagnetic radiation enhances the induction of rat mammary tumors in nitrosomethyl urea," *Cancer Lett.* 61, pp. 75-79.

Brainard, G.C., R. Kavet, and L.J. Kheifets [1999], "The relationship between electromagnetic field and light exposures to melatonin and breast cancer risk: A review of the relevant literature," *J. Pineal Research*, in press.

Brocklehurst, B., and K.A. McLauchlan (1996) "Free radical mechanism for the effect of environmental electromagnetic fields on biological systems," *Int. J. Radiation Biol.* 69, pp. 3-24.

Byus, C.V., S.E. Pieper, and W.R. Adey (1987), "The effects of low-energy 60Hz environmental fields upon the growth-related enzyme ornithine decarboxylase," *Carcinogenesis* 8, pp. 1385-1389.

Cadossi, R., F. Bersani, A. Cossarizza, et al. (1992), "Lymphocytes and low frequency electromagnetic fields," *FASEB* 6, pp. 2667-2674.

Cerutti, D.A. (1994), "Oxy-radicals and cancer," *Lancet* 344, 862-863.

Coulton, I.A., and A.T. Barker (1993), "Magnetic fields and intracellular calcium: effects on lymphocytes exposed to a condition of cyclotron resonance," *Phys. Med. Biol.* 38, pp. 347-460.

Davanipour, Z., E. Sobel, J.D. Bowman, et al. (1997), "Amyotrophic lateral sclerosis and occupational exposure to electromagnetic fields," *Bioelectromagnetics* 18, pp. 28-35.

Ekström, T., K.H. Mild, and B. Holmberg (1998), "Mammary tumors in Sprague-Dawley rats after initiation with DMBA followed by exposure to a 50Hz electromagnetic fields in a promotional scheme," *Cancer Lett.* 123, pp. 107-111.

Feychting, M., and A. Ahlbom (1993), "Magnetic fields and cancer in children residing near Swedish high-voltage power lines," *Am. J. Epidemiology* 138, pp. 367-481.

Feychting, M., U. Forssen, and B. Floderus (1993), "Magnetic fields, leukemia and central nervous system tumors in Swedish adults residing near high-voltage power lines," *Epidemiology* 5, pp. 501-509.

Feychting, M., G. Schulgen, J.H. Olsen and A. Ahlbom (1998), "Dementia and occupational exposure to magnetic fields" *Scandinavian J. Work Environ. Health*, in press.

Floderus, B., T. Persson, C. Stenbind, et al. (1993), "Occupational exposure to electromagnetic fields in relation to leukemia and brain tumors; a case-control study in Sweden," *Cancer Causes Control* 4, pp. 465-476.

Floderus, B., S. Tornqvist, and C. Stenlund (1994), "Incidence of selected cancers in Swedish railway workers, 1961-79," *Cancer Causes Controls* 5, pp. 189-194.

Galt, S., J.W. Wahlström, Y. Hammenius, et al. (1995), "Study of effects of 50Hz magnetic field on chromosome aberrations and the growth related enzyme OdC in human amniotic cells," *Bioelectrochem. Bioenergetics* 36, pp. 1-8.

Goodman, R., and A. Shirley-Henderson (1991) "Transcription and translocation in cells exposed to extremely low frequency electromagnetic fields," *Bioelectrochem. Bioenergetics* 25, pp. 335-355.

Graham, C., M.R. Cook, and D.W. Riffle (1997), "Human melatonin during continuous magnetic field exposure," *Bioelectromagnetics* 18, pp. 166-171.

Greene, J.J., W.J. Skowronski, and J.M. Mullins (1991), "Delineation of electric and magnetic field effects of extremely low frequency electromagnetic radiation on transcription," *Biochem. Biophys. Res. Communications* 174, pp. 742-749.

Grissom, C.B. (1996) "Magnetic field effects in biology: a survey of possible mechanisms with emphasis on radical-pair recombination", *Chemical Reviews* 95, pp. 3-28.

Guenel, P., P. Raskmark, J.B. Andersen, and E. Lynge (1993), "Incidence of cancer in persons with occupational exposure to electromagnetic fields in Denmark," *British J. Industrial Medicine* 50, pp. 758-764.

Harkness, T.T., and C.B. Grissom (1994), "Magnetic field effects on B12 ethanolamine, ammonia lyase: evidence for a free radical mechanism," *Science* 263, pp. 958-960.

Harrington, J.M., D.I. McBride, T. Sorahan, et al. (1997), "Occupational exposure to magnetic fields in relation to mortality from brain cancer among electricity generation and transmission workers," *Occup. Environ. Health* 54, pp. 7-13.

Jackson, J.D. (1992), "Are the stray 60Hz electromagnetic fields associated with the distribution and use of electric power a significant cause of cancer?," *Proc. National Acad. Sci. USA*, 89, pp. 3508-3510.

Kheifets, L., A.A. Afife, P. Buffler, et al. (1997), "Occupational electric and magnetic field exposure and leukemia," *J. Occup. Environ. Health* 39, 1074-1091.

Lagroye, I., and J.L. Poncy (1997), "The effect of 50Hz electromagnetic fields on the formation of micronuclei in rodent cell lines exposed to gamma radiation," *Int. J. Radiation Biology* 72, pp. 249-254.

Lai, H. and N.P. Singh (1997), "Melatonin and n-tert-butyl-alpha-phenylnitrone block 60Hz magnetic field induced DNA single and double strand breaks in rat brain cells," *J. Pineal Research* 22, pp. 152-162.

Li, C.Y., G. Theriault, and R.S. Lin (1997), "Residential exposure to 60-hertz magnetic fields and adult cancers in Taiwan," *Epidemiology* 8, pp. 25-30.

Lin, H., R. Goodman, and A. Shirley-Henderson (1994), "Specific region of the c-myc promoter is responsive to electric and magnetic fields," *J. Cellular Biochemistry* 54, pp. 281-288.

Linet, M.S., E.E. Hatch, R.A. Kleinerman, L.L. et al. (1997), "Residential exposure to magnetic fields and acute lymphoblastic leukemia in children," *New England J.Med.* 337, pp.1-7.

Litovitz, T.A., D. Krause, and J.M. Mullins (1991), "Effect of coherence time of the applied magnetic field on ornithine decarboxylase activity," *Biochem. Biophys. Research Communications* 178, pp. 862-865.

London, S.J., J.D. Bowman, E. Sobel, et al. (1994), "Exposure to magnetic fields among electrical workers in relation to leukemia risk in Los Angeles County," *Am. J. Industrial Medicine* 26, pp. 47-60.

Löscher, W., M. Mevissen, W. Lehmacher, and A. Stamm (1993), "Tumor promotion in a breast model by exposure to a weak alternating magnetic field," *Cancer Letters* 71, pp. 75-81.

Löscher, W., U. Wahnschaffe, M. Mevissen, et al. (1994), "Effects of weak alternating magnetic fields on nocturnal melatonin production and mammary carcinogenesis in rats," *Oncology* 15, pp. 288-295.

Lovely, R.H., R.L. Buschbom, A.L. Slavich, et al.(1994), "Adult leukemia risk and personal appliance use: a preliminary study," *Am. J. Epidemiology* 140, pp. 510-517.

Mandeville, R., E. Franco, S. Sidrac-Ghali, L. et al. (1997), "Evaluation of the potential carcinogenicity of 60Hz linear sinusoidal continuous-wave magnetic fields in Fischer F344 rats," *FASEB J.* 11, pp. 1127-1136.

Margonato, V., P. Nicolini, et al. (1995), "Biologic effects of prolonged exposure to ELF electromagnetic fields in rats: II. 50Hz magnetic fields," *Bioelectromagnetics* 16, pp. 343-355.

McCann, J., F. Dietrich, C. Rafferty, and A.O. Martin (1993), "A critical review of the genotoxic potential of electric and magnetic fields," *Mutation Research* 297, pp. 61-95.

McLean, J.R.N., M.A. Stuchly, R.E.J. Mitchel, et al. (1991), "Cancer promotion in a mouse-skin model of 60Hz magnetic fields: II. Tumor development and immune response," *Bioelectromagnetics* 12, pp. 273-287.

McLean, J.R.N., A. Thansandote, D. Lecuyer, and M. Goddard (1997), "The effect of 60Hz magnetic fields on co-promotion of chemically induced skin tumors on SENCAR mice: a discussion of three studies," *Environmental Health Perspective* 105, pp. 94-96.

Merchant, C.J., D.C. Renew, and J. Sevanson (1994), "Exposure of power-frequency magnetic fields in the home," *J. Radiological Protection*, 14, pp. 77-87.

Mevissen, M.A. Stamm, S. Buntenkotter, et al. (1993), "Effects of magnetic fields on mammary tumor development induced by 7,12-dimethylbenz(a)anthracene in rats," *Bioelectromagnetics* 14, pp. 131-143.

Mevissen, M., A. Lerchl, W. Löscher (1996a), "Study on pineal function and DMBA-induced breast cancer formation in rats during exposure to a 100μT, 50Hz magnetic field," *J. Toxicology Environ. Health* 48, pp. 169-185.

Mevissen, M., A. Lerchl, M. Szamel, and W. Löscher (1996b), "Exposure of DMBA-treated female rats in a 50Hz, 50 microtesla magnetic field: effects on mammary tumor growth, melatonin levels and T-lymphocyte activation." *Carcinogenesis* 17, pp. 903-910.

Mevissen, M., M. Haussler, M. Szamel et al. (1998), "Complex effects of long-term 50Hz field exposure in vivo on immune functions in female Sprague-Dawley rats depend on duration of exposure," *Bioelectromagnetics* 19, pp. 259-270.

Nafziger, J., H. Desjobert, B. Benamar, et al. (1993), DNA mutations and 50Hz electromagnetic fields," *Bioelectrochem. Bioenergetics* 30, pp. 133-141.

National Research Council Report (1997), "*Possible Health Effects of Exposure to Residential Electric and Magnetic Fields*," Washington, National Academy of Sciences.

NIEHS Working Group Report (1998) "*Assessment of Health Effects from Exposure to Power-Line Frequency Electric and Magnetic Fields*," Portier CJ and Wolfe MS. eds., Research Triangle Park, National Institutes of Health.

ORAU Committee Report (1992), "*Health Effects of Low-Frequency Electric and* Magnetic *Fields*," Washington, Oak Ridge Associated Universities.

Nordenson, I., K.H. Mild, G. Anderson and M. Sandstrom (1994), "Chromosomal aberrations in human amniotic cells after intermittent exposure to fifty hertz magnetic fields," *Bioelectromagnetics* 15, pp. 293-301.

NTP (1998), "*Toxicology and carcinogenesis studies of 60Hz magnetic fields in F344/N rats and B6C3F1 mice (whole body exposure studies).*" US Department of Health and Human Services, National Institute of Environmental Health Sciences, Research Triangle Park, NC.

Phillips, J.L., W. Haggrens, W.J. Thomas, et al. (1992) "Magnetic field induced changes in specific gene transcription," *Biochimica Biophys. Acta* 1132, pp. 140-144.

Preston-Martin, S., J.M. Peters M.C. Ya, D.H. Garabrant, and J.D. Bowman (1993), "Myelogenous leukemia and electric blanket use," *Bioelectromagnetics* 9, pp. 207-213.

Rannug, A., B. Holmberg, T. Ekstrom, et al. (1993), "Rat liver foci study on coexposure with 50Hz magnetic fields and known carcinogens," *Bioelectromagnetics* 14, pp 17-27.

Rannug, A., B. Holmberg, T. Ekstrom, et al. (1994), "Intermittent 50Hz magnetic field and skin tumor promotion in SENCAR mice," *Carcinogenesis* 15, pp. 153-157.

Reiter, R.J. (1993) "Static and extremely low frequency electromagnetic field exposure: reported effects on circadian production of melatonin," *J Cellular Biochemistry* 51, pp. 394-403.

Reiter, R.J. (1995), "Oxidative processes and antioxidative defense mechanisms in the aging brain," *FASEB Journal* 9, pp. 526-533.

Reiter, R.J. (1997), "*Melatonin aspects of exposure to low frequency electric and magnetic fields*," in J.C. Lin (ed.) Advances in Electromagnetic Fields in Living Systems, Vol. 2, New York, Plenum, pp. 1-27.

Reiter, R.J. (1998) "Oxidative damage in the central nervous system: protection by melatonin," *Progress in Neurobiology* 56, pp. 359-384.

Reiter, R.J., and B.A. Richardson (1992), "Magnetic field effects on pineal indoleamine metabolism and possible biological consequences," *FASEB Journal* 6, pp. 2283-2287.

Reiter, R.J., L. Tang, J.J. Garcia and A. Muñoz-Hoyos (1997a), "Pharmacological levels of melatonin in oxygen radical pathophysiology," *Life Sciences* 60, pp. 2255-2271.

Reiter, R.J., R.C. Carneiro and C.S. Oh (1997b) "Melatonin in relation to cellular antioxidative defense mechanisms," *Hormone and Metabolic Research* 29, pp. 363-372.

Rosenthal, M., and G. Obe (1989), "Effect of 50Hz electromagnetic fields on proliferation and on chromosomal alterations in human peripheral lymphocytes untreated or pre-treated with chemical mitogens," *Mutation Research* 210, pp. 329-335.

Roy, S., Y. Noda, V. Eckert, et al. (1995), "The phorbol 12-myristate 13-acetate (DMA)-induced oxidative burst in rat peritoneal neutrophils is increased by a 0.1mT magnetic field," *FEBS Letters* 376, pp. 164-166.

Sahl, J.D., M.A. Kelsh, and S. Greenland (1993), "Cohort and nested case-control studies of hematopoietic cancers and brain cancer among electric utility workers," *Epidemiology* 4, pp. 104-114.

Sasser, L.B., J.E. Morris, D.L. Miller, et al. (1996), "Exposure to 60Hz magnetic fields does not alter clinical progression of LGL leukemia in Fischer rats", *Carcinogenesis*, 17, pp. 2681-2687.

Savitz, D.A. and D.P. Loomis (1995), "Magnetic field exposure in relation to leukemia and brain cancer mortality among electric utility workers," *Am. J. Epidemiology* 141, pp. 123-134.

Savitz, D.A., H. Wachtel, F.A. Barnes, et al. (1988), "Case control study of childhood cancer and exposure to 60Hz magnetic fields," *Am. J. Epidemiology* 128, pp. 21-38.

Savitz, D.A., V. Dufort, B. Armstrong, and O. Theriault (1997), "Lung cancer in relation to employment in the electrical utility industry and exposure to magnetic fields," *Occup. Environ. Health* 54, pp. 396-402.

Savitz, D., H. Checkoway, and D. Loomis (1998a), "Magnetic field exposure and neurodegenerative disease mortality among electric utility workers", *Epidemiology*, in press.

Savitz, D., D. Loomis and T. Chur-Kit (1998b), "Electrical occupations and neurodegenerative disease: analysis of US mortality data," *Archives of Environmental Health*, 53, pp. 1-5.

Selmaoui, B., J. Lambrozo, and Y. Touitou (1996), "Magnetic fields and pineal function in humans: evaluation of nocturnal acute exposure to extremely low frequency magnetic fields on serum melatonin and urinary 6-sulfatoxymelatonin circadian rhythms," *Life Sciences* 58, pp. 1539-1549.

Scarfi, M.R., M.B. Lioi, D. Zenio, et al. (1994), "Lack of chromosomal aberration and micronucleus induction in human lymphocytes exposed to pulsed magnetic fields," *Mutation Research* 306, pp. 129-133.

Severson, R.K., R.G. Stevens, W.T. Kaune, et al. (1988), "Acute non-lymphocytic leukemia and residential exposure to power frequency magnetic fields," *Am. J. Epidemiology* 128, pp. 10-20.

Shen, Y.H., B.J. Shao, H. Chiang, et al. (1997), "The effect of 50Hz magnetic field exposure on dimethyl (alpha) benzanthracene induced thymic lymphoma/leukemia in mice," *Bioelectromagnetics*, 18, pp. 360-364.

Sobel, E., Z. Davanipour, R. Solkava, et al. (1995), "Occupations with exposure to electromagnetic fields: a possible risk factor for Alzheimer's disease," *Am. J. Epidemiology*, 142, pp. 515-524.

Stevens, R.G. (1987), "Electric power use and breast cancer: A hypothesis," *Am. J. Epidemiology* 125, pp. 556-561.

Stuchly, M.A., J.R.N. McLean, R. Burnett, et al. (1992), "Modification of tumor promotion in the mouse skin by exposure to an alternating magnetic field," *Cancer Letters* 65, pp. 1-7.

Tofani, S., A. Farrara, L. Anglesio, and G. Gilli (1995), "Evidence for genotoxic effects of resonant ELF magnetic fields," *Bioelectrochemistry and Bioenergetics* 36, pp. 9-13.

Theriault, G., M. Goldberg, A.B. Miller, et al. (1994), "Cancer risk associated with occupational exposure to magnetic fields among utility workers in Ontario and Quebec, Canada, and France: 1970-1989," *Am. J. Epidemiology* 139:550-572.

Tynes, T., A. Andersen, and F. Langmark (1992), "Incidence of cancer in Norwegian workers potentially exposed to electromagnetic fields," *Am. J. Epidemiology* 136, 81-88.

Vena, J.E., J.C. Freudenheim, J.R. Marshall, et al. (1994), "Risk of premenopausal breast cancer and use of electric blankets," *Am. J. Epidemiology* 140, pp. 974-979.

Verkasalo, P.K., E. Pukkalo, I, Karpio, et al. (1996), "Magnetic fields of high power voltage power lines and risk of cancer in Finnish adults: nationwide cohort study," *British Medical J.* 313, pp. 1047-1051.

Yasui, M., T. Kikuchi, M. Ogawa, et al. (1997) "Carcinogenecity test of 50Hz sinusoidal magnetic fields in rats," *Bioelectromagnetics* 18, pp. 531-540.

Wertheimer, N., and E. Leeper (1979), "Electrical wiring configurations and childhood cancer," *Am. J. Epidemiology* 109, pp. 272-284.

Wertheimer, N., and E. Leeper (1986), "Possible effects of electrical blankets and heated water beds on fetal development," *Bioelectromagnetics* 7, pp. 13-22.

SPECTRUM CONGESTION

Paul Delogne and Willem Baan

1. Introduction

Sharing the spectrum by dividing it into channels assigned to different users has always been on the mind of radio scientists and engineers. In reality, a radio system is rarely designed with an absolute square filter to make it operate only in its assigned band. Problems arise when the needs or characteristics of systems in neighbouring bands have been considered inadequately. Furthermore the same frequency channel can be reused several times on the surface of the Earth at distances that offer enough protection against interference.

Spectrum congestion relates to increasing difficulties in sharing the spectrum resource with a minimum risk of harmful interference. During recent years the use of the spectrum and its economic value have changed dramatically because of increasing demand. At the same time governments attempt to fill their coffers by trying to sell parts of the spectrum. This is not without danger. The use of the spectrum brings with it a "social responsibility" to use it wisely and to consider the needs of other spectrum users since it is public and natural resources. To take account of all spectral neighbours remains a task for all spectrum users.

Because the issues of spectrum congestion will continue to change and increase, new technical approaches need to be identified and new ways of thinking about existing challenges are required. URSI members must play a key role in this multi-disciplinary issue by providing scientific expertise and suggesting innovative solutions for an efficient use of the spectrum. This requires inter-commission work. On the other hand, URSI scientists are also spectrum users in areas such as radio astronomy and remote sensing and they need to defend their own interests. It is among the objectives of this paper to investigate how URSI could contribute to this exciting area.

2. Spectrum Management

Regulation of international spectrum use resides with the International Telecommunication Union (ITU) in Geneva, Switzerland, which is a part of the United Nations. Within ITU, the responsibility for radio communications lies with the Radiocommunications Bureau (ITU-R).

The ITU is an inter-governmental organisation. It has two categories of members, sometimes labelled with a capital 'M' and a lower-case 'm', respectively. The former category includes representatives of governments and hold the voting rights at Plenary Assemblies. The latter category includes "Sector Members" who do not hold voting rights, but are allowed to participate in technical studies performed within Study Groups. URSI and IUCAF (see below) are Sector Members.

Regulations or spectrum allocations adopted by the ITU have the status of *"international treaty"* that needs to be ratified by the member nations of the ITU. Subsequently each administration may want to implement the ITU adopted rules for its own territory and will normally do so. Therefore, the ITU-R Radio Regulations define the principles for operations of the radio communication services and for the protection provided for each of the Services.

2.1 The ITU-R Allocation Process

The spectrum has become a valuable natural resource and intensification of its use requires extensive planning and interference management. The allocation of frequency bands to various services is normally made during World Radio Conferences (WRC) which are organised by ITU-R at intervals of about two years. Preparatory work is performed in Study Groups and in particular in SG1 on Spectrum Management. This process requires careful consideration of technical, economical, and political arguments from all parties involved. Because of their global nature, the allocations for satellite services should be made particularly carefully.

Spectrum efficiency and *spectrum congestion* are very complicated issues within the ITU-R because they mean different things for different people. The allocation process itself causes some inefficiency because of many conflicting interests.

The ITU-R Radio Regulations provide the guidelines for operation and manufacturing of radio equipment. Often the regulatory process runs years behind the manufacturing standards. Whatever the case, frequency allocations have long-term consequences, as they are an essential condition for the economically secured development of systems. Questioning frequency allocations that have been granted years ago requires the study of economically acceptable evolution scenarios that also include development plans of all actors. Terrestrial television, discussed below, is a good example of a service for which modern technologies will allow for considerable progress in the area of spectrum efficiency.

In the past, some allocations resulted in spectrum fragmentation (or splintering) as the spectrum has been subdivided into small segments with sharply defined boundaries and no guard bands. As a result, many services with vastly different protection requirements have been placed in adjacent bands causing a number of problems. Improving the efficiency of spectrum use would require that services with similar protection requirements be allocated close together, in order to facilitate adjacent band coordination.

2.2 The Specific Case of Science Services

URSI does not only gather scientists who are experts on the efficient use of the radio spectrum for purposes such as communications. Other URSI scientists have as their primary interest the use radio waves for scientific investigation. Partly as the result of a long-standing collaboration between URSI and ITU-R, these applications have traditionally been recognised in the ITU-R structure, where they are considered by a specific Study Group (SG5) on "Science Services".

Science services include radio astronomy and the earth exploration by either active techniques (e.g. radars) or passive means (e.g. radiometers). In the frame of ICSU (International Council for Science), the interests of the scientific community as a spectrum user are represented to ITU-R by the IUCAF, which interacts as a Sector Member with other bodies such as the WMO (World Meteorological Organisation). The IUCAF (Inter-Union Committee on Frequency Allocations) is an ICSU Committee, sponsored by URSI, the IAU (International Astronomical Union) and COSPAR (Committee for Space Research).

In spite of this official recognition of science services, their specific problems are not always fully acknowledged. The Radio Regulations place all services on an equal footing and do not provide preferential treatment to the more vulnerable passive services as shown in the following articles from the RR:

S4.5 *The frequency assigned to a station of a given service shall be separated from the limits of the band allocated to this service in such a way that, taking into account of the frequency band assigned to the station, no harmful interference is caused to services to which frequency bands immediately adjoining are allocated.*

S4.6 *For the purpose of resolving cases of harmful interference, the radio astronomy service shall be treated as a radio-communication service. However, protection from services in other bands shall be afforded the radio astronomy service only to the extent that such services are afforded protection from each other.*

The specific case of passive services is thus not handled on an equitable basis. It is common practice in the design of active services that the power of emitted signals is raised to the point that the level of natural, additive noise onto the received signal is made negligible. In a such a context where active spectrum users may raise their transmit powers beyond that, spectrum management reduces to ensuring each user some required signal-to-interference ratio, i.e. handling relative signal power levels. Passive services are based on measurements of natural radiation, sometimes of very low levels. They need protection in absolute terms.

During the last years and still presently the radio astronomy service has suffered a number of harmful interference cases, mainly from satellites with inadequate protection in the adjacent band. Some radio astronomy operations in protected bands have been hereby made very difficult or even impossible for a number of years, although the ITU RR specifically allowed no harmful interference. How can the radio science community counteract such abrasive behaviour?

3. Spectrum Efficiency in Radio Communication Systems

The title of this section is very ambitious, since the concept of spectrum efficiency is extremely broad. In 1974, ITU-R issued its Question 47/1 entitled "Definition of efficiency and utility of spectrum use". The Question includes a number of quite interesting considerings, to finally conclude that:

The ITU Radiocommunication Assembly,

considering …

unanimously decides that the following Questions should be studied……

4. What are the appropriate methods of describing "spectrum utilisation" and how can such utilisation be compared for different radiocommunication services?

5. What is the best way of defining "efficiency of spectrum use" and how should this efficiency be calculated?

The Question remained unchanged for about 25 years. Recommendation [*ITU-R 1994a*] provides only the beginning of an answer, as it is restricted to some criteria for comparing similar systems. This state of matters is a good indicator of the problem complexity. It is by no way among the objectives of this paper to provide a real-world matched answer. This would necessarily involve numerous techno-economical and political considerations. The latter are extremely important, but they cannot be seen as more than a, frequently strong constraint for a scientific approach to the problem. By the way, this means that scientific approaches should devote enough attention to evolution scenarios. The main objective of this contribution is to try identifying those areas where the science community can contribute parts of the answer.

3.1 In-band Efficiency

3.1.1 Shannon's Limit

The existing telecommunication world is mostly based on a scenario whereby a communication channel is made available for transmitting information from some user T -say the transmitter- to another user R -say the receiver. This scenario remains compatible with most existing communication protocols and standards. The latter are extremely important for an efficient use of the spectrum. However, besides defining agreed syntactic and semantic communication rules, they remain based on the assumption that the basic resource of a telecommunications infrastructure is a physical channel made available between two users, perhaps for a limited time period and in a restricted spatial area. The users T and R are hereby not necessarily end users; they may be two arbitrary points in a telecommunication infrastructure, e.g. switches.

In this scenario, the basic element of the telecommunications world remains a physical channel, offering some bandwidth B, made available for transmitting information. At this level of the definition of spectral efficiency, the problem is restricted to the most efficient use of the channel. There is probably no better answer than Shannon's capacity theorem, which was formulated exactly 50 years ago [*Shannon, 1948*].

It was part of Shannon's genius to recognise that there is no other basic limitation to the channel capacity than the ability to separate the useful signal from additive, uncorrelated noise. Shannon anticipated that physical channel impairments such as dispersion, multi-path propagation, etc. were no fundamental constraints, in this sense that they can be corrected by suitable signal processing. Shannon therefore considered the simple case of a channel offering some bandwidth B, subject to additive white Gaussian noise (AWGN). He showed that the information transmission capacity of such a channel is given by

$$C = B \log_2 [1 + S/N] \quad \text{Sh/s}, \tag{1}$$

where S/N is the channel signal-to-noise ratio.

The beauty of this fundamental theorem lies in its simplicity. Its actual meaning and implications cannot be understood without some comments:

1) The proof of the theorem indicates that the channel capacity C is a theoretical limit, which can only be attained through thus far undefined signal processing and coding techniques. Reaching the limit necessarily implies an infinite processing delay. The delay which can actually be accepted in practice depends of the type of signal to be transmitted as well as of techno-economical constraints.

2) The assumed Gaussian character of additive noise represents the worst case. As indicated by the DeMoivre-Laplace theorem, the Gaussian character of a stochastic process generally results from the addition of a large number of independent contributions, of which none is dominant. This is the most unstructured type of noise. Additive, interfering, telecommunications signals may be very structured, i.e. internally correlated. The extreme case of a highly structured interfering signal is a sine wave: in the limit, such an interfering signal is not harmful when an infinite processing delay is available, since its characteristics (amplitude, frequency, phase) can be identified, and it can therefore be cancelled.

3) The channel capacity is expressed in Sh/s, where the *Shannon* (Sh) is the unit of information. This unit is defined as the information contained in a binary symbol, for which the two states, currently called 0 and 1, are equally probable. For a binary sequence, a further requirement for a binary symbol to contain one Sh of information is that successive symbols be statistically independent. It is important to note that this double condition is required for the elements of a binary sequence, currently called *bits*, to carry one Sh of information. When the condition is not satisfied, the binary sequence is *redundant*, which means that a bit contains less than one Sh of information, either because successive bits are not independent or because their two states are no equally probable.

The current practice stating that Shannon's theorem provides an upper limit to some transmission speed expressed in bit/s is a rather confusing language abuse. If a binary sequence with a bit rate R (bit/s) is redundant, its information rate C (Sh/s) is lower than R. This clearly indicates that the bit rate R (bit/s) which can be transmitted without any error through a channel can even be larger than Shannon's limit C as provided by equation (1).

Most natural signals are highly redundant. Redundancy offers some advantages, namely some protection against channel impairments. Natural redundancy is however not very efficient for this purpose. It therefore appears interesting to proceed in two steps: first removing natural redundancy, and then adding some artificial redundancy so as to provide the signal with a good robustness against channel noise.

It has been a part of Shannon's genial analysis to indicate that coding, a generic term for signal processing operations needed to make the best use of a transmission channel, includes two classes that are currently called source coding and channel coding.

3.1.2 Source Coding

Besides the more or less technical objective of defining universal communication codes -- e.g. the binary alphabet --, the main objective of source coding is to remove natural redundancy. "Compression", "bandwidth reduction", "bit rate reduction", etc. are terms which have been used in the literature over the last few decades to indicate this kind of operation. These appellations may be very inadequate, or even misleading. Once it is accepted that present transmission technology is binary, i.e. based on the binary alphabet, the basic question is: "which bit rate is needed to encode a signal with some accepted degree of degradation - not to say (content-dependent) distortion"? An alternative, equivalent formulation is: "which compression degree can be achieved"? The compression ratio is hereby defined as the ratio of bit rates resulting from, on the one hand, PCM (pulse code modulation) with some quantisation accuracy and, on the other hand, the proposed source coding algorithm for an equivalent degradation.

Source coding is obviously an essential ingredient for an efficient use of the spectrum. This kind of technology was nearly non-existent in the telecommunication landscape, ten years ago. It is presently used in a number of well-established techniques. Compression is currently used for commercial-quality speech (i.e. telephony) in cellular radio networks. Some standards are already defined for high-quality audio sounds, including speech and music. Image compression is part of established standards for digital television (DTV) and advanced (definition-enhanced) digital television (ADTV). Table 1 illustrates the performances of compression standards having reached the state of technical and commercial maturity.

It is quite clear from this table that source coding has played a fundamental role in the basic problem of an efficient spectrum use during the last decade. The basic question then comes: is this trend likely to continue? Should the scientific community continue investing efforts in this direction?

Table 1 *State of matters for source coding, according to established standards.*

Application	PCM bit rate	Compressed bit rate	Compression ratio
Commercial speech	64 kbit/s	4.8 - 9.6 kbit/s	7 - 13
High-quality sound	480 kbit/s	64 kbit/s	7.5
Digital Television	216 Mbit/s	2 - 10 Mbit/s	20 - 100
ADTV - HDTV	1 Gbit/s	20 - 30 Mbit/s	30 - 50

Our answer is definitely an enthusiastic "yes", although we do not feel qualified to show the way for further progress. As a matter of fact, fifteen years ago, a careful or timorous engineer would have considered the compressed bit rates shown in the table as unfeasible. That was indeed the right answer under the assumption that the complexity of acceptable algorithms would not lead to integrated circuits with more than a few kilogates in a chip. The present reality is that current compression chips contain a few hundreds kilogates. Further research on compression techniques remains essential, but it will not be successful if the search for new algorithms is excessively constraint by short-term hardware complexity considerations.

3.1.3 Channel Coding

Channel coding encompasses all signal processing operations aiming at configuring the signal for efficient transmission in a given channel. These operations are further ranked in two categories: forward error correction (FEC) and modulation.

In the most classical approach, FEC is performed independently of the modulation process. It consists of adding some calculated redundancy to the sequence of symbols to be transmitted, so as to allow the detection and correction of most transmission errors. The information rate of a code is defined as the ratio of symbol rates before and after coding. Linear block codes such as BCH and RS codes are most frequently used, at information rates which are quite close to one. Codes with different properties can furthermore be concatenated and/or combined with interleaving. In this approach, FEC is not directly related to the channel characteristics, beyond the possibility to match the code properties to channel error statistics, e.g. to errors occurring in bursts. Modulation is hereby restricted to the technique required to map the sequence of coded symbols into an analogue signal adapted to the channel characteristics, i.e. essentially some available bandwidth.

In a more advanced approach, frequently referred to as coded modulation, the error correction and modulation processes are combined, in this sense that the redundant code states are mapped into additional modulation states or symbols. Convolution codes with relatively low information rates (e.g. 1/2 or even 1/3) are currently used in these techniques.

It is outside the scope of this paper to quantitatively investigate properties of various channel coding/decoding techniques, as they are described in the literature [*Proakis, 1989*]. We will limit ourselves to a few general remarks. One or two decades ago, the channel bandwidth was immediately felt by most communication engineers to be a good indicator of the achievable bit rate for digital transmission. It is presently recognised that the channel bandwidth B is only a good indicator of the achievable symbol rate R_s through the approximate equation $B \approx R_s$. A given bit rate R_b·can always be matched into an arbitrarily small symbol rate R_s through the use of multi-level or multi-state symbols. If the binary information is mapped into symbols corresponding to binary words holding on n_b bits, the symbol and bit rates are in a ratio $R_s = R_b / n_b$. In this process, the size of the symbol alphabet - i.e. of distinguishable modulation states - increases exponentially, being given by $N = 2^{n_b}$.

Modulation techniques are currently characterised by a spectrum efficiency E defined as the ratio of the bit rate R_b to the channel bandwidth B :

$$E = \frac{R_b}{B} \qquad \text{in bit/s/Hz.} \tag{2}$$

It is clear that, for a given signal power, increasing the number of modulation states makes the signal increasingly sensitive to noise. Thus, additive noise appears as the fundamental limitation to increasing the spectral efficiency. Shannon's capacity theorem - equation (1) - provides the basic limitation: the spectrum efficiency cannot exceed the limit

$$E_{Sh} = \log_2 (1 + \frac{S}{N}) \qquad \text{in Sh/s/Hz.} \tag{3}$$

It is extremely interesting to compare the actual performance of existing channel coding techniques to this theoretical limit. Before proceeding, it should be observed that comparing the quantities defined in equations (2) and (3) implies that one bit contains one Sh of information, i.e. that the bit rate to be considered is the one at the output of the source coding process (thus before FEC) and that it is implicitly assumed that source coding is ideal. Furthermore, Shannon's limit corresponds to errorless transmission, whereas actual techniques can only guarantee some bit error probability $P_{e,b}$. Comparison of actual techniques with Shannon's limit makes sense when a very low value of $P_{e,b}$ is adopted. The latter was set to 10^{-7} for the purpose of comparisons.

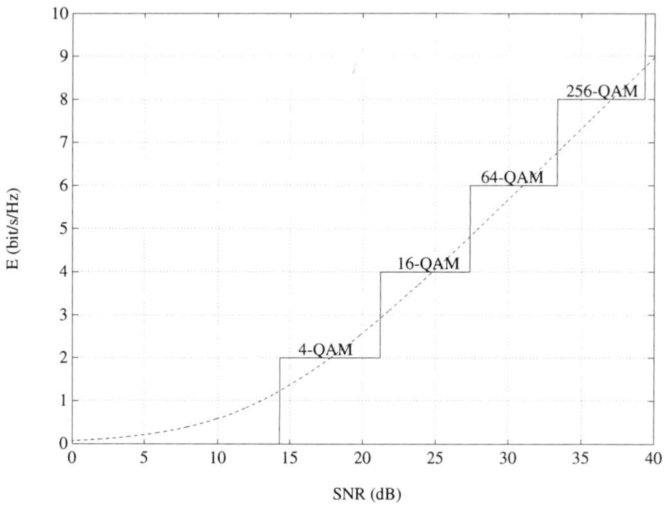

***Figure 1**. Achievable spectrum efficiency of the QAM family of modulations in function of the signal-to-noise ratio.*

The staircase line on Figure 1 illustrates the performance of uncoded QAMs (Quadrature Amplitude Modulations). This result is obtained by selecting, in function of the signal-to-noise ratio in a bandwidth $B \approx R_s$, the QAM level providing the highest spectral efficiency under the constraint of the specified bit error rate. The broken line curve illustrates a "practical" spectrum efficiency given by the equation $E_p = \log_2 \{1 + [S/(MN)]\}$: the parameter M can be seen as a margin, i.e. a penalty in terms of signal-to-noise, for some modulation against the Shannon limit. The broken line curve is drawn for $M = 20$ i.e. 13 dB; it represents well, the performance of uncoded QAMs. This indicates how classical modulations, without any error protection, perform compared to Shannon's limit. Repeating the exercise for more advanced channel coding techniques shows that the margin can be reduced to about 9 dB for QAMs combined with classical FEC techniques, such as BCH or RS. More advanced channel coding techniques, called coded modulations above, can further reduce the margin to about 6 dB.

3.2 Unwanted Emissions

3.2.1 Theoretical References

A basic assumption above has been that a channel with some bandwidth B is made available for transmitting information from say a transmitter T to a receiver R. Up to this point the reasoning does not include any consideration for technological issues or networking aspects. Let us, still for a while, remain in this frame.

It has been shown, several decades ago by Nyquist, that a baseband signal carrying a given symbol rate $R_{s,b}$ can be strictly confined in some (low-pass) bandwidth F_0, with no other limitation than $F_0 \geq R_{s,b}/2$. Strictly confining the spectrum of a signal unavoidably entails that the response to a transmitted symbol has an infinite duration. The related risk is that the response to a symbol be polluted by residuals from preceding symbols; this phenomenon is currently referred to as inter-symbol interference (ISI). It was part of Nyquist great merits to identify a criterion for transmitting baseband signals carrying a symbol rate $R_{s,b}$, without any ISI, with a strict limitation to a baseband $F_0 = (1+\alpha) R_{s,b}/2$, with no other constraint than $\alpha \geq 0$.

Nyquist's reasoning was later completed by Rice's representation of bandpass signals, which allows to represent any signal $x(t)$, strictly holding within some bandwidth B, centered on a frequency $f_0 = \omega_0/(2\pi)$, by:

$$x(t) = x_1(t)\ \cos(\omega_0 t) + x_2(t) \sin(\omega_0 t) = a_x(t) \cos\,[\omega_0 t + \varphi_x(t)], \tag{4}$$

where the signals $x_1(t)$ and $x_2(t)$ are baseband components, currently called I (In-phase) and Q (Quadrature), whose spectrum is strictly limited to $B/2$. Thus, it is perfectly possible to transmit a symbol rate R_s in a strictly limited bandwidth $B = (1+\alpha)R_s$ with $\alpha \geq 0$ as small as desired. "Strictly limited" means that the spectrum does not exhibit any side lobe outside this band. A given symbol rate can still be mapped into an arbitrarily high bit rate via increasing the number of modulation levels.

The QAM family of modulations appears as the simplest way of reaching this goal. It is obvious from Rice's representation that applying some low-pass filtering to the Rice components x_1 and x_2 is strictly equivalent to apply the equivalent bandpass filtering to x. Thus, the choice is left to the system designer between either building two perfectly low-pass filtered Rice components, whereby the modulated signal x does not need any bandpass filtering, or building the latter from imperfectly filtered Rice components, whereby the modulated signal needs to be strictly band-limited by a bandpass filter. The former approach appears as a first choice in the frame of presently available digital signal processing techniques.

3.2.2 Implementation Constraints

Whereas the previous section indicates ways to transmit in a strictly limited bandwidth, achieving this goal is frequently made difficult by various implementation constraints related to a specific application. The ITU-R terminology [*ITU-R, 1994b*] uses the generic term "*unwanted emissions*" for radiation outside what is called the "*necessary bandwidth*", and distinguishes between spurious and out-of-band emission. Roughly speaking, "*out-of-band*" (OOB) indicates spectral components, which are outside but rather close to the necessary bandwidth, i.e. within 250% of the latter, whereas "*spurious emission*" (SP) relates to unwanted emissions, which are more than 250% away from the necessary bandwidth.

Whatever the semantics behind the official definitions, these two types of *unwanted emissions* fundamentally differ in nature. *Spurious emission* generally indicates discrete spectral components resulting from frequency transpositions performed in the transmitter block-diagram or from parasitic spectral lines of frequency synthesisers. We will not elaborate on the reduction of spurious, because relevant issues are essentially related to technological choices, e.g. to a sound design of the block-diagram of equipment. *Out-of-band radiation* is a subject for contributions from a scientific community such as URSI are needed: it indeed involves some multidisciplinary considerations.

3.2.3 The Issue of Power Amplification

In economic terms, power amplification - i.e. the final stage of any transmitter - is a critical issue in the design of systems. The power efficiency of a power amplifier, defined as the ratio of the RF output power to the supply power, is an important parameter in most applications. For a given RF output power, a high power efficiency allows to decrease the supply power, hereby improving the weight and/or life time of batteries in portable transmitters, or the size of solar panels in satellites, etc. It also means less heat dissipation in the elements of the power amplifier, i.e. less cooling problems.

The power efficiency does not exceed 50 per cent for linear power amplifiers (e.g. push-pull class B), whereas it currently reaches 70 or even 80 per cent for class C amplifiers. The latter however are designed to work with nearly constant amplitude signals, whereby they exhibit an apparently linear external behaviour, although the internal working is basically non-linear.

Whereas the signal bandwidth can be perfectly controlled by signal processing means, e.g. via digital signal processing of Rice components, non-linearity in the final stage of a transmitter causes inter-modulation between spectral components of the signal, i.e. spectral broadening. This type of out-of-band radiation, which can only be removed via rather expensive bandpass filtering of the output signal, depends very much of the type of modulation being used. This implies that the choice of a modulation and the linearity requirements of power amplifiers are intrinsically related.

The literature contains a quite large number of papers elaborating on the spectrum efficiency of various modulations. Many authors take as a basis for comparison, the side lobe level of the modulated signal; such a criterion seems meaningless, since conditions for transmitting in a strictly limited bandwidth are well known, as shown above. Discussions around this topic frequently concentrate on a comparison between constant-envelope modulations, i.e. with $a_x(t)$ constant in Rice's representation in eq. (4). Candidate modulations in this frame are, for instance GMSK (Gaussian Minimum Shift Keying), TFM (Tamed Frequency Modulation), CPM (Constant Phase Modulation), etc. i.e. various kinds of angular modulation. The restriction to constant envelope-modulations probably contains the implicit assumption that class C amplification is used. Defining this as a reasonable context is excessively restrictive, not to say a fundamental aberration. The analysis of the Rice representation of signals - equation (4) - shows that, for a channel submitted to AWGN, the Rice components $x_1(t)$ and $x_2(t)$ of a modulated signal $x(t)$ can be seen as independent, i.e. submitted to independent AWGN components. This means that each of them is able to carry half of Shannon's channel capacity. Imposing that the instantaneous amplitude $a_x(t)$ be constant implies a unique relation between $x_1(t)$ and $x_2(t)$, and this means that 50 per cent of the channel capacity is wasted in the constant-amplitude scenario.

The latter appears as fundamentally questionable when we further consider the following. For a given supply power, the difference between Class (A-B) and Class C power amplifier choices, in terms of RF output power, does not amount to more than about 2 dB. Shannon's equation shows that, in a context of relatively high signal-to-noise ratios, this would not cost more than a fraction of a bit/s/Hz in terms of spectrum efficiency. Sacrificing half of the channel capacity by restricting to constant-envelope modulations cannot be justified, except maybe in a context where systems should work at very low signal-to-noise ratios.

There is a need for further research on the topic of linear amplifiers. The supply power P_s is probably a major constraint in most applications, and this obviously implies some saturation characteristic for the useful output power P_u. Basic questions include the following: (1) what are the most adequate parameters to characterise non-linearity effects (e.g. spectrum broadening) in the nearly-linear region of output power, (2) how can a linearity region be defined, as a specification of these parameters, and (3) how can the linearity region, expressed by the ratio P_u / P_s be extended. Such studies would obviously also benefit to cases where the signal is a multiplex of modulated components.

3.3 The Impact of Signal Processing

Digital signal processing techniques are used massively in recently developed communication systems, e.g. for source and channel coding, for adaptive channel equalisation, smart antennas, adaptive interference cancellation, etc. They contribute a lot to approaching Shannon's limit and are an essential tool for an efficient use of the spectrum.

A very impressive example of such progress based on digital signal processing is the advent of digital television. Classical terrestrial television uses analogue amplitude modulation and operates in 6 or 7 MHz wide channels. With about 60 channels, nearly one half of the spectrum below 1 GHz has been allocated to this service. This is probably the best example of spectrum wasting due to an obsolete technology, since the number of terrestrial television channels which can be received at a given location rarely exceeds 5 or 6. One of the reasons why so many channels are needed is that the analogue television signal, in addition to requiring a quite high signal-to-noise ratio of 40-45 dB, is very sensitive to co-channel and neighbouring channel interference. It is furthermore sensitive to multi-path propagation. Consequently, covering a country with one program requires a rather large number of channels, since the channel used by a transmitter can only be reused far away from the service area of this transmitter, either for transmitting the same program or another one.

Shannon's theory tells us that even with a comfortable margin in terms of SNR, a classical terrestrial TV channel is able to transmit digital data at a bit rate of about 40 Mbit/s allowing it to convey either four high-quality digital TV programs or one HDTV signal (Table 1). But this is only a small part of the story since, in addition, the type of modulation adopted for digital terrestrial TV (OFDM: Orthogonal Frequency Division Multiplexing) *[Tourtier et al. 1993]* allows to eliminate multi-path propagation effects even for very large delay spreads. This enables the concept of single-frequency network, whereby neighbouring transmitters use the same frequency to transmit the same program; receiving the signals of two such transmitters is indeed equivalent to receiving a single signal affected by a two-path propagation effect. As a result, a single TV channel will now allow national coverage for several programs whereas the traditional technique required several TV channels for national coverage with one program. The related gain, in terms of spectrum efficiency, is the range 20 - 30.

This kind of achievement is very impressive, but it is not an exception. The evolution from classical analogue mobile radio to digital cellular systems offers similar - or even much higher - improvement factors regarding spectrum efficiency. The key ingredient in this evolution is signal processing. This raises the important question: is this trend likely to continue?

Our answer to this question is again a very enthusiastic "yes". Technological evolution is by itself a justification for such a statement. Even recently designed systems can be re-questioned when spectrum efficiency is considered essential. Narrow-band channels suffer from non-selective multipath effects, for which deep, flat fading events are necessarily solved through an increase of the transmit power. This is rather catastrophic in terms of spectrum efficiency, since the channel capacity is restored at the expense of the interference environment. Selective fading is less harmful, since it does not reduce much the channel capacity, whereas the resulting effects (ISI: Inter-Symbol Interference) can be compensated by powerful channel equalisation techniques. A systematic use of wideband transmission and of spectrum spreading techniques therefore appears interesting in view of spectrum efficiency.

Similarly, the spatial dimension of communications needs better understanding. The cellular concept was an essential progress factor in terms of spectrum efficiency for mobile radio. However, it solves the interference problem via some spatial frequency reuse scheme which is probably not optimum. The technique was recently improved through the use of smart antennas, which performs some management and cancellation of interferences via adaptive antenna radiation patterns. Further progress could come from re-questioning the spatially frozen frequency reuse technique currently made in cellular systems.

Furthermore, the potential of signal processing leads to reconsideration of a number of basic principles in electrical science engineering. Several limitations and constraints (e.g. the performance of electrically small antennas, unavoidable loss in wide-band impedance matching, etc.) which are frequently considered as fundamental actually result from causality and can be removed by accepting some delay in suitable signal processing. This is a very interesting area for basic research.

3.4 System Related Issues

Above we concentrated on rather basic scientific aspects and techniques. Efficient use of the spectrum in practice requires to take into account peculiarities of each service being considered as well as system related issues. For instance, the multiple-access technique, power control and switching play an essential role for spectrum efficiency in cellular radio systems, but are not directly applicable in the broadcasting area. It is outside the scope of this paper to investigate such matters, as the subject is extremely broad. It is clear however that progress would benefit very much from a stronger interaction between radio scientists and engineers.

4. Scientific Spectrum Use in a Congested Spectrum

Already the scientific users are often hiding from man-made radio noise. Radio observatories have been located in remote sites, in deserts or on high mountains. However, terrestrial observatories cannot hide from satellite-based or airplane-based operations. In particular, certain satellite systems are already damaging or are planning to damage radio astronomy operations and many new satellite systems are being planned. Within the ITU Allocation Table about 80 percent of the radio astronomy bands have allocated satellite space-to-Earth operations in adjacent bands.

Operations in the bands specifically allocated to the radio astronomy increasingly encounter interference. In addition, new scientific research topics arise. The results of the Hubble Space Telescope and of large (8 to 10 m class) optical telescopes have caused a flurry of interesting radio projects at frequencies outside the allocated bands. Objects far away in the Universe having high redshifts will have their spectral lines shifted down in frequency and into bands allocated to other services. To operate in non-allocated bands on a "non-interference" basis requires an increased robustness of the observing systems to interfering signals.

4.1 Regulatory Procedures

Often in the past scientists have depended on their governments to be protected. Commercial interests currently prevent any preferential treatment for the science services. Active participation is needed at the local, national, and international level to improve the protection for stations in the science services. This participation is part of the *cost of doing business.*

4.2 On-line Data Processing

Interference robust receivers and on-line data processing are some of the possibilities for modern observatories. Currently test systems are being implemented that operate on narrow band signals in the time domain in order to perform blanking or whitening of the interfering signals. This objective is to have a more prudent approach to dealing with interference than that of throwing away the data. Further algorithm research is required to describe signal specific properties of the interference and the observational data and to do spatial filtering of the data. To some extent the science community is forced to re-invent some procedures and concepts that are well known in the communications area. This calls for more collaboration between science services and other spectrum uses.

4.3 Adaptive Interference Rejection

Designs are currently being made for the next generation radio telescopes called the Square Kilometre Array (SKA) which need to be build to be interference robust. One such design involves flat panel antennas with beam forming networks. Such antennas allow adaptive nulling of interfering sources such as passing satellites. A dialog and an exchange of knowledge on such technology between industry and the science community would be very productive.

4.4 Propagation Issues at Higher Frequencies

The radio astronomy and Earth exploration communities have great interests in frequency above the 40 GHz. Several radio astronomical observatories are already operating at frequencies above 1 THz within all the atmospheric windows. These observatories are indeed located in very remote areas such as mountain tops in the Sierra Nevada in Spain, the European or Japanese Alps, or high plateaux in the Andes mountains of Chile. Proper protection of such sites from terrestrial operations in the Mobile and Fixed Services by means of protection or quiet zones (of limited dimensions) will also require further study of propagation properties at frequencies particularly above 60 GHz. URSI members can make valuable contributions to such efforts. Proposals are currently being made for submission to the ITU World Radiocommunication Conference in the year 2000 (WRC00) to reorganise the spectrum above 71 GHz in order to re-align the spectrum and to minimise adjacent band conflicts between the passive services and particularly the satellite downlinks. Spectrum sharing between terrestrial fixed and mobile services and radio astronomy at isolated mountain tops is considered a plausible manner to provide access to the spectrum for passive spectrum use.

5. How Can URSI Help Solving the Spectrum Congestion Problem?

Our Union gathers scientists with a very wide range of interests in various areas of radio science, but with a strong common denominator. An efficient use of the spectrum is part of the common concerns to those who are spectrum users for scientific investigation, and those who are involved in scientific studies about the radiation and propagation of electromagnetic waves or the study of communication systems. URSI can contribute considerable scientific expertise in this area.

The multi-disciplinary character of the spectrum efficiency issue has been stressed several times in the previous sections. This calls for increased collaboration between URSI Commissions, but also with the external world. Interaction between scientists and engineers involved in the development of communication systems is important, because these two categories of actors hold different parts of the global truth. This can lead to fruitful cross-fertilisation effects.

Collaboration between URSI and ITU-R is a long-standing tradition, although there are presently no structural relations. A number of scientists and engineers are active in both organisations. URSI and IUCAF scientists are well present in ITU-R SG7 (Science Services) and in some task groups of SG1 (Spectrum Management), particularly in TG1-5 which carries out quite interesting discussions on out-of-band radiation. Several members of URSI Commissions F and G are active in SG3 (Wave Propagation) as representatives from their national administration.

This collaboration needs to be reinforced, but the problem requires a pragmatic approach because there are many differences in objectives, culture and time schedules between URSI and ITU-R. Many scientists do not know much about ITU-R, and the reverse is probably true from many people participating in the ITU-R work. A first step would be to increase mutual awareness and information, e.g. via news published in information bulletins. Much could be gained from creating links between URSI Commissions and ITU-R Study Groups, e.g. by appointing liaison officers. In addition to mutual information, their task could include suggesting topics and program committee members for URSI symposia, trying to co-ordinate and combine meetings of ITU-R working groups with such symposia, etc. Making access to the ITU-R documentation easier for URSI scientists would also be very positive.

Stronger URSI collaboration with ITU-R and helping to facilitate dialogues with the industry on all spectrum issues would greatly benefit all parties, including for a sound approach to problems such as the scientific use of the spectrum and the protection of science services. In addition, URSI members need to interact with their own governments and their national industries in spectrum related issues. The URSI series of *Commsphere* conferences have provided the correct international forum for dialogues.

Various topics considered in this paper have been intensively discussed during the last Commsphere, which was held in Toulouse (France) on 25-28 January 1999. Five sessions and four workshops were held. In the workshop on the Third Generation of Cellular/Personal Communications, the possible role of scientific contributions to resolve the disagreement between various CDMA proposals was discussed, as well as the need for an evolutionary, migratory approach from second generation systems. The workshop on the Electromagnetic Environment drew the attention on the fact that many existing standards on electromagnetic compatibility are obsolete, because they have not been adapted to the advent of digital systems and also because they do not cover recent developments such as the use of power lines for information transmission inside and outside homes. The evolution of global coverage satellite systems and projects and the problems of interference with radio astronomy were discussed in another workshop. Spectrum savings offered by digital terrestrial television, projects for the digitalisation of AM radio and the development of new services were discussed in the workshop on Future Spectrum Needs for Broadcast and Multimedia Delivery. Detailed conclusions of the workshops will be published in the Radio Science Bulletin.

References

ITU-R [1974], Question 47/1, "Definition of efficiency and utility of spectrum use," Geneva, Switzerland.

ITU-R [1994a], Recommendation SM.1046, "Definition of spectrum use and efficiency of a radio system," Geneva, Switzerland.

ITU-R [1994b], Recommendation SM.328.8, "Spectra and bandwidth of emissions," Geneva, Switzerland.

Proakis, J. G. [1989], Digital Communications, McGraw-Hill.

Shannon, C. E. [1948], "A Mathematical Theory of Communications," *Bell Syst. Tech. J.*, 27, pp. 379-423 and 623-656.

Tourtier, P. J., R. Monnier and P. Lopez [1993], Multicarrier modem for digital HDTV terrestrial broadcasting, *Image Communication*, 5, pp. 379-40